THE DREAM GENE

The Search for Genetic Memory
By Lee C. Bush

Printed and bound in the United States of America
First printing • ISBN 978-0-9991508-1-8
Copyright © 2017

All of the characters and companies in this book are fictitious and any resemblance to actual persons, living or dead, or companies is purely coincidental.

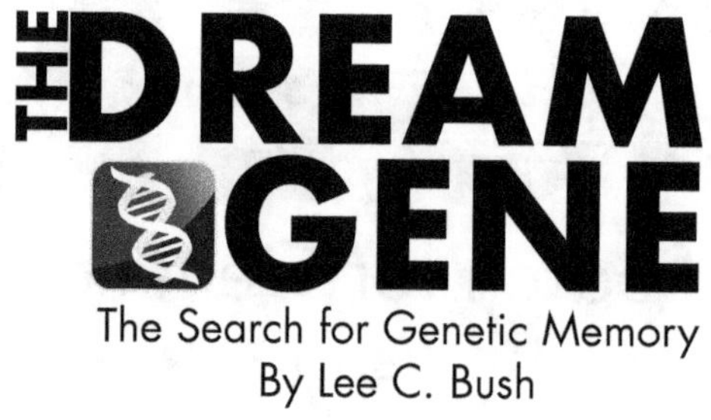

FOR MORE INFORMATION VISIT:
www.forwardobservermsg.com

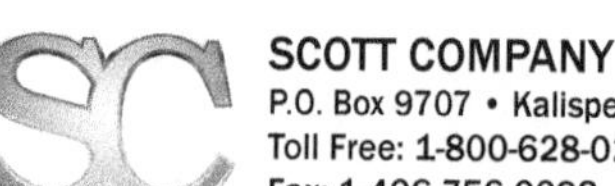
SCOTT COMPANY PUBLISHING
P.O. Box 9707 • Kalispell, MT 59904
Toll Free: 1-800-628-0212
Fax: 1-406-756-0098

Acknowledgements

The author would be remiss if he failed to recognize the many people and sources of information which made possible the writing of the book. Foremost is the information provided by the scholars of the Reasons to Believe ministry, especially Hugh Ross and Fazale Rana. The many recent books by Pastor Thomas Horn and recently deceased Cris Putnam have set the stage for the days we are now experiencing. Rabbi Steven Ben-Nun of the DeNoon Bible Institute has provided valuable information on the prophetic Scriptures and extra-biblical sources, including the recently discovered Dead Sea Scroll of Cave Twelve. The books and alternate media postings by Anthony Patch and S. Douglas Woodward concerning their revelatory research focused upon the Cern organization as it relates to the end times fulfillment of Bible Prophecy furnished invaluable information. Finally, the on-line information provided by the Universities of Johns Hopkins and Emory were most helpful.

Dedication

This book is dedicated to my wife Mildred.
She has been a support during the writing of this book.
She has stood by my side during the many changes in our journey
together.
Jeremiah 29:11

Table of Contents

1

The Journey Begins

*"For God speaketh once, yea twice, yet man perceiveth it not.
In a dream, in a vision of the night, when deep sleep
falleth upon men, in slumbering upon the bed."*
(Job 33:14–15)

Jacob's Dream

I am struggling to find an answer to a very simple question. What do you really want to do with your life Jacob Cahn? Like most kids growing up I wanted to be just like my dad. As the only son of Dr. Joel and Mrs. Rose Cahn of Omaha, Nebraska, my early aspiration was to be a medical doctor. As I draw nearer to the end of my pre-med studies, I'm having some second thoughts about this course for my life. I'm not sure whether this change of heart has emerged from my exposure to some of the university courses I've taken, or from some inner prompting. My early decision to pursue my father's career path came without any conscious thought on my part. It emerged over time from the unspoken expectations of my family, and no wonder. My dad is everything you hope to find in a general practitioner. He unselfishly serves his fellow man, healing the sick and diseased, and often just being there for people as they confront the hard issues of life and death.

Becoming a doctor is not an impossible task for me. I sprang from this mold with the required mental abilities. It has been a gift. In addition, I have a certain proclivity to view life as serious business. Taking life easy isn't part of my DNA. So, it was no surprise when my family assumed I would follow in my father's footsteps, attend Emory University in Atlanta, Dad's Alma Mater, and become a general practice physician. I accepted my personal destiny to take over my father's practice without a second thought. It was a given. So, why am I now feeling some uncertainty? I know I have the *chutzpah* to accomplish my family's expectations, but my heart just isn't in it. Yes, I often resort to a few Yiddish expressions. I'm a secular American Jew with a great Jewish heritage and a strong desire to succeed in life. Dad often recounts to me this strange destiny of the Jewish people, chosen to assist a transcendent God in bringing some degree of perfection into an imperfect world. *Mitzvah* is the command to serve the needs of our fellow man. It is at the center of the Jewish collective soul. But this commandment is often grievous because of the antisemitism Jews have experienced over the last three and a half thousand years. My dad very often expresses his strong opinions about the actual or perceived unrighteousness of Gentiles who, with demonic intent, have repeatedly tried to extinguish the seed of Abraham. Not infrequently my dad feels it necessary to vent his anger toward "that man" Jesus who challenged the central precept of Judaism, the oneness of the Almighty. His sudden outbursts of emotion invariably are followed by some anecdotal bit of Jewish humor. It comes with the script.

As I look back on my years growing up in Omaha, I can't recall experiencing even one incident of antisemitism remotely resembling what has taken place to the sons of Abraham in other times and places. I have never experienced even one slur or attack on my Jewishness. Perhaps if I had been born into an Orthodox family and wore braids in my hair and the *tefillin*, a box of special scriptures on my forehead, I might have encountered some form of persecution. At the least a bit of ostracism. No, persecution for being Jewish has never happened to me. I always have managed to get along well with my Gentile classmates in Omaha. In fact, many of my closest friends are Gentiles, even Christians. Here at college it has been the same. Hurrah for the Land of the Free and the Brave. But I wonder? If I had worn a tefillin on my head, would

Ha Shem have blessed my mind, my thoughts, my emotions, and my will even more? I did well in high school. It came easily for me. I did not reach the top of my class, but I always pulled good grades. When a particular course challenged my competitive spirit, I devastated the competition nine times out of ten. I have to admit, at times my aggressiveness has gone a bit too far. I often find myself apologizing for my bad manners. But then, who's perfect already? Early on, I made a conscious decision to give the feminine gender a wide berth.

I firmly decided to stick to business to fulfill my father's aspirations for me. Sure, I dated a couple of girls in high school, but I always carefully avoided any serious relationships.

You would think this combative quality of mine would serve me well in sports, but the truth is I had to work hard to develop those motor skills. Ever since I can remember I've had to work to overcome my poor coordination and clumsiness. I remember trying to learn to jump rope in the Yeshuva school. It took me a long time before I could do it. In time, my coordination improved. By my senior year in high school, I managed to earn letters in track and tennis. I even succeeded in making the tennis team here at Emory in my first year.

I met my best friend on campus, Allen Schaefer, on the tennis court. Our friendship is rather amazing. You ask, why so? It is because Al is an overly zealous, radical Christian. In our second year, Allen and I both decided to rent two apartments at the Clairmont Residential Center located a mile east of the main campus. Each apartment comes with a complete kitchen, laundry, private bathroom, bedroom, and living room. What I really like about my apartment is the view I get of the main campus. I have two large windows in the living room and one small window in my bedroom. The Center furnishes everything you could possibly want. The lower floor of our building includes a gymnasium, a computer lab, several classrooms, a café, and a large parking structure. Shortly after renting the apartment I brought Old Blue, my ancient '98 Ford, onto the campus. Another thing Allen and I like best are the seven tennis courts located right next to the building. There's also basketball and volleyball courts, a track, and a huge swimming pool. A shuttle bus on Starvine Way connects the complex with the main campus. I seldom use the shuttle. My daily schedule begins with a run between the Center and the main campus.

Today is Friday, the first week in May. I'm close to finishing the pre-med program, fulfilling the expectations of my family. Graduation is just eight days away. I should be looking forward to medical school, but I'm not. I'm still *kvetching,* i.e. complaining about this path of becoming a medical doctor. Questions keep running through my mind. I can't seem to make them go away. I keep trying to convince myself. After all, what's so bad about putting an MD after your name, Jacob? You could do worse. Why get so worked up about it? Don't get me wrong, I have a deep respect for the field of medicine, and my father's accomplishments, his caring attitude toward his patients, his mentoring of the young interns, and his absolute commitment to the practice of medicine. And I certainly appreciate the opportunity he has given me to attend Emory these last four years. But the fact is this future vocation is hanging over my head like a dark cloud. I'm not sure how well I will deal with the human side of medicine, the ability to relate to patients with compassion. There is another reason for my reluctance to sign on to the medical school here at Emory. Last semester's course in biology left a strong impact on me. Professor Adams included a brief survey of the expanding field of microbiology. I found the field very interesting, especially the topic of human genetics. The intricate design of life's DNA helix is totally beautiful and amazing. I asked Professor Adams if he thought I should change my major at this late date. He saw no reason why I shouldn't be able to change my major, but I would have to go elsewhere as Emory only offers a Post-Doctoral program in genetics. I asked him what schools he would recommend. He suggested the University of California at Berkeley and Johns Hopkins in Baltimore Maryland, but warned me gaining entrance to either school would not be easy. I thought, what have I got to lose? On the spur of the moment last semester I made the decision to throw my hat in the ring. I requested Emory to forward my transcripts to both schools and sent letters requesting applications. I have continued to struggle with this question of the future course of my life ever since last semester. Not so for my sister Emily. *Oui vey!* She just turned seventeen and already is convinced she's destined to become the best clinical psychologist in the world. No second thoughts for Emily. Well, good for her.

Tennis and track have helped me to largely overcome my lack

of motor skills and I have managed to earn a high grade-point average. To look at my undergraduate years here at Emory, most people would assume I have enjoyed four successful and carefree years. However, such is not the case. The truth is I have a problem. I'm not referring to this question about my future vocation in life. I have harbored a secret in my life from everyone, my family and my closest friends. Ever since my childhood I've experienced a recurring dream. It comes in the night, every week or so, completely unannounced and unpredictable, leaving no clue as to the timing of its arrival, or any insight into its meaning. At Emory, I hoped these night excursions would soon fade away. Such has not been realized. Instead the dream has intensified and proved to be relentless. It has invaded my nights here at the University even more frequently than before.

Lately I've begun to question my mental health. I've been trying some serious self-diagnosis. My weird dream and my lack of coordination have led me to believe I might in fact be an aspri, a person suffering from a mild form of the genetic disease Asperger Syndrome. So now you know the deep secrets in the life of Jacob Cahn. It's now twenty minutes to three and I'm jogging my way to the biology building. Dr. Miller, my virology professor, promised to post our final grades by three o'clock today. As I arrive in the hallway of the biology department, a group of fellow expectant classmates already are gathered in front of the bulletin board, waiting for Professor Miller to appear. We don't wait long before the door to his office opens and the Professor quickly posts the grades on the bulletin board.

Then turning around, he says, "I want to congratulate all of you. Virology is not the easiest of course offerings here at Emory. I am very pleased with your performance. Everyone has successfully survived this course. I wish you well as many of you move on to your graduate programs."

Professor Miller then quickly turns and walks back to his office. He locks the door and leaves the building. I am tempted to push my way to the front of the crowd, but I force myself to wait my turn. Holding my breath, I run my finger down the list to my name and across to the columns showing my final exam and course grades.

"All right, an A!"

My sudden outburst plainly annoys some of the other students who are scowling in my direction, but I don't care. This is the last grade I needed for the semester and I have aced the class. I hurriedly copy the grades into my notebook. This final semester has been awesome. I have earned all 4.0 or above. My name will again make the Dean's List and earn an invitation to attend the special honors commencement banquet. Tightly clutching my notebook, I set my course back to my apartment. My mind is racing ahead of my footsteps. I jog for the first mile, and then sprint the last hundred yards to the complex, all the while trying hard to contain the smile forcing itself across my usually serious face. A new sense of confidence is invading my thoughts. I hear myself saying, "Jacob, you have every right to feel good about this last year. You deserve a pat on the back. You're ready to move forward, perhaps to pursue a brand-new path."

Once back in my room, I slam the door shut with a bang and flop onto my bed. Staring at the ceiling, I begin to think about these last four years at Emory. No doubt about it, Emory University has been good to this secular Jewish boy from Omaha.

Memories of the last four years, race through my mind, spilling out like coins from a slot machine. I recall all of the professors who helped me, like Professor Rogers who turned me on to a love of basic research. Then there are the grad assistants who helped me get past those obstacles in the tougher courses. Best of all has been the students I now consider my close friends. I realize it's going to be hard to say goodbye to them, especially to Allen Schaeffer. It has been a full day. I decide to stick a pizza in the microwave and settle down to watch an old movie from my collection of VCRs and DVDs. Old movies are a fetish of mine. Some people collect old coins, stamps, or baseball cards. I collect old movies. I pulled out the weird movie *Tron*. After the movie and a quick shower, I decide to turn in. As I began to drift off into slumber, my dream is back again to torture me. I awake, breathing heavily and in a cold sweat. My digital alarm clock on the nightstand reads ten minutes after ten. What? I've been half-asleep for only two minutes? I lay there, half awake, breathing heavily. My ears are straining to hear some sound, anything to break the silence of the night. I hear the distant crack of lightning bolts followed by thunder rolling through the Chattahoochee Valley, breaking the silence. The pungent odor from the recent rain shower, seasoned by the smell of ozone, wafts through my open window. I continue to peer into the darkness. Short glimpses of rain clouds are illuminated against the night sky by flashes of sheet lightning. No stars or moon are visible. Can there be some connection between the thunderstorm and the onset of my dream? Not likely. It doesn't matter. The dream is over and I am safely back in the real world again. But the mystery remains. Why am I constantly reliving this same strange dream? I have kept my dream a secret. I have never told a living soul about what happens in my dream. It is just too bizarre. I will share just a bit about my dream with you. The dream always begins the same way. As I close my eyes and fall asleep, I have the sensation of falling, falling into some sort of dark passageway. I sense my soul is leaving my body, floating upward, bouncing along the ceiling like a helium balloon. At this point I feel no real anxiety. In fact, I rather enjoy looking down at my still form lying on the bed below. This short peaceful interlude is suddenly followed by a devastating second phase. In my dream, I am standing in front of a stone wall. There is a doorway in the wall. The door is locked, but there is a key hanging on the wall next to the door. Above the doorway is a sign in German. It reads "Arbeit Macht Frei" Translation: Work Makes Free. No doubt you know these words were displayed above the entrance to Nazi Germany's infamous Auschwitz Concentration Camp located in southern Poland where more than a million Jews were exterminated. This grim greeting can still be seen there today.

I will not tell you the rest of my dream. All I will say is my dream is quite detestable. It seems to be some sort of prophecy of the End Times. I do know I will be called to play a part in it.

I am now fully awake. I turn to look at my alarm clock. It's five minutes to seven, Saturday morning. Recuperating from this latest journey, I try to reframe the experience by imagining a garden full of beautiful flowers and well-trimmed trees, like the restored garden in the movie *The Secret Garden*. My attempt to revise the dream fails miserably. Will someone please explain this dream to me? How about you Ha Shem? You know everything. Can you explain it, or don't you hear me? In the early morning silence, I wait. Apparently, he prefers not to answer. Okay, I understand. You have bigger things on your Seder Plate these days.

Perhaps my dream has been passed down to me through my German DNA. I have a rare German VHS movie in my collection.

It is the propaganda film entitled *Triumph des Willens (Triumph of the Will)*. The film was produced in March of 1935. It won Germany's National Film Prize, and was awarded first place at the Paris International Exhibition in 1937. It has been named the best propaganda film of all time. Directed, produced, edited, and co-written by Leni Riefenstahl, it chronicles the 1934 Nazi Party Congress in Nuremberg, which was attended by more than 700,000 Nazi supporters. Albert Speer, Hitler's personal architect, designed the sets in Nuremburg and did most of the coordination for the work. I kept this film because it touches upon my family's history. My grandfather on my mother's side of the family, Ernst, perished in the camps. My grandmother Hilda emigrated from Germany after the war and rescued my mother from the holocaust of those dark days. *Triumph of the Will* seduced many men and women, persuading thousands from all over the world to support the Nazis, including my grandfather. Religion was a major theme in the film. It opens with Hitler descending god-like out of the skies in a plane. Many scenes contain church bells ringing, and individuals in a state of near religious fervor. There is prominent shot of the Protestant Bishop Ludwig Müller standing in his vestments among high ranking Nazis. In his final speech in the film, Hitler portrays himself as a messiah, and directly compares the Nazi party to a holy order. There are many other scenes with religious overtones. The film contains shots of the array of searchlights that made a powerful impression on the large crowds. In many scenes, the camera focuses on Hitler from below, as he stands on his podium, issuing commands to hundreds of thousands of followers. The audience happily complies in unison. Hitler was presented as the savior of the nation, asking only for the people of Germany to put their destiny in his hands.

I have in the past tried to get beyond the fear, to understand what is happening in my dream. Are these night journeys simply a figment of my imagination? Am I hallucinating? I have attempted to research the scientific understanding of dreams. I found there is a phenomenon occurring at the onset of sleep. The medical term is hypnagogia. Knowing the technical word for my dream doesn't furnish much help for me. I need to know where my particular strange dream comes from. Years earlier I resolved never to tell anyone about the dream for fear people would label me a complete *fachadick*, a nutcase. I took a deep breath and got out of bed.

Soul Brothers

With the dream of last night still weighing heavily on my mind, I showered and got dressed. On my desk are two unopened letters from Johns Hopkins and one from Cal Berkeley. I walked over to the desk and picked them up. I'm trying to decide. Should I open them now? No, I will save both of them for later, perhaps after breakfast. I put the letters onto the desk and headed downstairs. I am hoping Allen Schaeffer will be there.

As I enter the café, Allen Schaeffer is seated at his usual table near the window. Allen also is graduating this year. He has ordered another of his extreme breakfasts: ham and eggs, a stack of pancakes, a bowl of oatmeal, and a small glass of orange juice, topped off with a large cup of coffee. I grab a cup of coffee and a sweet roll. Allen is in the process of consuming a large forkful of Aunt Jemima pancakes as I sit down at his table.

"Good morning *akhi yedidi*."

Allen's His mouth is full of pancakes. I hear something completely undecipherable.

"What did you say Al? I didn't get that."

He manages to swallow the pancakes and asks, "What's this akhi yedidi thing? Should I be offended?"

"Not at all. It's a Yiddish greeting for a good friend, you know, like saying you're my Bro."

"Okay. For a while there I thought you came up with something worse than a *goy*."

"No way, Al. Say, I hear the grads have elected you to give the address at the Class Day Reception and the Dean of Students has asked you to give a brief talk during commencement as well. That's quite an honor Allen. Have you finished your talks yet?"

I wait as Allen washes down another mouthful of pancakes with a swig of coffee.

"I've written a few notes. The Class Day talk won't take long, probably ten minutes or so, and the commencement talk even shorter."

"You seem a little edgy this morning. Are you nervous about standing up in front of the whole student body?"

Allen just shrugs as he continues chewing his pancakes. I answered my own question.

"Well, not to worry. I've heard you speak in class. You are a good public speaker. I'm sure you'll do just fine. Of course, no body's going to remember what you say an hour after the reception."

Spearing a piece of ham on his plate Allen manages a smile at my bit of encouragement.

"Hey, one sweet roll won't take you very far, Jake. Have a piece of my ham."

"How many breakfasts have we shared together, Al? Have you ever seen me eat ham?"

"Oh yeah, I forgot. Sorry."

"You forgot? Like you're willing to give me your last piece of ham? I don't think so."

I wasn't offended. I have to endure Allen's dry humor many times. He enjoys pulling my Jewish leg.

"So, tell me Al, what possibly could you say meaningful enough for these young *shlemiels* to remember?"

Allen briefly stopped eating as he considered my question. "Jake, you are right. It's hard to come up with something memorable, something helpful to be remembered, but I need to try. After all, this day will mark a major turning point in all of our lives. What do they call it, a Rite de Passage? I want to encourage our class, to stoke up their fires, to urge them to pursue their dreams. God created each of us as unique individuals, one of a kind. The Almighty has a purpose and destiny for every one of us. He has preordained a path for our lives. We must seek our own unique and divine calling, Jake. Unfortunately, few realize their true potential."

"I agree with you on one thing, Al. Several in our class are certainly unique."

"Jake, it's our task to determine what God's plan is. Unfortunately, many will never realize this because of the barriers they erect in their lives. Fear, pride, and the love of pleasure in this world all too often get in the way."

"Al, I'm not sure many even know how they are supposed to discover these special callings? I don't know myself."

"There are a number of ways we can determine our calling. We are gifted with reasoning abilities, self-awareness, the ability to forge relationships with others, and just by making use of our creative juices, to name a few. The best way for us to zero in on our destiny is to develop a close relationship with the Lord. He promises to guide us.”

“Al, I’m not sure how this happens. Sometimes he tells us in very direct ways, such as in a vision or a dream.”

Al’s reference to visions and dreams makes me uneasy. It is evident he believes dreams can have a supernatural origin and important meaning. Dreams to him are messages direct from God. Even though Al is my best friend, I don’t feel I can share my bizarre dream with him. No, I can’t go there. I’m concerned about where this line of conversation is taking us. Fear is rising up from the pit of my stomach. I wonder if Al can see fear in my eyes? Have I blown my cover? I need to devise a strategy to avoid further discussion about dreams and visions.

“I think most dreams are meaningless Al, especially the weird ones. Even if the same dream comes up over and over again, it doesn’t mean there is some hidden message or truth behind it. I’m convinced dreams are the result of eating too many slices of pizza the night before.”

As Allen takes another bite of pancakes, he looks up from the table and stares steadily at me for several minutes. His blue eyes are boring right into my soul. Finally, he answers.

“Jake, I can tell you I often experience some really strange dreams, real winners. They seem meaningless to me at the time, but later often prove to be extremely accurate and useful. Once I dreamed I was in a small boat and suddenly this storm comes up and—”

I need to interrupt him. “Perhaps your dreams are special, but not mine, Al. God may give you meaningful dreams, but I’m convinced dreams for most people just don’t mean a thing. Our subconscious mind just takes some memory, mixes it with another unrelated experience and twists it into some nonsensical mental video blog.”

The look on Allen’s face tells me he is not buying my argument.

“Jake, I can’t agree with you. Everyone should consider their dreams as important. Often only when we are asleep can God get his message into our busy minds. I know my dream is meaningful when it touches something deep down inside, like my boat dream. The Bible teaches we should expect God to speak to us in dreams and visions, especially as we enter into these last days. When a dream happens over and over again, or is confirmed to us in some other way, we know it has meaning for us.”

I paused to consider Allen's words. Does my recurring dream actually contain some hidden meaning? If so, what is its message? My strategy to discount the value of visions and dreams is not working. Sooner or later Al is going to realize I'm hiding something from him. He will zero right in on my dream, and I don't want him to go there.

"Jake, there are many signs telling us we are living in the last days. God's prophets announced thousands of years ago events now taking place in our generation."

Thankfully Allen decides to turn our conversation in a slightly different direction. I know where he is headed. In our conversations, Allen often succeeds in planting a seed of Christian faith in me.

"What signs, Al?"

"You know a good example is one found in the Book of Isaiah. Isaiah prophesied about the last days in Chapter Sixty-six and verse eight. He prophesied the nation of Israel would be reborn in a single day. What do you think the probabilities are for such a prophecy to come true given the many years since Israel existed? The prophecy seemed utterly improbable, yet it happened, just as Isaiah said it would. On May 14, 1948, the State of Israel was reborn in a single day."

Allen's reference to Isaiah's prophecy passed in one ear, but not quite out of my other one. I have to admit, Al certainly knows Scripture a lot better than I do. Sure, I attended Hebrew School as a youngster, but I don't remember much from those lessons. I tried again to steer Allen away from this topic of dreams.

"Are you still planning to study theology at the same time you're in medical school? Seems like a pretty heavy load to me, Al."

Al's casual demeanor is replaced by an intense, purposeful expression. He now is sitting on the edge of his chair, ignoring the remaining food on his plate.

"Yes, it will be hard, but I want to start working with Doctors Without Borders just as soon as I finish my medical training and residency. Jake, I want to minister the Gospel at the same time as I use my medical skills. I plan to work in the underdeveloped world, perhaps in Africa. The need is so great there, especially in areas like Northern Nigeria and the Sudan. The people of these African countries are extremely poor. The curse of poverty is always hanging over their heads. Rising food prices are turning their economies upside down. The region is woefully short of medical doctors and there are few medical facilities. Outbreaks of diseases like Ebola and Malaria happen frequently. Do you know HIV now has infected over thirty percent of the population of Kenya? The economies of the majority of countries in North and West Africa and the Middle East. These areas are nearing collapse. Their people are caught up in riots against the repressive entrenched regimes, demanding greater freedom. Unfortunately, these protests are met by violent reactions of governments. The people end up far worse than before, in slavery to radicalized Muslim tyrants practicing Sharia law. In many countries, Islamic Jihad is persecuting Christians, Jews, and other faiths. As a result, thousands have abandoned their homelands, attempting to make their way to someplace where they can find some measure of security and stability. Unfortunately, these emigrants often bring seeds of violence along with them. Jake, my total passion is to help the people of the third world, not only giving them the benefits of modern medicine, but reaching them with the good news of Jesus Christ."

I'm impressed as I listen to my friend reel off his future plans. I cannot but admire the fire and religious fervor of this goy. At the same time, I am feeling very uneasy, even dreading what dangers he may encounter in his life as he follows this dangerous calling. Allen finishes his animated description and returns to his plate of pancakes. I watch as he takes the last bite, lays down his knife and fork and becomes very quiet. I don't need to be told what is taking place now. I've seen this all before. Al is praying for me, most likely asking God to bring Jesus the Messiah into my life. I am aware of how little I really know about the Christian faith and this God-man, the one Dad always refers to as "that man" Jesus.

His silent prayer finished, Allen asks, "I haven't heard anything about your plans for next year, Jake. You will be entering medical school with me here at Emory, right?"

I take time to choose my words carefully. "Well, Al, I'm going to let you in on a little secret. I'm contemplating embarking on a different course. I have pretty much convinced myself not to enroll in medical school here next semester, or at any other medical school."

Allen is visibly shocked. "Why, Jake? I thought sure you are planning to work with your dad in Omaha?"

"Well to tell you the truth Al, I really don't want to become a medical doctor. I just can't picture myself treating the same illnesses over and over again, pushing the pills, constantly on call, never free to enjoy some peace and quiet in my life. You should see my dad's schedule. Every day of the week he begins at six in the morning making hospital rounds, then gets to his office by nine. He arrives home about seven or eight in the evening. Throughout the day and night, he's constantly interrupted by emergency calls from the hospital. Talk about slavery, my dad is a prisoner to his practice. Besides, I don't like the direction our national medical system is taking. There's no way people are going to be able to pay for the rising cost of medical care in the near future. In my opinion, it's going to result in an absolute disaster. Unlike you, I'm not looking forward to serving humanity on some foreign mission field. No, I don't want to practice medicine, not in Omaha or anywhere else. I am deciding whether to change my major."

"Wow, surprise, surprise! So, what are you planning to do after you graduate?"

"I want to work in microbiology, specializing in genetics. If I can earn the doctoral degree, I hope to land a research position at a major university or with the National Institute of Health, working on the human genome."

"Human genetics? Wow, that's interesting. Is your dad on board for this change of heart?"

"Actually, I haven't told him about my decision yet. He won't be thrilled about it. He's looking forward to me helping him in Omaha. I think he'll come on board once I explain the reasons for my decision. Al, there are so many exciting opportunities opening up in this field. Really, the sky's the limit. I have a whole lot to learn, but I think I can handle it."

"Yes, I'm sure you can, Jake. Just think, you might be blessed to find a genetic cure for some serious disease. What a great vision. I assume you've sent an application to the Yerkes National Primate Research Center here?"

"No, Al. The Yerkes is purely a research center. I would have to earn a degree in microbiology before I could work there. I have two unopened applications sitting on my desk right now, one from the Microbiology Department at Cal Berkeley, and one from Johns Hopkins in Baltimore."

Although Allen's response is encouraging, I detected a note of disappointment as the reality of my leaving Emory sank in.

"Can you afford to pay for one of those university programs?"

"Al, students accepted for these programs receive some very large stipends from the National Institute of Health. Also, I may be able to earn additional funding by working as a genetic counselor or lab assistant."

I was not surprised when Allen's interest cooled. I knew he would not look with favor upon any process of selecting marriage unions or planning families according to the compatibility of a couple's DNA. There seemed to be something else bothering him as well.

"So, what are you thinking, Al?"

"Are you sure this is the course for you? Alice and I are looking forward to struggling through medical school with you. She will be very disappointed if you run off to Baltimore."

"Yes, I certainly will miss you two as well."

"By the way, Alice and I have arranged a little surprise for you a bit later today."

"A surprise?"

"Yes, a surprise."

"Now you have me curious. I'm sure I will like it, whatever it is."

"Oh, yes, you definitely will like it Jake. What we came up with may not be as viable now with you leaving for parts unknown. We will just have to see how this all plays out."

I wondered. Perhaps Al has proposed to Alice and they are planning on inviting me to be the best man at their wedding. But why would my leaving Emory be an issue? It will be hard, but I'll have to be patient and wait until our tennis match later today.

"Are we still meeting on the tennis court today Al?"

"Definitely. I reserved the court for ten o'clock."

"Okay. I'll see you then. I have to go upstairs now and take a look at my mail."

Al is still working on his breakfast as I finished my coffee, sweet roll, and left.

An Open Door

It is a few minutes after eight when I arrive back in my room. It was an introductory course in biology from Professor Adams first leading me to consider straying from my father's chosen profession. The course opened a door for me. I became aware of the exciting things taking place in the fields of microbiology and genetics. The initial report of the completion of the human genome project was reported first in 2003. In May of 2006, the journal Nature first announced the complete sequencing of human DNA. I instinctively knew this was just the beginning. This explosion of knowledge will certainly open many doors of opportunity for a properly motivated researcher like myself. In my spare time, I began reading up on genetics. Already new fields have emerged, such as the using genetic markers to trace the historical lineages of mankind and other species. Practical applications such as the genetic counseling and advances in preventive medicine also have emerged. Researchers are now discovering the relationships between susceptibility genes and disease. Even the so-called hoc genes now have been shown to be regulators of gene expression. I realize such knowledge inevitably will lead to entirely new methods of treatment.

Actually, the field of gene susceptibility closely relates to my father's interest in preventive medicine. I plan to use this to convince him of the correctness of my decision to enter the field of genetics. I could use the master's degree in human genetics as a stepping stone to reach my primary goal, the doctoral degree. The final step would be to secure a post-doctorate research position. I don't see any reason why you can't walk through one or more of these open doors, Jacob Cahn. I am facing a crucial decision, selecting the best school to attend. My first choice is the McKusick-Nathans Institute of Genetic Medicine, offered by Johns Hopkins. The institute coordinates with the other Departments of Pediatrics, Medicine, Molecular Biology, Oncology, and other departments to provide a comprehensive program in human and medical genetics. What appeals to me about this program at Johns Hopkins is it does not require a student first to obtain a medical degree. I could avoid the hospital residency requirement and can move directly into a research position. While the training program is geared toward the Ph.D. The M.A. degree is available either with or without a thesis. This program seemed to fit my goals exactly. There is another plus to this path. The school works closely with the National Institute of Health in Washington DC. My second choice is based on the recommendation of Professor Adams, my biology professor. He favors Berkeley because of the talented group of micro-biologists working on cutting edge research, literally developing methods of splicing whole gene segments into existing DNA/RNA strings. I am anxious to open the two letters I received last Thursday. They are lying on my desk next to the still incomplete application form for the Emory Medical School. I decide to open the letter from Cal Berkeley first. The letter is from the admissions office. I'm disappointed. The envelope contains a standard form letter. What's more, they have ceased processing additional applications to the genetics program for the upcoming year due to the large number of applications. The letter suggests I contact them further if interested in applying to one of their other related majors. There is a printed signature from a Dr. Joan Harbaugh, Supervisor of Admissions.

What a bummer! Why did I take so long in making up my mind to change majors? Well, too late now. I stared at the second letter on the table from the McKusick-Nathans Institute of Genetic Medicine at Johns Hopkins. This is the school I really want to attend. My hand is shaking as I pick up the envelope and open it. The envelope contains a detailed application form and a personal cover letter signed by the Director of the Institute, Dr. Jonathan Hobbs. The letter requests applicants to forward their transcripts and grades to the Institute. In addition, they request three letters of recommendation from those familiar with the person's character and suitability for undertaking a career in human genetics. The completed application form and references are to be sent directly to the Office of Admissions. The Office of Admissions has enclosed four prepaid postage-free envelopes addressed to them. I have the option of submitting the completed application to them over the internet. After submission, applicants are to contact a Ms. Roberta Capoletti, the Program Administrator. She will schedule a meeting with the applicant in Baltimore. Her telephone number appears at the bottom of the letter. The envelope includes a very encouraging note from the Director congratulating me on my successful completion of the undergraduate degree at Emory, and says he looks forward to personally meet with me to discuss the graduate program at the Institute. The note is signed by Dr. Hobbs, Director of the Institute.

A queasy feeling is rising in my stomach as I consider several questions. Do I have the confidence to press on in this direction? Will the three references I furnish help or hinder my acceptance to the school? Will I be able to meet the substantial cost of grad school? I do know there are stipends available for students accepted to the program, but will my dad financially be on board now with my change of plans? I'm fairly sure he will be there for me to provide additional funds if necessary. There's always the student loan program, but I don't want to dig a financial black hole I will have to climb out of later.

I put the three recommendation envelopes into my backpack for safekeeping. After scanning the application form into my computer, I began entering the requested information. It's the usual required information: my full name, social security number, age, sex, race, religion, marital status, and questions about my work experience. *Lekherlekh*, just how many times do we have to tell people this stuff? Their system probably already knows what kind of shorts I wear, even where I bought them. Marital status? No entanglements there. Race, what does race mean in today's culture? I place a 'X' for Caucasian. Religion? How relevant is a person's faith to entering a university program? I enter 'Jewish.' As I type the word, I am aware I only dimly understand the significance of being Jewish. I grew up attending a reform synagogue in Omaha. I even attended Yeshiva, the Jewish elementary school, and attended the Hebrew school long enough to pass my bar mitzvah when I was thirteen. Although our family kept the Jewish holidays, my faith has been largely irrelevant to my life. How can an ancient 4,000 years old culture possibly be of use in today's world? Hebrew, what can you use it for? After all, this is the twenty-first century. If I had been born into a Catholic family, Latin would have given me a leg up in pre-med school and beyond, but not Hebrew, and certainly not Yiddish.

I took a philosophy class in my second year at Emory. With no small amount of gusto, the professor stated he is a firm atheist, a real *apikoros* skeptic. I remember thinking at the time, I'm not buying it. To be an atheist I would have to believe man is merely the fortunate end of a naturalistic process of evolutionary mutation and random selection, from the goo to you. No Professor, I'm not ready to rule God out of life's equation. Atheism doesn't fit my view of the world. The professor wasn't finished with us yet. He suggested another alternative those of us who didn't agree with his atheistic view. Since science has yet to prove God's existence, he argued, honest seekers of truth should at least embrace agnosticism. After all, man has no way of knowing whether God exists. I weighed his second argument, but found it equally unconvincing. I realize my viewpoint is politically incorrect, but I'm no lemming. God has never been a prominent part of my life, at least thus far. Perhaps this reluctance is in my genes. Just the same, I will follow my gut instinct and assume God does exist somewhere out there. I realize the question demands a more specific answer. I don't want to stay sitting on the fence for the rest of my life. I just can't deal with it now. The bottom line for me is this; I'm a secular Jewish boy trying to survive in a Gentile world. Maybe someday I will try to understand what it means to be spiritually Jewish, but not today.

I pulled my thoughts back to finishing the form. Yes, I am going to Johns Hopkins, without trepidation. I carefully folded the application, made a copy of my semester grades, wrote out a check for the registration fee, and stuffed everything into their self-addressed envelope. I carefully put the letter back on my desk. I'll mail the completed form later on my way to the commencement rehearsal.

After posting the amount of the check in my checkbook my balance is a little over five hundred bucks. If I am frugal during this last week at Emory, it should be enough to get me to Baltimore for the interview. Another bout of panic is coming over me. Here I am a few days from graduating and I haven't told Dad about my defection from the family plan. I'm angry at myself for dragging my feet on this. I can't postpone discussing it with him any longer. It is way past time to do so. I'll call him today. I'm sure he'll understand my decision once I explain everything.

It's a quarter past nine, but an hour earlier in Omaha. Dad probably is at home having breakfast after his early rounds at the hospital. My heart is pounding as I dial the number and wait. It's ringing. Perhaps he won't answer and I can just leave him a short message. No such luck. Dad's voice is on the other end of the line, loud and clear.

"Hello Son. Nice of you to call. In fact, I was just about to give you a jingle. Your mother and I finished making our reservations for next week. We will be staying at the Omni Hotel at the CNN Center downtown. Emily is coming earlier. In fact, Emily should be arriving just about now. The Admissions Office has scheduled an orientation tour for incoming students this week. They offered to put her up in one of the dormitories, but she wanted to stay at the Omni. I said you will watch out for her. I want you to give her a quick tour of Atlanta. Please keep an eye on her for us. She is convinced she can handle any situation, but I'm not so sure. If all goes as planned your mother and I will be arriving on Delta flight 240 at 10:35 on Friday morning. Can you pick us up, or shall we take the hotel shuttle instead?"

"Dad, I think it would be better for you to rent a car at the airport. Thursday is Class Day and I have a busy schedule all day. I have an appointment with a librarian in the morning. The Senior Class Reception is at seven in the evening, and I plan to spend some time with my friends afterward. I won't be getting back to my apartment until late in the evening. You will need a car to get around in Atlanta. Many of the functions are off the main campus."

"Oh yes, I remember. Your advice sounds good. I will reserve a car at the airport."

"Dad, you should have plenty of time to check in at the hotel and touch base with Emily on Friday. I can meet you there for lunch. Did you receive your invitations to Friday night's Senior Honor's Reception and the Torch and Trumpet Soirée I sent to you? The reception at the Miller Alumni house starts at eight o'clock. The Soirée at the Emory Conference Center isn't over until midnight."

"Yes, we received the invitations and the map. I will rent the car at the airport and check in at the hotel. We will wait for you at the hotel with your sister."

"Dad, I have some concerns about Emily. I won't have time to spend a lot of time with her given my tight schedule. I certainly can't watch over her twenty-four-seven."

"I understand. Do the best you can, Son. At least take her to one or two social events. You know she loves to be the center of attention. Show her around campus when you have time. I assume you already have registered for next semester. Did you look up my friend Dr. John Schroeder yet? He said he would be glad to give you some pointers and help you get settled in at medical school."

"Dad, I haven't contacted Dr. Schroeder, and I haven't registered for medical school either. In fact, the main reason for my call today is to talk to you about this coming semester. I've been thinking a lot about my future, you know, what I really want to do with my life, and the future of the medical profession isn't—"

Before I get any further, Dad interrupts me.

"What? You are still thinking about what to do with your life? What's to think about Jacob? This is not the time to consider changing our plans. Those slots in the Medical School are limited. You need to get registered in the medical program right away. What are you planning, to take a semester off to see the world? Surely not! Do you remember the kid, a couple of years ago, who graduated from Emory and ended up starving to death in the Alaska wilderness? Jacob, your career as a successful physician starts next semester."

"Dad, let me explain. I'm not planning to take any break from schooling. Last semester I took this course in microbiology, and I really liked it. The course is interesting and challenging. The professor encouraged me to look into the emerging field of human genetics. He suggested I might want to become a member of The American Society of Human Genetics located in Bethesda, Maryland. I have been reading the research articles published by the Society. They are just incredible. As a result, I am seriously considering changing direction, enrolling in a school to become a micro-biologist. In fact, I hope to become a human genetics research scientist someday."

"Don't be ridiculous, Jacob. No doubt biologists are discovering many things about the cell, greatly affecting medicine in the future, but this will never replace the sense of fulfillment medical practice brings to you. You work with people, not test tubes and electron microscopes. When you see a baby's life preserved, or a person healed of some chronic disease, you touch the heartthrob of life."

"Dad, I understand what you are saying, but my gifting is not the same as your own. You know the handicap I struggle with. I don't relate to people easily. Besides, I really am interested in this new field of human genetics. It's the basic stuff we're all made of. There's great potential for the development of cures for diseases having plagued humanity for untold centuries. Human genetics research comes close to the medical field. In fact, it's really just a separate part of the same profession. I can't imagine anything more exciting."

"Exciting perhaps, but this won't happen overnight. It will take many years of research before such cures become available. Research must be followed by animal testing, then human trials, and who knows if the drug companies will view the results as economically feasible."

"Dad, several treatments have been approved already. For example, third degree burn victims sprayed with stems cells taken from the patient's healthy skin, heal in days instead of months. With all respect, my bottom line is I am going to change my major to pursue a career in microbiology rather than becoming a physician."

The revelation of my decision is followed by a long period of silence on the other end of the line. I held my breath, waiting for his reaction, wondering what would come next. Finally, Dad broke the silence. His voice is slow and deliberate. There is no mistaking the import of his voice. It is the tone he used whenever I crossed the invisible line between parent and offspring, ruler and ruled. "Jacob, do I hear you saying you are definitely not going to enter medical school next semester? If so, I must say you have made a bad decision, a decision I absolutely do not approve of. I cannot and will not accept your decision."

His response is devastating. It pierces me like a sword thrust into my belly. I realized my decision is not going to fly with him. It's not going to get off the ground.

"*Klog ist mir.*"

"What? Did I hear you say? 'Woe is me?' Son, you need to rethink this whole crazy idea of yours."

"Dad, let me explain."

"Jacob, I don't want to hear anything more from you except your intention to enter medical school at Emory next semester."

"Dad, hear me out. I know this must be a great disappointment for you, but I am convinced it is the right path for me. I do not want to enter medical school next semester. I've done well in my studies here at Emory. In fact, I pulled straight A's this semester. Our family has received not one, but two invitations to attend honors receptions on Saturday. I really think microbiology is my calling and I know I can succeed in this field. By taking this concentration, I will be able to immediately work in this field, to get my Ph.D. without having to spend time in residency as some hospital. Please understand and go along with me on this."

"Understand? I understand you are throwing away a promising career as a medical doctor. This plan is not new to you, Jacob. Our family has discussed it many times in the past. *Saichel*, its only common sense! As far as the intern requirement, I can arrange an internship right here at Creighton. You know I plan to hand my practice to you once you gain the experience. I can't believe what I'm hearing. Do you realize how many young students would give their right arms for opportunity you have? *Zerr schlect!* You are throwing a golden opportunity away? Think about the expectations your family has for you. We have made a considerable sacrifice to send you to one of the best universities in the country, one with a highly respected pre-med program and a medical school par excellence. Just because you have pulled a few good grades as an undergraduate does not mean you can excel as a research scientist. What if you fail to cut the mustard in genetics? Will you then decide to pursue some other career, or will you just become a professional student like so many young people are doing today? Are you assuming we would be willing to continue to support you financially as you bounce from one career path to another, always learning and never accomplishing anything worthwhile? Really, it makes no sense, Jacob. You need to reconsider your plans. Think this over, Jacob. Think it over carefully."

Dad's objection to my decision completely shakes me. Did I hear him say his financial support for a career in genetics might not be forthcoming? Oh boy. Without Dad's financial help, money might become a big problem. I decide to mount one more attempt to win him over.

"Dad, I know we certainly need to be on the same page on this. I fully understand how you have looked forward to my medical career, a son to follow in your footsteps. But I've already given this matter a lot of thought. I know genetics is my calling. I'm sure I have the ability to succeed in this field. In addition to my scholastic record, I have included a list of the honors I have received here at Emory in my application. I also requested the school to send my records to a number of other schools. Dad, there is one school which has already responded quite favorably to me, encouraging me to apply for admission to their program. I will work just as hard to succeed at the graduate level as I have done here in the pre-med program. I need your support Dad. I promise. If you will back me on this, someday I'm going to make you proud of me."

"Jacob, this is not a matter we are going to resolve in a single phone call. We will discuss this decision of yours on Friday and resolve it as a family, just as we always have done in the past."

"Okay Dad, I will meet with you to discuss the whole thing on

Friday afternoon at the hotel. I have to go now. Allen and I have reserved the tennis court for this morning at ten o'clock."

"Very well, Jacob, your mother and I will see you on Friday."

My dad skipped his usual *schmaltzy* goodbye and hung up the phone. I put my cell phone back in my pocket. I stared at the envelope lying on the desk in front of me containing the completed admission form. My earlier confidence in Dad's support is shaken badly. The outcome of Friday's meeting now is highly uncertain. Doubts are assailing my mind. I can't decide what I should do at this point with the letter. Is this a hill to die on? Should I mail it now, or wait until after our meeting? I know what I want to do, but I'm not sure I have enough chutzpah to pull it off, to see it through on my own. Do I dare take this road without dad's support? Where is your life heading, and just who are you anyway Jacob Cahn? What are you here in this world to accomplish? Do you have enough gray matter to succeed as a geneticist? What kind of a brain experiences such weird dreams, dreams which I cannot understand the meaning of? Why wasn't I born into an Orthodox Jewish family? Those *schmucks* know who they are in this world. An Orthodox kid would have the faith to press on. He would know Ha Shem would turn it out for the best.

I began pacing the floor in my apartment. Suddenly I'm doing something completely unexpected. I hear myself praying, praying out loud for help from the One I really don't know at all. I'm praying the only passage I can recall in Hebrew. For my bar mitzvah, the rabbi required me to memorize Psalm Twenty-three, it is the Shepherd's Psalm. For some reason, it stuck in my head all these years. I feel better after praying. Somehow, I know my prayers are heard. I picked up the letter to Johns Hopkins from the desk. I put it right back down again. I'll mail it, but not right now. It is time to forget my troubles, meet with Allen and play tennis. I quickly put on my ragged tennis shorts and shoes, grabbed my bag of equipment, and left for the courts.

2

First Love

"And Jacob served seven years for Rachel: and they seemed unto him but a few days, for the love, he had for her."
(Genesis 29:20)

Love's Challenge

I took the elevator down to the ground floor. In addition to the team practice sessions held during the week, Allen and I usually play a game of singles together at ten on Saturday mornings. Besides sharpening our skills for competition, these Saturday morning games helped both of us release the tensions of the previous week of classes. When I arrive, Allen has two girls in tow, dressed and ready to play tennis. Alice Winters, Allen's girlfriend, is one of them, but the other girl I've never met before. So, this is his big surprise. Alice is all smiles as she introduces her friend to me.

"Jacob, this is Rachel Kaufman. We shared a class together last semester. Rachel, this is Allen's friend I told you about, Jacob Cahn. They both are on the tennis team."

This is quite a surprise. Apparently, Allen and Alice have planned to do a little match-making. I'm not disappointed. This girl is a real beauty. She's about 5 feet 8 inches tall, auburn hair, liquid dark brown eyes, and a slim but muscular figure. She is wearing a neatly pressed pair of yellow tennis shorts with top and shoes to match. I am embarrassed and very uncomfortable. Here I am sporting a scruffy beard, wrinkled tee, and tennis shorts with a rip on one side. Al should have at least warned me to shape up a bit. I manage to flash a quick smile toward the girl.

"Please excuse my appearance, girls. Al didn't tell me we are going to play mixed doubles today."

The girl returns my smile and says, "Think nothing of it. Your grungies don't bother me. We're here to play tennis. Last term, Alice and I discovered we both love tennis. We are here and ready to play."

My angst about meeting new people, especially young females, immediately surfaced. Still her smile is reassuring. I reached out and took her hand. A current of electricity shot up my arm, literally taking my breath away. The look on her face tells me she is feeling the same thing. I don't want to let go of her hand. She blushes slightly and pulls her hand away. No doubt about it, this girl is absolutely captivating. She has completely trashed my gender-avoidance policy in one brief moment.

"Jacob, Alice told me all about you on our way to the courts. She said you and Allen are on the tennis squad. I just love the game of tennis. It seems we have something in common."

"Yes, we do. It's really a great game, especially when you win."

Winning? I would not experience a winning game this day. At least I have an excuse. My partner completely distracted me. I couldn't take my eyes off of her. Several times I missed returning the easiest of lobs. What really embarrassed me were the times when I failed to get my racket on the ball, Rachel effortlessly returned it. Her movements are so graceful. She reminded me of a film in my collection of old movies. I watched it again just the other night. It is about the Serengeti Plains, narrated by James Earl Jones. Yes, she definitely reminds me of a gazelle bounding across the Serengeti. After losing the first set, we took a break. I keep telling myself, Jacob, you gotta be cool. Gazelles keep their distance from young lions. I decide to break the ice by asking about her family.

"Rachel, is your family Jewish like mine?"

"Yes, my family is very Jewish. Our family is Orthodox. My Papa, Samuel Kaufman, is the editor of Atlanta's weekly Jewish newspaper, The Jewish News. Is your family Orthodox, Jacob?"

"No Rachel, I'm not. Our family attends a reformed congregation in Omaha, but not regularly. If you don't mind me asking, how is it you can play tennis on Shabbat?"

"Jacob, you must not understand the deep meaning of Shabbat. Ha Shem gave this commandment out of his love for us. We are to take time from life's daily tasks to recharge our batteries. When I play tennis, I am not working, I'm being restored. Anyway, Papa doesn't have to know everything I do. I lit my Shabbat candle last night and prayed the Modeh Ani this morning. So, I'm covered with the Man upstairs."

"I'm sure Ha Shem has taken your love of the game of tennis into account. I should have taken time to say my prayers this morning. Then we would not have lost this first set."

When all was said-and-done, there was more said-than-done on our side of the net. We lost all five sets. Talk about embarrassment. There's little hope this girl will agree to play another game of tennis with me.

I am quite surprised when Alice says, "Jake, I thoroughly enjoyed playing tennis with you two this morning. I think Rachel enjoyed it as well. Why don't we do this again soon?"

"No wonder you want to, Alice. You and Al won every set. Since you both had such a great time brutalizing me, I'm ready to take you on again next Saturday if Rachel is."

The two girls briefly confer with one another. Even though next week is going to be a busy one for everyone, the girls agreed to join Allen and myself for a rematch on Saturday at ten o'clock. I often wondered, can two people fall in love at first sight. Could it happen during a tennis match? I seem to have fallen in love with this girl Rachel. Is this destiny? Will I spend the rest of my days married to this woman? I am going to marry Rachel Kaufman if it takes every bit of Jewish chutzpah I can muster.

I realize I must take advantage of the opportunity, to immediately move this relationship to a deeper level.

"Rachel, will you be attending the commencement rehearsal at six this evening?"

"Yes, of course."

"I have a car on campus. It's old, but it runs well. Can I pick you up for the rehearsal? This will give us time to get better acquainted. Where do you live?"

Alice interjected, "Jacob, I have already arranged to pick Rachel up. You can meet her at the rehearsal."

I realized my ploy has been shut down, but I persisted. "Alice, this will allow me a chance to see where Rachel lives and give you more time to get dressed for the rehearsal."

Rachel is smiling as she gives me the go ahead. "You can take me home today so you will know where I live. You can pick me up for the rehearsal, but don't be late."

I quickly agree. "That sounds good to me."

Allen is quietly listening to our conversation with a wry smile on his face.

Alice turns to Al and adds, "Well, since Rachel has a way home, perhaps you would like to take me to lunch, Al?"

"Sounds good to me Alice. I'm hungry."

When I turn to look at Rachel she is staring at me, deep in thought. I can tell her wheels are spinning. I wonder, what is going through her lovely head? I'm feeling quite uncomfortable, standing there in my grungy shorts. Then she looks away.

I decide to ask her what she is thinking. "Rachel, you seem to be lost in thought. Is there something you want to share with us?"

"There is something I want to share with you, Jacob. We have just met today."

"Yes. We have known one another for a little over two hours."

Then comes another long pause. She is choosing her words very carefully.

"Jacob, I have a question to ask you. Is it your intention to court me?"

Her question takes me completely off guard and I hesitated, not knowing exactly what to say. My brain is spinning, but I manage to blurt out, "Yes, Rachel, I believe I am."

"Can you give me a more definite answer?"

"Yes, I definitely do want to court you, Rachel."

"Very well then, I feel I should stop addressing you as Jacob and start calling you Jake."

"That's fine by me, Rachel, but I really prefer to use your given name. It reminds me of another Jacob and Rachel."

"That's a nice thought, Jake. You have my permission to call me Rachel. Now, since you are not Orthodox, are you aware of the rules for courting Orthodox girls?"

"I don't recall reading about any dating rules in the Torah?"

"That's true, Jake. The rules are not written down in the Torah. The rules are in the traditions established by the rabbis long ago. Papa, my father, insists I strictly adhere to these traditions. You know, like in the movie *Fiddler on the Roof*, traditions, traditions, traditions. Shall I briefly go over them with you?"

"Sure, fire away."

"Very well. There are four rules Orthodox young people must follow when they first begin a courting relationship. Rule number one: The boy can greet the girl by taking her hand, but he is not permitted to hold it for any prolonged period of time, like you did with mine earlier today."

"I'm sorry, I didn't realize."

"I understand. You are forgiven. Rule number two: A boy does not embrace his girlfriend except on very special occasions, and always with the utmost decorum."

"Such events as graduation ceremonies?"

"Perhaps, but don't count on it Jake. Rule three: Young people of the opposite gender do not dance together. Orthodox girls dance with other girls, and Orthodox boys dance with boys."

I recognized the opportunity to pull Allen's chain.

"Say Al, can I have the first dance with you at the graduation party?"

Al scowls at me as Rachel continues to expound on the rabbis' rules of courtship.

"And finally rule number four: Under no circumstances are boys allowed to kiss their girlfriends until later in the courtship."

I'm realizing courting this girl is going to be a major undertaking.

"And how long is it before we get to this later stage, Rachel?"

I expected to get a smile, but none came. Instead she answers my question with a stern voice.

"Jake, couples must keep these rules until they are formally engaged, until the girl's father has signed their *ketuba*, the formal covenant agreement of marriage presented to him by the prospective groom. Do you think you can keep these four rules, Jake? If so, you have my permission to court me."

I considered it well worth my while to abide by the good rabbis' rules. In addition to the four rules, I learned another thing from this conversation. This girl is definitely a strong-willed, no nonsense kind of gazelle.

I raised my hand and officially swore an oath. "I hereby agree to keep these rules until a later time, yet to be determined, when Rachel Kaufman and I become officially engaged."

The rules agreed to, Rachel said she is ready to go home. As we drive toward her house located at 1520 North Peachtree Battle NW, strong emotions are welling up inside of me. The intensity of my emotions is amazing. My heart is overflowing with love for Rachel Kaufman.

Rachel and I said very little during the journey to her home. Both of us are trying desperately to control the strong emotions welling up inside. I am finding it very hard to keep my oath, to at least reach for her hand. I try to reach over to her, but she moves further away from me, and opens her window to let in the cool afternoon air. I pulled up into the driveway of her home. She once again reminds me to be sure to pick her up on time. Then she quickly exits the car, runs to her front door, and waves goodbye to me. It is plain she is having as much difficulty holding in her emotions as I am. After returning to my apartment, I have just enough time to get a shower and a change of clothes. I picked up my letter to Johns Hopkins. I will mail it on the way back to Rachel's home.

I realize I need to move this courtship forward quickly. I have only a few days remaining on campus. I will take her to dinner at Bernie's Delicatessen after the rehearsal. Bernie's is located just off campus, and all my friends will be there. I will have another opportunity to get to know Rachel even better. If all goes well, I might broach the subject of plans for our future marriage. The thought of a future life with Rachel is making my head spin. All of this is working out so well. Soon I will be engaged to marry Rachel Kaufman. Rachel is waiting for me on the front porch when I arrive in front of her home. I quickly get out and open the door for her. She briefly smiles as she gets into the car. As I pull away from the curb, she is staring straight ahead, saying nothing. I break the uneasy silence, "Hey, girl, why all of this silent treatment?"

Tears are filling her eyes. She continues looking straight ahead. I'm not sure if she is sad or very angry. I conclude she is feeling a lot of both. I remain quiet as she tries to regain her composure. After a few minutes, she manages to share why she is so upset.

"Jacob, this is the worst day of my entire life. I told Papa about our day together and your intention to court me. He said he absolutely refuses to even consider it."

I struggled to find the right words. "Rachel, I know we are meant for one another. Once he gets to know me I'm sure I can find a way to resolve this disagreement with your father. After all, I have not even had the opportunity to meet him yet. I don't know right now how or when, but I'm sure I can win him over. Don't let this problem overwhelm you. When you look back at this day, it hasn't been all bad. In fact, it has been a wonderful day for both of us."

Rachel retreated into silence again. I didn't know what else to say. I continued to drive slowly back to the campus. As we neared the campus, she gradually began to share her thoughts.

"Of course, you are right, Jacob. It has been a wonderful day. We met only this morning, but I feel like I have known you all of my life. I'm certain you are the love of my life. I will always love you, whatever happens. But you do not know my Papa. Unless Ha Shem hits him between the eyes, he will not change his mind about our courtship or his permission for us to marry, not today, not tomorrow, not ever. The painful truth is I have a stiff-necked Jewish father who has made up his mind against our marriage."

"Rachel, I don't know him yet. People can change over time."

I feel the heat of anger rising up in me. I'm hot under the collar.

"Unfair! What's the big problem? Rachel tell me his objections so I know how work things out with him?"

"Papa has his reasons. He will not accept our courtship because … Oh, I can't talk about it now Jake."

She retreats into silence again, leaving my question unanswered. Somehow, deep down, I know she is right. Her father is not going to accept me as his future son-in-law. Our beautiful dream of a life together has just morphed into the impossible dream. It is as if our love boat has run aground onto some rocky shoal, and we're drowning in the sea without life preservers. The analogy suddenly becomes crystal clear to me. These are no ordinary shoals. Our love boat has run aground on the hard rocks of Orthodox Judaism.

Rachel and I arrive at the main campus at five thirty in the afternoon. I parked Old Blue and we walked to the Quadrangle near the Admin Building where the rehearsal for the commencement service is to be held. The area already is filled with students. As I expected, total chaos and confusion reigns. The few faculty members present are trying vainly to gain some control over the excited perspective graduates. Rachel and I have been assigned seats apart from one another.

Before we separate, I asked her to meet me at the north end of the Quad after this is over. Then I made my way to my assigned seat. After several minutes, a semblance of order finally is achieved and the rehearsal begins. Stand-ins for the dignitaries and faculty begin filing in, taking their seats on the grandstand according to the prescribed pecking order. Next, the graduates selected for special honors take their seats in the grandstand. The rest of the graduating class, take their seats in front of the stage. The director for the graduation ceremony introduces the Dean of Students and the crowd quiets down as the Dean begins to speak.

"Ladies and gentlemen, first let me congratulate you on your matriculation from our fine university. At this time, I need to review with you the busy schedule of activities planned for the forthcoming week. I cannot possibly review with you all of the many activities by our various schools and community groups. I must refer you to the information packets you received this evening. Our celebration begins on Thursday afternoon with Class Day. This is for graduates only and will be held in the Glenn Memorial Auditorium on campus. Everyone must assemble at 5:30. Following the ceremony there will be a banquet held at the Emory Conference Center Hotel in town at seven o'clock. There will be shuttle transportation available. On Friday evening, a reception for honors students will be held from eight to nine at the Miller-Ward Alumni House. Honor students, I assume you know who you are by now. Family and guests are included. Following the reception, the Torch and Trumpet Soirée will be held at the Hotel. Again, transportation from the campus will be provided. The Soirée will begin at nine o'clock and continue until midnight. Then on Saturday morning those graduates who have alumni family members are invited to attend the Commencement Legacy Brunch with their family members. This will be held in the Glenn Memorial Auditorium from nine until eleven o'clock. You must assemble by eight thirty. On Sunday, all of you will again return to the Glen Memorial Auditorium for the Baccalaureate Service. The Service begins precisely at nine in the morning. Your attendance for the Baccalaureate Service is required. Please do not be late. Our University President will give the key address. Following the ceremony, there will be a brunch at the Dobbs Center.

Monday morning is the Commencement Service. You are here today to practice this gathering. The Commencement Service will be held right here on the Quad beginning at eight o'clock. All Arts and Science graduates will remain after the exercise to receive their diplomas. Other departmental groups will then be assembling at their respective school locations. I need to admonish all of you of your obligation to arrive on time for all of these scheduled events. Let me again refer you to your information packets for the many other activities being held during these commencement days. I thank you in advance for your patient attention. At this time, we will begin the practice preparation for Monday's commencement service beginning with the presentation of honor students."

The prospect of enduring all of these commencement activities is putting me in a bad mood. I wish they would just hand my graduation certificate to me and free me to take the road to Baltimore. The idea is tempting, but quite impossible with my family coming. The Dean's instructions completed, each of the honor students came forward to the podium and introduced themselves to the graduating class. I walked to the podium, stated my name, and quickly returned to my seat. Next, those students selected to give addresses to the class are asked to come forward and recite a few lines from their prepared talks. As each student finished their talk, attempts at humor bursts forth from the crowd. Allen walked to the podium and quickly said a few words.

As Allen finishes his talk, a clown in the crowd shouts, "Hey Al, I understand there is a job opening for a preacher in Antarctica. I think it would be a really cool opportunity for you!"

Without skipping a beat, Allen replies, "I understand the Southern Cross is clearly visible from there. I'm ready to go wherever God sends me, even to the South Pole."

Painfully, the rehearsal drags on well over two hours. At 8:25 we finally are given back our freedom to go our separate ways. I manage to find Rachel at the north end of the Quad and we walk together to the car. Then I headed for Bernie's Delicatessen.

The Controversy

As we arrive at Bernie's, I see the place is packed. I don't know what I was thinking. My plan to spend time alone with Rachel is not going to happen tonight. We spotted Allen and Alice at a table near the back of the room and they waved us over to join them.

As we sit down at the table I asked Allen, "Well Al, are you considering taking the ministry position in Antarctica?"

Allen's face lights up in a broad grin, "Jake, you never know. It might work out. Alice could teach me how to talk to the penguins down there."

Alice adds, "Well preacher, you will have to learn killer whale language as well since those are the real sinners in the Antarctic."

Rachel looks a bit confused, "Alice I understand you are studying linguistics. Do they also teach you how to communicate with animals?"

I shake my head in disbelief, "Rachel, I can't believe what I'm hearing. You've been watching too many Disney movies? Alice is studying human linguistics, you know, the language Homo sapiens-sapiens speak."

Alice chimes in. "Jake, you might be surprised to learn I know quite a bit about animal communication. Although animals lack the physical abilities to speak as we do, they can and do talk to one another, and they can communicate with us. Many have the ability to express emotion, such as love for instance."

This line of scholarly conversation is cut short by the arrival of Hank Bowman, a member of Allen's church and one of Atlanta's native sons. Hank shares Allen's dream of serving on the mission field as a doctor. He wants to become a pilot with Mission Aviation Fellowship. Hank pulls up a chair and sits down. He launches into a spiel using his best southern drawl.

"Hi y'all. What I have to say will just take a minute or two. I'm invitin' y'all to a party tomorrow night at our church over on the corner of Elm Street and Appomattox Avenue. It starts at seven. We've lined up a groovy band. The ladies of the church are bringing plenty of victuals and drinks, non-alcoholic of course."

Alice says, "Sounds good to me, Hank. Al and I will be there."

Looking at Rachel and myself, Hank asks, "How about you two, Jake?"

Courting Rule number three crosses my mind, the rule against dancing with the opposite gender. I tell him, "Well, Hank, we aren't sure. I wouldn't count on us."

With a shrug, Hank says, "You two will be missing a great party.

If you change your mind, you know where to find us. Oh, by the way Allen, I enrolled for flight lessons yesterday. I start my ground school over the internet next week."

Al tells him, "Well Hank, be careful when you land not to crash your hard drive."

"No way, man! I'll land those birds as smooth as a feather on a still summer's day in Georgia. I'll see y'all tomorrow night."

Hank moves on to the next table. Everyone at our table decides to order the specialty of the house, Rueben sandwiches. Al adds a bowl of clam chowder to his order. As we wait for our meals to arrive, I seize on the opportunity to ask Alice a question.

"I'm not sure how you will take this Alice, but I'd like to ask you a rather personal question. After all, we are all good friends here, right?"

"Sure, go ahead and ask, Jake. If I don't want to answer, I'll let you know."

"Well, Alice, it's a question of some concern to both Rachel and myself. You are Jewish like us, and you and Al have been going together for quite some time. I would guess you must have at least considered making your friendship more permanent, so to speak. I wonder, Al being a Christian and all and you from a strong Jewish background, have you discussed your differences as far as the compatibility of your two faiths?"

Alice is eager to answer my question. "Yes, we have talked about it, Jake. You will be happy to know we don't have a problem when it comes to our faith. We both are believers in Yeshua, Jesus. He is our Messiah. Over a year ago, I accepted Yeshua as my Lord and Savior. I came to understand he is the promised Messiah described in the Torah, the Psalms, and many other prophesies in Scripture."

I am taken back by Alice's answer. What on earth is she saying? It takes me several minutes to digest it.

"I don't understand what you mean, Alice. Did you convert to Christianity? Are you no longer a Jew? Aren't we Jews born Jewish, whatever faith we hold, or lack thereof?"

"Jake, you need to understand. I'm not a *shishka*, someone who has rejected being Jewish, and I didn't, as you say, convert to Christianity. I was born a Jew and I will die a Jew. God chose the Jewish People to be the vehicle to bring salvation to all of mankind. When I accepted Yeshua as my Lord and Savior, I became complete, a completed Jew. You certainly must know Jesus is Jewish, don't you? Jesus gave his life to redeem everyone, Jew and Gentile alike. Jesus broke down the wall of separation between Jew and Gentile and established the New Covenant. Did you know the veil separating the Holy of Holies was torn from top to bottom when Jesus died on the cross? The veil was decorated with Cherubim. No longer do the Cherubim stand guard at the Garden of Eden with their swords drawn to keep mankind separated from their God. No longer do we Jews require a priest to intercede for us with God. Now believers in Yeshua have direct access to God. Al and I are a perfect example of the way the Lord has broken down the wall of separation between Jew and Gentile."

"Alice, this is all new to me. I've never heard the term, a completed Jew. Do the rabbis teach this from the Torah? Is this a new branch of Judaism?"

Alice is smiling. "Well the rabbis do if they have accepted Yeshua as their Lord. In Scripture Yeshua is referred to by the prophets as The Branch. The vast majority of rabbis strongly object to recognizing Jesus as the Messiah. The rabbis have been blinded by their own sense of tradition. They depend more upon the *Midrash*, the rabbinical teachings drawn from the Scriptures and Jewish traditions, than they do upon the actual written words of the *Tanakh*."

Rachel can no longer remain silent. "Alice, we observant Jews do not accept the books written by the followers of Jesus. They are not part of our Tanakh. These writings are not the words of Ha Shem. They are not part of our Holy *Cepher*.

Alice answers, "Rachel, I know your rabbis do not accept the New Covenant books, but why do they overlook the scriptures about Yeshua in your Hebrew Bible? There are many places in the Tanakh identifying Yeshua as the promised Messiah. Does your rabbi ever read Chapters Fifty-two and Fifty-three from the Book of Isaiah, or Psalm Twenty-two to your congregation? I think not."

I can see a violent storm gathering. Sparks are about to fly. Rachel is bristling with righteous rage. Unaware of the nerve she has touched, Alice does not see what is about to come her way.

Alice continues, "Rachel, do you know no holy rabbi, Orthodox, Reform, or Conservative, reads these passages to their congregations because they describe in great detail the sufferings Yeshua bore at the hand of the Jewish and Roman leaders."

Rachel fires back, "Our Rabbi Schneider reads through all of the Tanakh and much more from the Talmud, and the commentaries. He plainly explains *Adonoi's* word to us. What is more, my father reads the Tanakh every day and he has never mentioned any such passages describing Jesus as the Messiah."

It is time to stop them both from going any further. Raising my voice above the two of them I interject, "Ladies, I think we should table this discussion for now. It's getting late. We can continue it at another time."

After my plea, Alice resorts to a more conciliatory approach.

"Perhaps both you and Jake can find time to attend our Messianic congregation before next semester, Rachel. This way you can learn firsthand the truth about Yeshua and what it truly means to become a completed Jew."

Rachel is not placated. She retorts, "My father would never permit me to go to your congregation Alice."

Alice is not willing to back off. She proceeds to suggest another alternative.

"I'm willing to loan you some of my books to help you understand this, Rachel."

Rachel's next response resembled a blast of icy wind on a cold winter's day in Omaha. "Thank you for the offer, Alice, but I will not be borrowing any of your books. My father has an entire library of books on our Jewish faith, including many books describing the evils done to Jews by Christians over the past two thousand years. These persecutions continue today in many countries. I pray the *Shema Yisrael* three times each day, declaring the Lord is One, baruch Ha Shem. Perhaps you do not understand, Alice. Our Scriptures state plainly. Ha Shem is One, not three as Christians believe. As far as that man being born of a virgin, I don't think so."

Allen did not weigh in on the matter. Rachel's fiery response abruptly ended any further conversation on the subject or any other topic. There followed an excruciating few moments of uncomfortable silence. It's clear the evening has ended in a major disaster. It is over!

Glancing at my watch I tell everyone, "Wow, ten o'clock. Rachel, we have to go."

Everyone nods their agreement and we all say our goodbyes. I paid our bill. Taking Rachel by the hand, we head for my car. As I pull out of the parking lot onto the boulevard, Rachel says, "I'm sorry Jake, but I just could not sit there and listen to Alice spouting off about her new-found Christianity. I hope you understand."

"Yes, I understand, Rachel. Let's set it aside for now. I'm sorry I ever began the conversation. Let's just forget the entire thing. I do need to say this to you. I don't think you gave Alice much of a hearing. After all, she was just sharing her new faith with us. My logic professor shared something I believe is very useful in situations like we experienced tonight. A person's world view most always overrules the conflicting facts presented to them, even when those facts are clearly true. Ha Shem encourages us to come to our studies with a blank slate and with an inquisitive mind. Pre-conceived positions set in concrete condemn us to remain in the status quo."

"You don't just reject your moral upbringing because someone suggests something new, Jake."

"I'm not asking you to accept her ideas. I'm just saying you need to realize she is sharing a position she strongly believes to be true. We all need to have the freedom to do that and the grace to listen to others."

"Thank you for your unswerving support, Jacob."

"Rachel, I just want you consider logically what others share with you. Can we agree to do that?"

"Okay, I will try to be more understanding in the future."

Although my words calmed the storm, I instinctively knew the words spoken tonight had eroded a deep chasm between the two girls. Neither of us said anything more for several minutes. I decided to bring up the problem of Papa Kaufman again.

"Rachel, what possible strategy do you believe can I best use in winning your father over? How are we going to convince him to approve of our courtship and sign our ketuba? I really do not know how to proceed?"

"Jake, I don't know. This has been a terrible experience for me. I just don't know what to suggest to you."

"Well for starters, we need to find out exactly what your father's objections are. I need to know exactly why he doesn't want us to marry. I want to schedule a meeting with him right away."

"Jake, meeting with him won't solve anything. He is no match for you. He'll probably bite your head off and hand it back to you on a silver Seder plate. He might even politely ask you to leave our home and never come back again."

"Perhaps so, but I still have to meet with him, even if it is painful. Do you understand?"

"Yes, I understand. Since you insist, I will ask Papa to meet with you. Just remember, I warned you. It is not going to be pleasant. What day is best for you?"

"I would like to meet with him as soon as possible."

"Well, Papa is very busy during the week. Would you want to meet with him tomorrow morning?"

"Sunday? Yes, I can meet then. How early does he start his day?"

"Papa is up very early, usually by six or so. Would eight be too early for you?"

"No, I can meet with him at eight o'clock."

"Okay, I will ask him when I get home and let you know. Is there anything else we need to discuss?

"Yes, Rachel, there is. I need to be perfectly honest with you. I have something important I need to share with you."

"Oh, so now you are going to tell me you are already married."

"No, I am not married. I need to tell you I may be a bit of a fachadick, a nuthead. There is this weird dream I experience from time to time. The dream is persistent. I first began experiencing it when I was very young, about five or so years old. It has caused me to doubt my sanity at times. I have never shared this with anyone, not even with my family. You need to know all about it before we agree to marry."

"Okay. So, tell me about this dream of yours."

"As I said, I never know when it will show up. The dream always begins the same way. As I close my eyes and fall asleep, I have the sensation of falling, falling into some sort of dark passageway. I sense my soul is leaving my body. I seem to be awake, but not really. I am still asleep, looking down at my body from the ceiling of my bedroom. I begin to bounce along the ceiling like a helium balloon. At this point I feel no real anxiety. In fact, I rather enjoy looking down at my still form lying on the bed below."

"I have never experienced an out-of-body experience. I believe many of the prophets experienced similar dreams when Ha Shem needed to get their attention."

"Yes, I remember the story of Job when Ha Shem spoke to him out of a whirlwind. Getting back to my dream, this first short peaceful interlude suddenly changes. I see a flash of light. Fear now washes over me in successive waves. I realize there in no escape. I try to force myself back down into my body again. Sometimes I am able to do it, but usually I cannot. Last night, the dream advanced to the next totally frightening phase. I am catapulted through the ceiling into another dimension. This is when I really begin to panic. Then I am standing in front of a stone wall, covered with vines. I try in vain to gain some measure of control over the dream. I attempt to look downward, to see myself lying on the bed again. I realize I am a prisoner in this strange place. There is a small door in the wall. Above the door is a sign. It reads "Arbeit Macht Frei." Do you know where those words have been seen before? They were written over the gate of the Auschwitz death camp. Is it my sentence of death? Then slowly the doorway opens and I can walk into a garden. Like the abandoned garden in the movie *Secret Garden*, it is filled with gnarled trees, tangled fallen branches, tall weeds, and half-dead plants, remnants of a vegetable garden. The gnarled limbs of trees above me seem to be looking for an opportunity to crash down and bury me in the ground below my feet. Unlike the movie, I sense this dark garden is home to grotesque and evil entities. Like the pictures in a horror film, I see all sorts of demonic creatures. I'm going to spare you their description. The most terrifying fact of this dream is these entities are alive. Yes, alive! I grab a dead limb off the ground and frantically attempt to defend myself. Just when I am convinced there is no hope of ever returning back to my bed, a voice rings out, echoing off the stone walls of the garden. The voice proclaims, "It is finished. Time is no more."

"The effect of this pronouncement is immediate. All activity of these apparitions suddenly ceases. Their forms seem to melt away, as if they are being sucked down into Hell's bottomless pit. I slowly float down, passing through the ceiling of my room and into my body again. Then I wake up from my dream."

"This is really bizarre Jake."

"It's more than bizarre. So, there you have it. I'm not sure you want to marry this crazy dreamer, Rachel. If not, I can understand."

"First of all, Jacob, I think you are perfectly sane. You may be suffering from an attack by the enemy of your soul. Your dream sounds very much as if you are experiencing something called 'hypnagogic state' where the mind does strange quirks as you fall asleep. Does this always occur right after you go to bed?"

"Yes, it always comes then."

"I'm impressed. I've read about the hypnagogic state.' Are you familiar with this phenomenon?"

"Yes, but just knowing the term doesn't help much."

"I have to agree with you. Jacob, you should not keep this a secret any longer. You need to talk to someone about the dream."

"Do you know someone who might be able to help me learn about this affliction?"

"Yes, matter of fact I do. A good friend of my father is the head librarian at the Woodruff Library on campus. His name is Dr. Michael Stein. He has spent many years studying such unusual things. You need to take time to pay him a visit."

"I shall as soon as I can. Commencement begins with Class Day next Thursday."

"You might just stop by the Library during one of your morning runs and get acquainted."

"Yes, I will. In the meantime, I'll wait for your call."

"One more thing, Jacob. I do not want to play tennis with Alice next week. I'm sorry, but I can't handle it."

"I can understand how you feel. Perhaps sometime later you two can manage it."

"Perhaps, but I rather doubt it."

That's the way we left it as I pulled up in front of her home. Rachel said goodnight, quickly jumped out of the car, and bounded up the front steps. There was no opportunity for me to see her to the door. Driving back to the campus I thought about the events of the day. The letter! I checked my pocket. Yes, it is still there. I stopped at the post office and dropped it off. The pick-up would be at ten on Monday morning. I made a mental note to contact three people to furnish references to send to the genetics department at Johns Hopkins as soon as possible. I arrived quite late at my apartment, mentally exhausted. Rachel left a message on my cell. She says her father is able to meet with me at 9 o'clock tomorrow morning. It is too late to return her call. I will check back with her first thing in the morning.

I cannot get to sleep. I toss and turn for what seemed like hours, thinking of the myriad challenges I will face in the coming weeks. Mostly I wonder if Rachel's father ever will agree to our engagement. What are his objections? It is after two when I manage to fall asleep.

3

An Uncertain Future

*"For I know the thoughts that I think toward you, saith the Lord.
Thoughts of peace, and not of evil, to give you an expected end."*
(Jeremiah 29:11)

Wisdom of a Jewish Father

The sound of the carillon calling the faithful to morning service woke me out of a sound sleep at eight o'clock. It is Sunday and I felt great. It has been several days since experiencing my dream. Perhaps my dream is a thing of the past. I showered and dressed. From my almost bare cupboard I pulled out a box of Wheaties and made a cup of instant coffee on the stove. Then I made a quick call to tell Rachel I'm on my way to meet with her father. Driving to her home I rehearse how best to greet this man. I scour my brain trying to find a strategy to change the mind of Papa Kaufman. The more I think about it, the less hopeful I am. I arrived at the home five minutes early. I don't want to appear too anxious, so I parked the car around the corner and waited. I turned it off my cell as well. I don't want to be distracted by a phone call during our meeting. My mind is still frantically searching for a winning solution to this problem. At two minutes before nine, I started the car and pulled up in front of the Kaufman home. I quickly walked to the front door and rang the bell. This is my first time to set foot inside Rachel's home. I wondered … will it be the last? There is an elegant *mezuzah* hanging at an angle next to the door. It contains the Shema Yisrael verse from Deuteronomy. The tilted mezuzah reminds all Jews of the chaos in the world, a chaos which will remain until the return of the Messiah. I kissed my fingers and touched the mezuzah. This is the traditional way of blessing a Jewish household. Hannah Kaufman, Rachel's mother, opens the door. Yes, I remembered briefly seeing Mrs. Kaufman one time before, the day I took Rachel to the commencement rehearsal. The woman standing in front of me is short and slightly overweight. The smile on her round face immediately tells me I am a welcome guest in her home. There is nothing pretentious about her. She is wearing a plain house dress. She impresses me as completely outgoing, warm, and friendly. Perhaps she may prove to be a potential ally. I certainly can use one.

Mrs. Kaufman extends her hand to greet me, "Hello, Jacob. Exactly nine o'clock. My, aren't you the punctual one."

I took her hand. "Yes, I always leave a little early to be sure I arrive on time. You never know what the traffic will be like here in Atlanta."

"Yes, that's true. Come in, come in. Welcome to our home."

As I enter the entryway I catch a glimpse of Rachel, handkerchief in hand, bounding up the stairs. No doubt she is taking refuge in her room. The home is elegantly furnished. A large crystal chandelier hangs over the entryway. The great room has a high vaulted ceiling. As I enter the great room, there are a number of oil paintings on the walls, many depicting historical events from the history of Atlanta and the Civil War Period.

Mrs. Kaufman asks, "Will you be staying to have lunch with us later, Jacob? I have a chicken in the oven."

"Oh, no thank you, Mrs. Kaufman, not today. I must say, your home is very elegant and the historical paintings are interesting? Atlanta has such a rich history."

"So, you recognize the subject of the paintings. I so enjoy them. The city's history is very interesting to me."

"I am somewhat of a history buff myself. I enjoy taking a few side trips around the city when I have time. I see a Jeremiah Leavitt has painted most of your paintings."

"Yes, Jeremiah stayed with us last year while teaching an art class at the Jewish Studies Center on the campus. He now lives in New York. He is becoming quite well known as an artist, specializing in American history. He intends to open an art gallery there soon."

I wondered ... just what is Rachel's connection with this successful artist, Jeremiah Leavitt? I will need to find out more about him, that's for sure.

"Mrs. Kaufman, the design of your home resembles an antebellum home. Is it actually an original?"

"Oh, my gracious no. Unfortunately, General Sherman made the decision to have his army destroy many of the antebellum homes in Atlanta during his Atlanta Campaign. Those mansions the Union Army missed, soon disappeared with the rapid growth of the city. No, our home is only about 30 years old. I am a member of The Atlanta Conservation Agency. We have managed to save only a half-dozen homes. We can discuss Atlanta's history at another time. I believe my husband will be free very shortly. Can I get you a snack or something to drink before your meeting?"

"I appreciate the offer, but no thanks, Mrs. Kaufman. I'll just wait here in the great room until Mr. Kaufman has time to see me. I can reschedule for another day if he is too busy this morning?"

"He's just putting his editorial to bed for tomorrow's paper. He will be free very soon. In fact, why don't we check up on him and see if he's finished. He is in his study."

Mrs. Kaufman leads me to the study. Knocking on the door she asks, "Sam, have you finished the article? Your young visitor, Jacob Cahn, is here to meet with you."

"Yes, Hannah, I'm finished. Show him in."

Mrs. Hanna Kaufman opens the door and I enter his study. Hannah excuses herself and closes the door behind her. I am now alone facing this man for the first time. Mr. Kaufman is seated behind a large oak desk, engrossed in reading from the Atlanta Constitution, Atlanta's daily newspaper. He lays the paper down and stands up to greet me. His appearance is what I expected. He is a short man, in his forties, sporting a full beard and wearing horn-rimmed glasses. He has a yarmulke on his *kop*. He seems to have an honest face. I detect a bit of a twinkle in his eyes.

"Ah Mr. Cahn, I see you're very punctual. Please, pull up a chair son and sit down."

I tried to voice the greeting I rehearsed earlier in the morning, but despite my best efforts, not one word comes out of my mouth. I cleared my throat and tried again. Still nothing. Mr. Kaufman sits back down and continues to read his paper. I placed a chair in front of his desk, and also sat down. There is a lump in my throat. A sense of foreboding begins to well up inside of me.

Mr. Kaufman looks up. He points to the paper on his desk and says, "I am trying to understand why this paper gives so few lines to our charitable events, Mr. Cahn. We held a benefit for the Children's Center yesterday, and they gave it only ten lines. Not much of an endorsement would you say?"

"No Sir. Certainly not."

"So, let's get down to business. I assume you have something rather important to discuss with me."

"Yes Sir, I do."

"Perhaps I already know what you want to discuss with me, the possibility of a courtship with my daughter. Is this the reason for your visit Mr. Cahn?"

"Yes Sir. This is the matter I have come to discuss with you this morning."

"You know, Mr. Cahn, marriage is like a three-ring circus. First comes the engagement ring, then the wedding ring, and then the suffering. But don't get me wrong. I love my wife. I even still hold her hand. Of course, when I let go, she goes shopping."

The previous twinkle in his eyes is now joined by a broad smile. His humor succeeds in breaking the ice.

I manage to recall a similar quip. "My father says when he asked for my mother's hand in marriage, he didn't realize it would always be in his back pocket."

"Yes, this has been my experience as well, Mr. Cahn."

I am discovering the fact this man is quite human. We both share an appreciation for Jewish humor. Encouraged, I decide to begin the presentation of my case.

"Sir, I appreciate this opportunity to speak with you and I will try not to take too much of your valuable time this morning."

"Well, don't feel pressured. I have ample time to discuss your concerns. Take as much time as you need, Mr. Cahn. Since we may be discussing some personal matters, can I call you by your given name, Jacob?"

"Yes Sir, please do."

"Jacob, let me begin by saying I am very aware you and my daughter Rachel already are quite good friends. I believe you first met her on the tennis court just last Saturday?"

"Yes Sir. I have known her for just eight days."

"You met on Shabbat, no less."

"Yes Sir. It is amazing how quickly we have become close friends. As far as the Shabbat issue, I must explain it wasn't my idea to meet her on a Saturday."

"There's no use wasting our time discussing *bupkis*, these small things. I understand your father is a physician in Omaha and teaches at Creighton University."

"Yes sir. He has his own practice and also is on the staff at the Hospital. My father, Joel Cahn, has roots in Georgia. He grew up in Valdosta. His father was also a doctor. Dad attended Emory University and completed his internship in Buffalo, New York. My mother Rose's grandfather, Ernst Stein, was born in Germany and died in the camps. His wife Hilda and her young daughter Rose, my mother, emigrated to Buffalo after the war. Hilda worked on the janitorial staff at the hospital where Dad interned. She introduced him to her daughter and my mother Rose. My parents moved to Omaha where Dad established his practice. I have a younger sister Emily who will be attending Emory this Fall. I attended public school. I did briefly attend Hebrew School long enough to pass my *Bar Mitzvah*. Our family attends a reformed synagogue in Omaha. I have done well here at Emory. I will be attending the Honor's Reception next week. I guess it about sums up my life to date."

"Yes, you have done well, Jacob. Not every student gets invited to the Honor's Reception. Rachel did well in school, but not quite as well as you have. I think it's time to get down to business, to address the reason you are here today. I wonder, do you know the Hebrew meaning of your name, Jacob?"

"I believe it means a supplanter."

"That's correct. Jacob, are you a supplanter? Do you wish to supplant my relationship with my daughter?"

I now have discovered the source of Rachel's directness. She has her father's DNA for sure. A cold chill runs up my spine. My earlier composure, such as it was, quickly evaporates. I'm unsure of the reason for his last comment. Is he simply outspoken, or is he attacking my character? I certainly don't want this meeting to erode

　　　　　　　　　　　　　　　　THE DREAM GENE

into a war of words. Still I need to insert a bit of resistance to the man's question. I decide to give him the benefit of the doubt and assume he is merely testing me.

"Sir, no way do I intend to supplant your relationship with your daughter. As far as the implications of my name, I must admit I do wrestle with Ha Shem at times. I seem to remember Jacob's father-in-law Laban was a difficult man to deal with at first, but they finally overcame their differences. I believe Jacob and Laban ended up making a covenant together."

The twinkle remained in his eyes. I first thought Mr. Kaufman received my response favorably, but his next comments gave me pause to consider just what kind of situation I am facing.

"I believe the Torah does not give us the end of the story, Jacob. My name is Samuel. He was the one who heard from Ha Shem and received wisdom. Perhaps you should be concerned with the intentions of a Mr. Jerome Leavitt. He could present an even worse problem than Eli's sons for Samuel."

I set the Leavitt comment aside for now. I need to assure this Samuel of my strong commitment to preserve the unity of his family.

"Sir, I have absolutely no intention of causing any breach in your relationship with your daughter. In fact, when Rachel and I discussed our relationship this morning, I told her the very same thing. I can assure you I will never come between you and your daughter."

"Jacob, your decision to meet so promptly has made a favorable impression on me. It is plain you are willing to wrestle with the issues in life head on. Perhaps we can come to some common understanding. Do not let this opinion of mine raise your hopes too far. At this point I am not going to change my mind. I definitely am not willing to sign a ketuba for the two of you. If I do decide to sign one, it will not happen any time soon. Still you display a refreshing degree of wisdom for a lad of your few years and I appreciate the difficulty my decision is presenting to both of you. The bottom line is this, Jacob. I do not believe either of you are ready to enter into an engagement."

"I understand what you are saying Sir, and I certainly welcome this opportunity to fully discuss this matter with you. I never expected to fall in love with a woman so quickly. My bottom line is this; I am hopelessly in love with your daughter. I don't have adequate words to express it to you. We both feel we have known one another, not for a few days, but forever. I am not Orthodox, but I have agreed to faithfully follow your rabbi's rules for courtship. I'm not even averse to consider changing my religious persuasion and embrace Jewish Orthodoxy."

"Jacob, you must realize this is my only daughter and I also love her dearly. I want the best for her, to see her enjoy a happy and fulfilling marriage. Perhaps I should list a number of my concerns. How do you as a graduate student intend to provide for her? What would you two do if she became pregnant while you both are in school? I understand you intend to attend Johns Hopkins in Baltimore, and she has entered the graduate archaeology program here in Atlanta. How would you deal with living apart from one another, you in Baltimore and she here in Atlanta? But these issues are of minor importance. My greater objection lies elsewhere."

I am surprised by the conciliatory, even compassionate tone in his voice as he lists his objections. I wondered what his ultimate objection could be?

Mr. Kaufman continues. "Setting aside these more mundane issues, I feel inclined to share a bit of Jewish wisdom with you, if you have time to listen."

"I am more than willing to hear whatever you want to share with me, Sir. Take all the time you need."

"Jacob, the Hebrew language as Ha Shem expressed in the Torah is wonderfully rich in meaning. It never ceases to amaze me what a person can find when one devotes himself to its study. Let me begin by asking you a simple question. Are you at all familiar with the Hebrew language?"

"Not really Sir. As I said earlier, I did not fully apply myself in Hebrew school."

"Jacob, the story I am going to tell you helped me in my own life, and perhaps it will help you in yours as well. Let me begin by stating that the lifeblood of every marriage is the union flowing from the spiritual covenant sealed by the Almighty, baruch Ha Shem. This spiritual component is the basic foundation of any successful marriage. Do you possibly recall the Hebrew words for a man and a woman?"

"No Sir, I do not."

"The Hebrew word for man is *iysh*, spelled aleph, yod, shyn, and the word for woman is *ishah*, spelled aleph, shyn, hei. When you examine these two Hebrew words for man and woman you find they share two letters in common, the alef and the shyn. The name for man has an additional letter, the yod, and the name for woman contains an additional letter, the hei. Together these two Hebrew letters, yod and hei, have a special meaning, a kind of shorthand if you will. Do you know what they signify, Jacob?"

"No, Sir, I do not."

"Let me enlighten you. They are the abbreviation for Ha Shem's personal name, the most holy name of our Lord, baruch Ha Shem. We never pronounce the personal name because it is so holy. The original Hebrew did not use letters, points or marks for vowels. In fact, it has been so long since the name was pronounced no one is certain how it should sound. This is why we use the term Ha Shem. Translated into English it means 'The Name.'"

"I do remember we never speak or write the personal name of Ha Shem."

"It is true. So, once you remove those two letters from the names of man and woman, two identical Hebrew letters remain, the aleph and the shyn. They spell *esh*, the Hebrew word for fire. Our rabbis have interpreted this as referring to the passion existing between a man and a woman. No doubt you are experiencing some of this passion in your relationship with my daughter Rachel. Now we are getting into the heart of my story Jacob, and it is this. The fire of passion is not strong enough to base a marriage upon. A marriage between a man and a woman, devoid of this spiritual dimension, all too often crashes and burns. In other words, without the yod and hei, the spiritual component in a marriage relationship, there will be no lasting union. Do you understand what I am telling you, Jacob?"

I paused to consider the impact of his message. It is plain what Mr. Kaufman is telling me. Without a common faith, my marriage with Rachel will not last.

I quietly answer him, "Yes, I understand the point of your story, and I appreciate the wisdom you have shared with me. I do understand I must take a hard look at the spiritual foundation our relationship rests upon. This is a hard thing for me."

"Yes, Jacob, it is. You must examine carefully the basis of your relationship with Rachel and with Adonoi. As you do so, I am sure you will arrive at the right decision."

"I promise to give this my careful consideration, Sir. You can consider it an oath."

"Well, no need to erect a pile of stones. Your word is enough. There is another bit of advice I want to give to you. As the son of a medical doctor, you must know wounds take time to heal. If we reinjure a wound, it takes an even a longer time to heal. To use this analogy, I think the two of you need some time for your fresh wounds to heal. It will be wise for you to break off contact with Rachel for a period of time. Perhaps six months or so might be necessary. Do you think you can agree to such a time of healing Jacob?"

"Yes, I believe we both need some time to heal. I don't see how I will be able to break off contact with her right away. First of all, I need some time to discuss these matters with her, plus all of the graduation ceremonies will involve both of us. I will need some time to sort things out before I leave for Baltimore."

"Yes, of course, Jacob. Take some time to talk these matters over with Rachel. When we first meet a special person, we feel like we have known them all of our lives. We are sure we know everything about them. You have expressed just such a feeling. Later, we often find ourselves saying *vos ist dos*? In other words, just what is going on? It takes time to really know another person. I have a humorous story to tell you. There was a *shadchen*, a matchmaker, who goes over to a *buchur*, a student at the yeshiva school and asks him if he wants to meet a girl. The boy says he's not interested. But the shadchen says 'she's beautiful!' 'Really?' says the buchur. 'Yes, and she's rich too.' 'Really?' 'And she's from a fine family?' 'Sounds great,' says the buchur. 'But why would a girl like her want to marry me? She'd have to be crazy.' 'Well,' said the shadchen, 'You can't have everything now, can you?' Now I'm not saying my daughter is crazy, but she is complicated."

"Sir, I realize your daughter is, as you say, complicated, but aren't they all?"

"Yes, that's true. Women are not easy for us to understand. Do you feel we have covered what you came to discuss with me today, or are there some other things we need to address?"

"I think we have covered everything I needed to discuss with you. Thank you for your time and advice. Again, I promise to seriously consider your suggestions."

Smiling, Mr. Kaufman rises, reaches over his desk, and shakes my hand. "Very well, Jacob. That's all I expect from you. I wish you every success in your future endeavors as you pursue the path Adonoi has chosen for you, baruch Ha Shem. And now, unless you object, I would like to pray the Priestly Blessing over you. It is taken from the Bemidbar, Chapter Six and verses Twenty-three to Twenty-seven. Perhaps you can recall the better-known name for this fourth book of the Torah, the Book of Numbers?"

"I believe the *Bemidbar* also is commonly referred to as the Book of the Wilderness."

"Yes, it is known as the Wilderness Book. Shall I pray this blessing over you in Hebrew as well as English before you leave?"

"Yes, I would greatly appreciate your blessing, Sir."

I watch as Mr. Kaufman raises his hands over his head, and makes the sign of the letter shyn, signifying the name of the Almighty. He begins to sing the blessing over me in Hebrew, repeating it in English.

"May the Lord bless you and keep you. May the Lord make his face shine upon you, and be gracious unto you. May the Lord turn his face toward you and give you peace. Amen."

As I leave the Kaufman home, I feel like a heavy weight has been lifted from my shoulders. Still, my change in attitude toward Mr. Kaufman is a bit puzzling to me. In some mysterious way, my earlier resentment toward this man has changed. The words of this Jewish sage have the ring of truth in them. Has this meeting birthed something quite profound, perhaps a spiritual bond between us? We each hold very different views of the world. We do share one thing in common. It is a love for Rachel, and a desire to cherish and protect her from the disappointments of life. This is something I can hang my hopes upon. Receiving the Priestly Blessing, the raising of the hands, also is very special to me. It seems enough for now. I am beginning to understand the reality of Sam's objection to our marriage, our different spiritual backgrounds. I realize we will not resolve this situation overnight. That said, ending contact with Rachel, even for a short time, is hard to imagine. What did he say, six months apart from Rachel? The level of pain will be a 'ten' for both of us. Also, what about my strange recurring dream? Will she decide our problems are too great and I am too weird to marry? I do not know the answers to these things. I drove Old Blue slowly back to the campus turning them over in my mind.

The Decision

I arrived back at the campus at twelve noon. I decide to have lunch at the dorm lunchroom. As I enter the game room, Allen Schaeffer and Terry Johnson are sitting at a table, talking. Terry looks up and says, "Hi, Jake. We are just hanging loose this morning."

Allen nods in agreement and asks me, "Have you talked to Rachel this morning?"

"I just returned from talking with her father. We talked for over two hours. I told him about our desire to be engaged."

Pressing me for more information, Allen asked, "Has he agreed to allow you two to marry Jake?"

"Well, not exactly. He has not ruled out a future approval of our engagement, but he feels Rachel and I have to work out some issues first."

Terry asked, "What kind of issues, Jake?"

"Guys, I don't feel I should mention the issues we discussed this morning. I haven't even had time to discuss them with Rachel yet. I hope you understand."

No doubt Terry is thinking of his own similar situation. "We understand, Jake. Given time, I'm sure you both will find a way."

Allen turns to Terry and asks him, "Terry, did I tell you Jake is tentatively planning on changing his major? He doesn't want to go to medical school. He wants to enroll at Berkeley or Johns Hopkins in Baltimore. He is planning to become a microbiologist, a geneticist no less."

"Wow, what a surprise. Have you made a decision on which school you want to attend, Jake?"

"Allen, Berkeley is out of the picture. I am going to Baltimore."

Terry is excited. "Jake, do you know I grew up in the Baltimore area? My father's thoroughbred farm is located in Timonium, just north of the city."

"Really, Terry. How did you end up down here in Atlanta?"

"Well, my high school girlfriend liked the look of the Emory campus."

Allen asked, "I haven't heard you say anything about a girlfriend, Terry. What happened?"

Terry hesitates before he answers, "She informed me she has found someone else, Al."

I can see the hurt in his eyes. Now I know why he earlier felt so sorry for me. I did my best to ease his pain. "I'm sorry Terry. This must have been very painful for you."

"Yes, but I'm okay with it now. As the saying goes, there are more fish in the pond. I'm going to return to Baltimore and enter vet school at JH. Say, if you have some time, let me get a map of Baltimore from my room. I'd like to clue you in on a few things about 'Ole Balmore.'"

"Sure, Terry, but I have only a few minutes. Make it quick."

Terry left the room and returned shortly with two maps, a highway map of the US and a map of the City of Baltimore. He spread the US map out on the pool table.

Pointing to the map Terry says, "Jake, the best route for you to take is Interstates 20 and 95 up to Baltimore. It's a trip of 675 miles, and it will take you a whole day to get there. My advice is for you to leave early and watch your speed, especially in Virginia. The cost of a traffic ticket in Virginia is obscene. Interstate 95 no longer is a toll road in Virginia, but you never can tell what the politicos may do next. I would be sure to take some change along with you just in case."

Then Terry spread out the map of Baltimore and began pointing out various locations to me.

"The Medical School complex has one location here on Broadway next to the Hospital and another on 33rd Street on the east side of town. Students also attend classes on the Homewood Campus located on North Charles Street located a bit north of the CBD. The design is interdisciplinary and works well. Most undergrad students change their majors at least once. The McKusick-Nathans Institute of Genetic Medicine is located in a new building right near the hospital, the Broadway Research Building. Everyone refers to this main building as the BRB. Medical students usually rent rooms in one of the brownstone row houses on the streets south of the East Campus. Baltimore is a swinging city at night. East of the CBD and the Orioles Stadium is where the night life is, but I don't suppose you will have time for partying."

"No, probably not Terry."

"Now if you want to find a place in the Jewish section of Balti more, you have to travel to the northwest side of town, to Pikesville. There's quite a large Orthodox Jewish population there. Pikesville is located right next to the Pimlico Race Track. We stable horses there."

"Terry, I won't be frequenting any race tracks either. I plan to spend my time outside of class making some money, possibly working as a lab assistant in the Biology School."

Terry seems disappointed. "Knowing you, I'm sure your nose will be on the grindstone, but you really should take time to at least see the Preakness Stakes in the Spring."

I don't want to deflate Terry's expectations. "I probably can take time to go to the Preakness. Perhaps your dad can point out a winner for me."

Terry is lost in thought for a few moments. "Jake, it suddenly occurs to me that your plan to become a geneticist might just come in handy. Who knows, perhaps someday you will find a way to help the racing industry. The biggest problem my dad and other horsemen have is dealing with the weakness in the legs of thoroughbreds. Their lower legs are too fragile to stand up under the grueling strain racing puts on them. It's a major problem. When horses break down they often lose the will to live and must be put down, disposed of. The strain also causes them to bleed, or hemorrhage inside."

I doubted I would be working on thoroughbreds, but I didn't want to discourage Terry. "You never know what microbiologists will learn next Terry. This is an exploding field. Perhaps a solution will be found someday for such thoroughbred ailments. I'll keep the problem in mind. I believe I read somewhere artificial insemination is not allowed when breeding thoroughbreds. This would rule out any altering of their DNA at the germ level."

Terry nodded, "That is true, but if a way was to be found to genetically solve the problem, I think that rule could be changed."

A suggestion came to my mind. "Terry, the Department of Animal Sciences at Cornell in Ithaca New York has an on-going research program into proteomics and transgenic animals. You might want to have your dad contact them."

I can tell from the frown on Terry's face he has no idea what I just mentioned. "Terry, proteomics is simply the study of proteins in the cell, and transgenics is genetic research crossing DNA across species in animals or plants. Both fields attempt to introduce improvements in the genome. For example, one application might be to strengthen the legs of thoroughbred horses."

Terry took a piece of paper and wrote down the name of the research group, then shook my hand and thanked me, "This is great. I really appreciate the information Jake. I had no idea these research programs are doing work like this. I'll have Dad look into it. This new major of yours really sounds exciting."

Terry reaches into his wallet and hands me one of his dad's business cards. We all exchanged email addresses and agreed to keep in touch with one other.

Terry adds, "I'll send you an email after you get settled in at graduate school. Remember, Timonium isn't very far from the center of Baltimore. You're welcome to come visit the farm any time. I plan to return home right after graduation."

"Thanks, Terry. I'll be sure to look you up when I get to Baltimore."

After lunch, I returned to my room. I gained a lot of useful information from Terry about 'ole Balmore,' and it took my mind off my immediate problems for a little while. I began to look forward to my visit with the Genetics Department at Johns Hopkins. I even felt a fresh sense of confidence in my ability to take on a career as a human geneticist. The problem still weighing on my mind is the dream. I decide to pay a visit to the Woodruff Library soon, but first I need to call Rachel and give her the results of my meeting with her father. I am about to pick up the phone when it rings. It is Rachel.

"Why haven't you called me, Jake?"

"Rachel, I just returned from lunch with Al and Terry."

"So how did the meeting with Papa go? Papa wouldn't tell me anything."

"I believe the meeting went well.

"When I asked Papa how the meeting went, he just said, 'We had a meeting of minds. We understand one another.' Do you know what this means, Jake? I don't. He did say he enjoyed meeting with you. He said you are a bright young man. That's all I could get out of him. What's the bottom line? Are we soon going to be engaged, or not?"

"Well, Rachel, as I say, our meeting went well. We talked for almost two hours. It was a good conversation. In fact, I learned a lot. I hope you know we both love you very much and want the very best for you."

"I know you both love me. Give me the bottom line. Is Papa in favor or our courtship, or not?"

"Although your father did not give his approval for our engagement at this time, he didn't close the door to a future engagement and marriage. He felt we need more time to consider all aspects of our relationship."

"All aspects? What kind of aspects Jake?"

"Well, he felt there are some issues we need to work through."

"Really. What issues?"

"Your father raised some valid questions we need to consider. I can't tell you everything on the phone. We need to talk face-to-face. How about discussing it over dinner, say tomorrow night?"

"Tomorrow night? Why wait until tomorrow night, Jacob? Why not tonight?"

"Something has come up. I thought Emily was coming with Mom and Dad on Friday. Dad called earlier today and informed me she's already in town. She came early to attend a freshman orientation. Dad asked me to show her around Atlanta."

"I see. Obviously, I will be playing second fiddle to your sister Emily."

"This is a complete surprise to me Rachel. I remembered Hank's party is scheduled for this evening. I thought, why not take Emily to the party and show her a good time? I can take both of you."

"Jake, I can't go if Allen and Alice will be there. Go on and take your sister to Hank's party. I'll just sit here in my room and cry."

"Rachel, I don't know what else to do with Emily. We three could take in a movie?"

"No, go ahead and take her to the party. Just remember I will miss being with you. When do you plan on picking me up tomorrow evening?"

"How about six?"

"Six is fine."

"Where shall we go, back to Bernies?"

"No, I don't want to go there. You pick the spot Jake, but not Bernies."

"I'll find a nice quiet place where we can enjoy dinner together, just the two of us. I'm sorry about this Emily thing, Rachel."

"Okay, I understand. You would rather be with your sister than your girlfriend."

"You know that's not true. I wasn't planning to go to the party, but I need to take advantage of the opportunity. This is not just about

Emily. I need to keep on the good side of Dad. Do you know what I'm saying?"

"Yes, I fully understand. Go ahead and take your sister to the party. I'm sure she will have a great time."

"Are you sure you don't want to tag along?"

"Yes, I'm sure. Just don't let me hear you've been dancing with any of Alice's girlfriends."

"I'm keeping the rules. I won't even be dancing with Allen."

"Okay, Jake. Have a good time with your friends and I'll see you tomorrow evening."

"Thanks, Rachel. I love you."

My talk with Rachel over, I put a call in to Emily on her cell phone.

"Hello, Emily. Welcome to Atlanta."

"Hi Bro. How did you find out I'm here in Atlanta?"

"Dad told me. He asked me to keep an eye on you."

"I'm sure he did, but you don't have to. I'm perfectly capable of getting along without an assigned guardian."

"Well, I'm sure you are, but Dad asked me to show you around town. So, I called to invite you to a party tonight with some of my friends."

"Oh, I see. You need a date and can't find anyone, so your inviting your sister. Pretty lame Jake. When will you get out of this shell of yours and find a girlfriend?"

"I have a girlfriend."

"Really? Well, I will look forward to meeting her at the party."

"Her name is Rachel, but she won't be attending the party."

"Why not? I want to meet her."

"I can't explain why. It's kinda complicated. Do you want to go with me to the party or not? It's casual dress. There will be a live band."

"A live band? Well, I suppose I can help you out. I'm staying at the Omni Hotel here downtown. What time do you plan to pick me up?"

"It's now one o'clock. I'll pick you up in front of the hotel at six. Can you be ready by then?"

"How about 6:30?"

"Okay, Emily. I will pick you up at 6:30, but don't be late."

"I'm sometimes delayed, but never late, Bro. Will I have to ride in your wreck of a car?"

"You expect a limo no less? Unlike some people I know, Old Blue is very dependable. Please don't be late. I will meet you in front of the Hotel at 6:30."

With those parting words, Emily hangs up. I have to smile. As soon as I mentioned a live dance band, I knew she would be hooked. I don't really expect her to be on time. I probably will wait another 30 minutes or so before the gabby little *klip* shows up.

Although it is pushing it, I decide to take a quick trip to the Woodruff Library. I quickly dressed in my running shoes, put on my backpack, and set out for the Library.

4

Realm of The Spirit

"Trust in the Lord with all thine heart
and lean not unto thine own understanding.
In all thy ways acknowledge him, and he shall direct thy paths.
Be not wise in thine own eyes; fear the Lord and depart from evil.
It shall be health to thy navel, and marrow to thy bones."
(Proverbs 3:5–8)

Game of Life

It is twenty minutes after one when I arrive at the Woodruff library located on Asbury Circle. It is a Sunday in the semester break. I fully expected I would find the library closed. I wanted to give it a try anyway. I really need something or someone to help me understand the underlying meaning of this dream of mine. I arrived at my destination sweating profusely. An older man is just opening the heavy front door of the library. Out of breath I managed to run up the steps and introduce myself.

"Good afternoon, Sir. My name is Jacob Cahn. I wonder if the library is open this week?"

"On Sundays, we usually close the library at two o'clock. Today the library will be open until five. I have a few things to catch up on. I'm the only one on duty today. I was just returning from lunch when you arrived."

"I wonder if you happen to know a Dr. Michael Stein?"

"Yes, I know him quite well. You are looking at him. I don't remember seeing you here before, Jacob. Are you a history major?"

"No Sir I'm not. I'm pre-med and graduating this semester. Are the archives limited to students taking specific majors or can other students use them as well?"

"The library is open to all students on the campus, Jacob. By the way, I'm in charge of the MARBL, the Manuscript Archives and Rare Book Library here at the Woodruff. This afternoon I need to review a few new books we received yesterday. Perhaps I can put you to work moving a few of these heavy books for me?"

"Sure, I'll be glad to help, but I will have to leave by about five."

"I don't think this will take us that long."

The man ushers me into the library and asks, "So, what brings you to the library this afternoon Jacob?"

"Well sir, my girlfriend Rachel recommended I check it out. She says MARBL contains a goldmine of information about medical phenomena, as well as Jewish history."

"So, she is familiar with MARBL. Do I know this girlfriend of yours?"

"You probably know her father, Samuel Kaufman."

"Ah, Sam Kaufman. Yes, he and I are long-time friends. I occasionally write a column for his newspaper. I have known Rachel since she was just a baby. She's a beautiful girl, a real *tsatskelah*, and smart too. Does Rachel think our library might have some useful information for you?"

"Yes, she recommended I pay you a visit. I am interested in finding out about dreams, weird recurrent dreams. She said you are familiar with the subject and possibly can help me. You see I have experienced a recurrent dream for a very long time. I've never told anyone in my family about the dream. I did tell Rachel part of it, but I left out the most disgusting parts."

"You have come to the right place, Jacob. The library has quite a number of references on dreams and unusual phenomenon, but I must say we don't have many young people often asking for this kind of information. Do you perhaps have an interest in astrology and the Kabballah?"

"No. Astrology in my opinion is so much *bubkis*, a whole lot of nonsense. I don't believe the stars or occult séances control our lives. I have never read the Kabballah. I'm interested in gaining insight into what the medical field calls hypnagogia."

"Yes, I am familiar with the term. Besides being the head librarian here I have a doctoral degree in psychology. Since you and I have time, we can embark on a *chavruta*."

"I'm afraid my knowledge of Yiddish is limited, Sir."

"Chavruta means to do something together, to embark on a joint effort. You want to learn about the hypnagogic state in three or four hours? You are expecting too much. Young millennials move too fast for me. It's quite sad. What with ipods, texting, kindles, and tweets, I fear our culture is drowning in technology. I feel more and more like a dinosaur every day. Let me start by giving you a brief overview to the subject. Then I'm going to ask you to help with my review of some new books the Library has just receive. After our work is out of the way, we can delve into this dream of yours. How does this plan sound to you?"

"It sounds fine. I appreciate the help, Dr. Stein, but I need to warn you. Even though my dad is a general practitioner, I'm starting from zilch, ground zero on knowing about dreams. I'm just a secular Jew trying to understand why I continue to experience a very troubling dream."

"Not a problem, Jacob."

"I want to respect your position. Should I be calling you Professor or Doctor Stein?"

"The position I value most these days is just being vertical, able to get out of bed in the morning. My students all call me Dr. Mike."

"Dr. Mike, are you also a professor here at Emory?"

"Yes, I teach one or two psychology courses each semester. I want to emphasize, my field is psychology, not psychiatry. I don't hold psychiatrists in very high regard."

"Why not?"

"It is because the great majority of psychiatrists are looking at every case with secular eyes. They are blind to the spiritual realities present in our world, Jacob. Unlike the fields of psychiatry, psychology is concerned with your overall being, not only your mental health, but also your comprehension of the meaning of life, your world view. Let me ask you another question. If you are to go to one of these credentialed psychiatrists, what do you think they will ask you first, Jacob?"

"Well, I suppose they will want to know about my symptoms."

"No, they would not start there. They would first ask if you have medical insurance covering mental illness. Do you understand what I am saying?"

"Yes, I fully understand. I have held the same opinion for some time now."

"I have clarified my position on psychiatry, let me start with a brief overview of this common, but unusual state of mind called hypnagogy. This technical term hypnagogy was originally coined by Alfred Maury. Sometimes the word is used in a restricted sense to refer to the onset of sleep, contrasted with hypnopompia, the term for waking up. However, hypnagogia is also used in a more general sense, covering both falling asleep and waking up. There really is no need to have separate terms. Indeed, it is not always possible in practice to assign a particular episode to one or the other state. The person's experiences are generally the same. They may drift in and out of sleep. We will use the broader sense of the term. I won't attempt to describe this other terminology with you, since they describe the same basic phenomena. Perhaps after you tell me what you are experiencing in your dream state, we may be able to narrow down the specific kind of dream you are struggling with, to relate it to an underlying cause. In general, this phenomenon of threshold consciousness is commonly called 'half-asleep-half-awake' or 'mind awake-body asleep' states. These other terms simply describe the mental state of someone who is moving towards a period of deep sleep or fully awake periods. We are examining the interval of falling asleep or waking up as a transition. Such transitions are usually brief, but can be extended by sleep disturbance or deliberate induction, such as occurs during meditation. Does this sound like your predicament, Jacob?"

"Yes, it does."

"Very well then. Let's get to work."

Dr. Mike proceeds to outline my task. He points me to a cage across the room marked "New Books."

"If you don't mind Jacob, bring me one of the boxes of new books from the cage."

Each box contains some 20 books. As the Doctor examines the new books, he begins showering me with a barrage of questions. It is like a full-blown inquisition. He wants to know about my college experience: the subjects I have taken, the professors I most admire, and my grade point average. He follows this up with more personal questions. He asks who my friends are on campus, what life in the Clairmont Center is like, and about my experience playing on the tennis team. He asks me what local synagogue I attend and how often I darken their door. His questions reach all the way to Omaha. How many brothers and sisters do I have, and how did I decide to attend Emory University? I tell about my father's high regard for the University.

There is one question I noted he carefully avoided, the status of my relationship with Rachel Kaufman and her father. It is pretty clear he understands the pickle I'm in with Papa Kaufman. I answered his questions as completely as possible. I gave him a complete rundown on my experience at Emory and my decision to apply for admittance to Johns Hopkins to pursue a career in human genetics. I postponed describing my dream at this point for fear he will conclude I am a complete mental case. It takes him only a few minutes to review the books, separating them into two stacks. One group he intends for the library to purchase and the other he will return to the publishers.

"Jacob, from what you have shared, I am of the opinion you have a promising future ahead of you. You say you do not know much about dreams. Everyone has dreams at one time or another. I'm left wondering why your dreams are so important to you?"

"I have a simple answer for you. This recurrent dream is scaring the living daylights out of me."

"I see. If this dream of yours has a dark side, and I'm guessing it does, you are going to have to tell me everything about your dream, Jacob."

"If I do, you are going to conclude I have a *loch-in-kup*, a hole in my head."

"It is evident to me you are quite normal Jacob, at least as normal as any of us are. Can you agree to describe it in detail for me please?"

"Yes, I can, but as you say, it has a dark side. As I said earlier, I have experienced this same dream for a long time, but have never even shared it with my family. It's too weird. I did share it with Rachel, but I omitted the horrible parts for her. I don't want her to know just how terrible the dream is. If I tell you about the details of my dream, can you agree to keep it confidential?"

"My lips are sealed, Jacob."

I began to share the details of the dream with him, my initial exit from my body and my journey into the evil garden. He listened intently, often stopping me to ask me for greater details about what I experienced. He waited until I finished describing the dream before making any comment. "Jacob, your dream is very interesting. I understand now why Rachel sent you in my direction. You say these creatures appear to be alive? Do they look the same each time you take this journey?"

"Yes, they're the same, and they are definitely alive."

"Jacob, you are going to discover this topic is a lot deeper and more mysterious than you ever imagined. You are not dealing simply with hallucinations or the aftermath of a bad movie. No, you are dealing with a dark entity using your transition from wakefulness to sleep to attack your spirit, attempting to corrupt your very soul. I will have to round up a few books for you to read. Part of your dream is from what is called the *Mazzaroth*. The Mazzaroth originally was a pure revelation to early man of God's plan for redemption of all mankind. It was written into the stars by Ha Shem, but then it was corrupted by Satan down through history, especially during the Greek and Roman periods. It would take us a very long time to review the history of the Mazzaroth, Jacob. I can loan you some books about it."

"I need to let you know my time here at Emory is rapidly drawing to a close. I will be leaving for Baltimore right after commencement. Regarding the library materials, I won't have much time to read them once I begin my classes in Baltimore."

"Not a problem. We will work out a way for you to make use of the library. Just today I loaned out some reference books to a scholar visiting from China. He is here to attend a conference organized by the Society for Neuroscience. The Society is holding their annual meeting here at the Omni Hotel this afternoon and evening."

"This sounds very interesting. I will try to attend the meeting, even if I have to sit out in the hallway."

"When you pass from wakefulness to sleep you experience hynagogia. Often you will have a sense of falling. I want to describe some of the history of hynagogia. Early references to it are found in the writings of many ancient scholars, Aristotle, Iamblichus, Cardano, Simon Forman, and Swedenborg. Romanticism brought a renewed interest in the subjective experience of the edges of sleep. In more recent centuries, many authors have referred to the phenomenon. For example, Edgar Allen Poe wrote about the 'fancies' he experienced. He said they occurred only as he was on the brink of sleep. In Chapter Thirty-four of *Oliver Twist*, Charles Dickens' writes an elaborate description of his hypnagogic state.

"Scientific inquiry into sleep began in the 19th century with Johannes Peter Müller, Jules Baillarger and Alfred Maury. There is a continued history of the investigation stretching into the 20th century. The advent of electroencephalography (EEG) has supplemented the introspective methods of early researchers with physiological data. The search for neural correlations with hypnagogic imagery began in the 1930s and continues with increasing sophistication. There has been a revival of investigations of hypnagogia and related altered states of consciousness in recent years. It has played an important role in the emerging multidisciplinary study of consciousness. Nevertheless, much remains to be understood about these strange experiences and their neurology. The topic has been treated as questionable science. Some have described the field as a well-trodden and yet unmapped territory.

We do know this state is related to memory, creativity, and the revelation of new knowledge. Professor Einstein often said he received new concepts and new knowledge during this transition period between wakefulness and sleep.

Now I want to begin Part Two of our discussion. It is clear you are dealing with a second, more serious problem, namely your world view. Let's take a little rabbit trail down through history. How well do we understand the causes of historical events? Just what is the motivation behind those events? No doubt you have taken at least one course in Political Science. Have you read *The Prince* by Machiavelli?"

"Yes, I've read it in my political science class."

"If you recall Machiavelli describes how the leader seeks to elevate his power to bend the will of others to his will. The book clearly describes the immoral way the strong leader uses every method at his disposal to secure power, wealth, fame, and control. Setting aside those naturally occurring events such as earthquakes, volcanic eruptions, floods and the like, most regard historical events as originating from the individual decisions of leaders in a society. In other words, this mindset views history as simply a record of human decisions and the consequences resulting from those decisions. We call this human process natural history. These decisions can result from moral or immoral objectives, but most notably the result of godless, immoral pursuits. But there's another way to view the history of the world. We call it' conspiratorial history.' Man's preconceived view of natural history has blinded him to the reality of what has taken place in the past and what is now playing out in the present. You see, there is a spiritual dimension to history. The enemy of our souls has blinded us to the existence of an ongoing war in the heavens. Most are unaware of the reality of this war between good and evil. They are unknowing pawns in what I like to call The Cosmic Chess Game."

"How is this view of history relevant to my dream, Dr. Mike?"

"An excellent question, and I will answer it if you answer my question."

"I'll try to. What is your question?"

"I believe you called yourself a secular Jew. Does this mean you have rejected the belief in the Creator of the Universe?"

"Not at all. I believe the heavens, this world, and humanity are created by Ha Shem. As you mentioned in respect to the Mazzaroth, it did not take Satan long to corrupt this pure evidence of Ha Shem's eternal plan for mankind."

"Your answer confirms my earlier favorable impression of you, Jacob. We must view history as a chess match. There are invisible entities at work to achieve the desired goals of our enemy, Satan. This conspiratorial view recognizes the reality of these unseen malevolent forces at war against God. They attempt to control worldly events, seeking to overturn the divine plan of the Creator. Most of these conspiratorial events of history are hidden from public view. It can be compared to an iceberg. Only ten percent of an iceberg extends above the level of the sea. Ninety percent is hidden from view, but this is changing as the Almighty unseals the knowledge of the past. Let me give you a modern-day example. The culture of our young nation was birthed by a small ship carrying some fifty people who landed on the rugged shores of New England. These Puritans sought to be free from the autocratic rule of King James. Their culture birthed the rule of law. Our culture matured a hundred years later under the hand of our Founding Fathers. Not all of these brilliant framers of our Constitution were Christians, sharing the vision of these Puritans. A good number of them were deeply involved in the Masonic and Rosicrucian secret societies. These men were not free to divulge their beliefs because they would have been persecuted by the dominant Bible-believing population. They buried their beliefs below the surface of the water, so to speak. Nevertheless, they participated in building a new form of government, a constitutional republic declaring the rights of the people. The government placed power into the hands of the common people, the best form of governing the world has ever experienced. The hidden religion of prominent men who occupied the highest levels of secret societies was anything but Christian. Some were deists. Others are deeply involved with the occult. These secret organizations based their beliefs in the Roman, Greek, Chaldean, Egyptian, and Sumerian gods. Although the principal gods bore different names, in actuality they are the same. I am referring to the cult of Osiris, Horace, and Isis. This occult belief looks forward to a day when Lucifer will be reborn to rule the Earth. The timing of his rebirth has been prophesied by Old and New World peoples. Many believe this final challenge to Ha Shem may be close at hand. I can't really go into any more detail today. You will have to come to your own conclusions about this dark side of history. As I indicated before, I will select several references for you to check out and examine. Right now, I would like to tell you a story. It's more like a parable. There are two men who happened to meet one another and fast became friends. The first man was a well-known scientist. The second man was a famous artist. One day, the artist invited his scientist friend to visit an art gallery in their city. The artist began describing each of the paintings as they toured the gallery. Along their way, they came to a large painting of two men playing chess. The artist made a few comments about the painting and then continued walking, not noticing his friend has stopped, and is staring at the painting and the title on the brass plate below, which reads "Check-mate." The player on the left of the chess table appears to be proudly gloating over his triumph, while the man on the right is quietly examining the chess board. The artist walked back and asked his friend, 'Why are you so interested in this painting?' The man firmly answered, 'Because, the man on the right has one more move.' Do you understand the meaning of the story, Jacob?"

"I'm not sure I do. Does this story have some bearing on human history?"

"Yes, it certainly does. The painting depicts the history of the world to this point in time. You see, we are involved in a contest between good and evil forces for control over this planet. This contest began in the Garden of Eden with a great deception. Satan convinced Adam and Eve they were able to pursue their own independent path and did not need to depend upon their Creator. The game will continue until Satan and his cohorts ultimately are destroyed by the Lord. He then will declare 'Check-mate.'"

"Are you saying natural history is played out only in the physical realm whereas the conspiratorial view reaches beyond the physical into the spiritual dimension and is responsible for wars, pestilence, famine, and much of the misery we see in our world?"

"Exactly. With the good Lord's help, good and righteous men and women are waging a war against Satan's Evil Empire. The sad fact is the majority of people are blinded to the existence of this spiritual war. The parable of the painting tells us our God has one more move."

"This view of history is quite new to me. You have me wondering whether my dream is possibly an attack to derail my ultimate destiny."

"Yes, absolutely. Your dream is to delay or derail your destiny, Jacob. You have been thrust into the game. I believe you have been selected to serve as a front-line player, a pawn if you will, in this grand conspiratorial chess game. The pawns must endure the early attacks by the enemy."

"This is an awesome thought. I have another question for you."

"What do you want to know?"

"How do I contend with these repeated dream attacks on me?"

"You cannot deal with them single-handedly, Jacob. You are going to need the help of your Creator. You must read his Word and pray for deliverance. He is there for you, even now."

"Yes, I believe he is. I certainly appreciate your help today, Dr. Mike. This parable is awesome. Can I also ask your help on another smaller thing? Johns Hopkins has asked for three recommendations, and I need them to be in the hands of their Department very soon. I know we have just met, but I wonder if you would consider sending them a recommendation for me?"

"Jacob, based upon the information you have shared with me today, I am very comfortable with recommending you. Do you happen to have their address with you?"

"As a matter of fact, I do. I have come prepared."

I opened my backpack and gave him one of the self-addressed envelopes to Johns Hopkins.

"Thank you so much, Dr. Mike. I really appreciate your willingness to help me with my application process."

"Yes, of course. I'm here to help."

"I just happen to running short on time today. I promised to take my sister Emily to a party this evening. Also, there is the meeting you mentioned with the Society for Neuroscience this afternoon."

"Before you rush off Jacob, I need to have you fill out this request for a visiting scholar library card. The card gives you full access to all of our reserve holdings."

As I fill out the form, Dr. Mike printing out a care giving me access to the reserve library holdings.

"Since I don't know where I will be staying in Baltimore, I will give you my family's address in Omaha."

"That's fine."

Dr. Mike went into the library stacks and soon reappears. He has selected four books for me to read.

"Jacob, you can mail these books back to the library as you finish them. By the way, you need to purchase a few helpful books for your personal library. You will need a good study Bible, one containing a Greek and Hebrew/Chaldean dictionary. You also will need a good concordance to help you quickly locate specific passages in the Bible. You also need to purchase a few good commentaries on selected topics in the Bible. You also can find some excellent commentaries on the internet, and they are free. These references should get you started."

"I will try to purchase a study Bible and a concordance before I leave Atlanta. Should I purchase only a Hebrew Bible, or one including the New Covenant Scriptures also?"

"Get the complete Bible as well as some of the apocrypha books. There is a new Bible by the Cepher Publishing Group containing these apocrypha, but it is expensive."

"I'm not sure what those are, but I will ask about them. Well, I really must leave now."

"Jacob, I wish you success in your graduate program and your relationship with Rachel. She is a wonderful girl. I hope and pray you can convince Sam to approve your courtship."

"Thank you, Dr. Mike. Any help you might be able to furnish along those lines will be very appreciated. I will keep in touch."

I put the four books Dr. Mike loaned to me into my backpack and headed back to my apartment. I didn't arrive at my door until half-past eleven. I would have to cut the visit to the conference short in order to pick up Emily for the party. I made myself a peanut butter sandwich and drank a leftover cup of coffee. Then I quickly changed into my suit. If I tell the attendants at the conference I am a student, I may be able to wrangle a free admittance to the conference.

5

The Message

"And the Spirit of the Lord will come upon thee,
and thou shalt prophesy with them,
and shalt be turned into another man.
And let it be, when these signs come unto thee,
that thou do as occasion serve thee, for God is with thee."
(1 Samuel 10:6–7)

A Word of Knowledge

I am leaving for the conference when the phone rings. It is Emily.

"Jake, where have you been? I've been calling you, but you haven't returned my calls."

"Yes, I know. I went to the library to look into something. What's your problem?"

"I don't know what to wear for this party. Is it a fancy smancy event or informal?"

"It is rather informal Emily. Just wear a nice dress and you will be fine. I am leaving here just as soon as I change clothes. There is a conference downtown I want to go to before I pick you up."

"What? You are going to a conference? How do I know you will pick me up on time? Where's this conference going to be held, anyway?"

"Not to panic, Emily. It is located right next to the Omni in the Georgia World Congress Center. After the meeting, I will pick you up for the party. See you at six thirty."

"Okay, I will be waiting for you in the lobby."

"Fine, I'll see you then."

I hung up and left for downtown. Since Emily is staying at the Omni, I can park there and walk a block to the Center. I arrived at the hotel and walked to the registration counter. I may have to use my chutzpah to get permission to enter the auditorium. There is a young girl manning the ticket table. I walked to the sign labeled "A to G."

"She is smiling as she welcomes me. Hello, Sir. May I have your name?"

"My name is Jacob Cahn. I am a student here at Emory University. An instructor friend told me about the conference and recommended I attend this afternoon's lectures."

"Do you have a reservation? I don't see your name on my list of attendees."

"No, I did not know about the conference until an hour or so ago. I didn't have time to purchase a reservation. I understand Emory students suffering from poverty often are allowed to attend local conferences without charge."

"No one has informed me about any such policy, Jacob. You must pay for admittance."

"Perhaps they failed to let you know about this policy. Can you contact your supervisor and ask him about it?"

"No, he is not here at the moment. Perhaps I can ask one of the other workers here."

"No doubt this will take you some time and I would very much like to hear this particular person now speaking. Would it be a problem for you to let me just slip into a seat in the back row?"

"Well, I don't know."

"I see your name is Nancy. Nancy, I have always believed rules are made to be broken in unusual situations. As I say, I would very much like to hear this particular speaker. Can you make an exception, just this once?"

"Well, I guess it will be alright, Jacob. Here, let me give you the conference materials and a lanyard. You will have to write your name on the badge. Please, don't tell anyone what we are doing."

"No. My lips are sealed."

"Yes, mine are sealed as well, Jacob?"

"Yes, I see. You have beautifully sealed lips, Nancy. Thank you for cutting me some space."

"You are welcome. The conference today is almost over, Jacob. I am going home in a few minutes, so I won't be here when you leave. Take care."

I took the package of conference materials and shook her hand. Then I walked into the auditorium. The speaker was just beginning his presentation. I opened the agenda handout. There is something written on it. Ah, yes. Nancy has her own agenda. She has written her phone number, drawn a heart, and neatly signed her name at the top. I found the title of the current lecture. It is The Response of Neural Pathways to Magnetic Fields. I took careful notes. One fact stuck me. The lecturer said these neural pathways can be improved by the introduction of nano copper or gold into the blood stream. I remained to listen to two other presentations before it was time for me to walk over to the Omni to retrieve my car and Emily.

To my surprise, Emily is waiting for me when I arrived in front of the hotel. She's wearing a brand-new outfit, a yellow dress with a bright orange sash around her waist. I have to admit, she looks very stunning. She quickly gets into the car and we leave for the party. Emily is talking my ear off the whole way.

The sign on Elm Street reads Willow Creek Christian Fellowship. As we pull into the parking lot, the sound of loud music meets our ears.

"Emily, do you recognize the song? It's an old one by Delirious. It's not the kind of song you expect to hear at a church, but then this is the first time I've darkened a church door."

"It's the first time for me as well, Bro."

Signs at the front of the church point to the location of the Fellowship Hall. The Hall is filled to overflowing with noisy grad students as we enter. It takes us a few minutes to locate Allen. He is sitting with Alice and some of their friends at a table near the back of the room. I guided Emily through the crowd to their table and introduced her.

"Hi everyone, this is my younger sister Emily from Omaha. She's here for the freshman orientation and the commencement. I brought her along tonight."

Hank Bowman is quick to offer her a seat next to him. "Emily, why don't you sit here next to me? Jake can find his own seat."

I am feeling a twinge of brotherly responsibility. I located an empty chair and placed it between Hank and Emily. Hank simply sits down in my chair, spoiling my strategy.

I firmly tried to discourage Hank's advances. "Hank, Emily will be graduating High School this coming June. She is just 18."

Emily flashes a ferocious glare at me and kicks my shin really hard. Then she turns and addresses everyone at our table.

"Hi y'all! It's so nice to be here in Atlantah. I trust I am using proper Southern pronunciation."

Her comment elicits an immediate response from the guys at the table. Lifting their pop cans, they shout in unison, "Here, here. May the South live forevah!"

Emily responds, "Oh my, is this a rebel yell I'm hearin?"

It's clear Emily is in her element. Hank quickly escorts Emily to the dance floor.

Allen taps me on the shoulder and points at a table across the room.

"Do you see the man standing at the table near the bandstand, Jake?"

"You mean the tall one facing our table?"

"Yes, that's the one."

"Is he a friend of yours?"

"He's Bill McGuire, the associate pastor of our church. He leads our young adult class."

The pastor appears to be in his early thirties. He's casually dressed for the occasion, wearing a yellow Hawaiian shirt with big blue flowers and a pair of white trousers. I watch as he individually greets each of the grads sitting at the table. After saying a few words to each of them he places one hand on their head and raises the other hand, palm upward into the air. No doubt the good pastor is dispensing a graduation blessing on each of them. I doubt he's praying the Priestly Blessing I received from Samuel Kaufman early this morning.

I decide to find out whether my assumption is correct. "Can someone explain what Pastor McGuire is doing? Is he praying a special blessing over those guys, or just trying to mess up their hair?"

Terry Johnson, who has been quietly sitting at our table, suddenly comes to life. "Jake, you're right. Pastor McGuire is praying a blessing on each of them, no doubt giving them the benefit of his special gift of knowledge."

I'm puzzled. "gift of knowledge? What do you mean Terry?"

"Jake, Christians believe the Holy Spirit sometimes gives you a supernatural understanding, a word of knowledge about a particular situation. This is called a gift of knowledge. Another gift given is the awareness of spiritual entities or forces. This is called the gift of discernment. In both cases, the Lord reveals things only he knows."

"I hear what you are saying Terry, but I still don't understand it. It sounds to me like something from the outer limits."

Terry's face breaks into a broad smile. "You are right, Jake. It's something from the outer limits."

I watch all of the guys at the table. They are nodding their heads in agreement with Terry and begin chuckling. I wonder what these guys finding so humorous? Just perhaps Terry and the others know a lot more about this gift of knowledge than they are letting me in on. I continue to watch the pastor make his rounds from table to table. It doesn't take long for me to realize this pastor will soon be arriving at our table. I'm feeling a bit uncomfortable over the idea. Just then, Hank returns from the dance floor without Emily. I searched the dance floor and found her dancing with a guy I've never seen before. Harmless enough I suppose, but I still don't like it. When I turned my attention back to our table Pastor McGuire is standing next to Hank. Hank introduces the pastor to everyone. Pastor McGuire proceeds to move around our table, shaking hands with each of us, and sharing a few kind words. He congratulates each grad and shares some conversation with the girl friends present. He asks the grads about their future plans. He concludes each of these conversations by offering to pray for them. Suddenly I realize I am next in line, and there's no place to hide. I envy Emily, out on the dance floor. As the pastor shakes my hand his blue eyes are boring holes right through to my soul.

"It's a pleasure to meet you, Jacob. Allen tells me you plan to attend graduate school at Johns Hopkins in Baltimore."

"Yes, Pastor, I have applied and hope to be accepted. I plan to study microbiology and human genetics. Perhaps I might even return with a Ph.D. in hand to work here at the Yerkes."

"That's very interesting, Jacob."

Suddenly the pastor seems distracted from our conversation. He pauses and takes a small notebook out of his shirt pocket.

"Excuse me for a moment, Jacob. If you don't mind, I need to make a few notes."

"No problem. Go right ahead."

A few minutes seems like an hour. He then closes the notebook and slips it back into his shirt pocket.

"Jacob, I needed to write something down before it escaped me. You are Jewish I assume?"

I know it would be foolish for me to try to convince him I am a devout Jew.

"Yes, Pastor, I'm Jewish. I attended Beth Simca Reform Synagogue on Oak Street here a few times. Actually, I only attend on High Holy Days. When I get to Baltimore I plan to be more faithful."

"Do you know you will be in the center of a large Jewish community in Baltimore?"

"Yes, Terry Johnson gave me a rundown on the Baltimore area, including the Jewish community located in Pikesville."

I tried to ease my nerves by taking a drink of soda. The pastor wasn't finished with me yet. I suspected our conversation is heading into deeper waters. His next question confirmed my suspicions.

"Jacob, can we take a short walk outside? I would like to talk to you privately for a few minutes. I have something quite interesting to share with you."

Now I know this pastor is not going to let me escape with a simple handshake, a greeting, and a blessing. I seized on Emily as my escape route.

"Pastor, I really need to keep tabs on my sister, Emily. Right now she's out there on the dance floor dancing with someone I've never seen before."

Always helpful, Allen offers a solution. "Not to worry, I'll keep an eye on her for you Jake."

Alice does not approve of his offer. "Al, your attention needs to be placed in my direction. I'm sure Hank won't mind keeping an eye on Emily. He has been doing so all evening."

Hank says "I'll go cut in on them for you, Jake."

Pastor McGuire tells me, "Problem solved Jacob. I will take only a few minutes of your time."

I had no way to escape, so I smiled and agreed to follow the pastor outside. As we walked toward the door, Pastor McGuire asks, "What field did you say you are going to pursue, microbiology?"

"Yes, I hope to concentrate on the human genome. I would like to do research into the process of precursor neurons in the preborn. I am also interested in the location of human memory."

"I don't really know anything about precursor neurons or the location of human memory, but I think you will be a good researcher."

Once outside, both of us stood for a few moments, quietly looking up at the night sky. The city lights failed to dim the brightly shining Orion Constellation in the Southern horizon.

Finally, Pastor McGuire broke the silence. "No doubt you are wondering why I have asked you to take this walk with me. You see, the Lord sometimes reveals things to me about people. As I shook your hand at the table, he told me something quite interesting about you, Jacob. Before I share it with you, I have to confess I don't understand what this message means. God has never given me a message quite like this one. Given time, I am sure it will become clear to you. The message is quite detailed. That's why I took out my notebook and wrote it all down. I wanted to be sure I passed all of it along to you."

"Sort of like when God told Moses to write down the Big Ten?"

"No, God did not dictate the Torah to Moses. He actually engraved the Big Ten on those stones himself. Tonight, he has seen fit to dictate it to me for you."

I watched as Pastor McGuire pulled the notebook out of his pocket and began to read what he had written.

"Jacob, the Lord said you are a prophet. He is calling you to the minister in these last days. He said you will be sealed as one of the one hundred forty-four thousand sons of Israel, that you are from the lineage of Levi. Then he told me to go outside with you. He promised to give me yet another revelation about you."

"Really, this is very strange, Pastor."

Suddenly the Pastor's face takes on a strange, frightened look. "Jacob, the Lord just did something even more remarkable. In an instant, he has given me a very scary vision. It is so vivid in my mind. Let me first read to you the exact words he spoke to me at your table: 'The word is very nigh unto thee Jacob, in thy mouth, and in thy heart that thou mayest do it. Son of Levi, I have set before thee this day, life and goodness, not death and evil. I am calling you forth to produce fruit and life-giving water. I will open your eyes and reveal knowledge to you hidden from the beginning of the world. Jacob, my Israel, if you will wait on my Spirit and remain faithful to the Spirit's calling, I will bless and exalt you. You must remain faithful to the mantle I am placing upon you. The Tempter desires to deceive you, to tempt you to misuse this gift. He desires to cause you to walk in the ways of Balaam into sorcery and false dreams. Already the enemy has troubled you with a dream. Verily I declare this dream will never trouble you again. Behold, I am the alpha and omega, the beginning and the end of all things.'

These are the words I heard and wrote down. Then as we were standing here, the Lord showed to me this horrible dream you have been experiencing for most of your life. After the vision ended, the Lord asked me, 'What did you see William?' I told him there is a strange walled garden. It is abandoned. There are fallen branches from the trees, untrimmed plants, and a large number of strange animals of many kinds, snakes, spiders, and dragon-like lizards. They seem to be bent on attacking me. The Lord said: "You have seen correctly William. Neither with you only do I make this covenant and this oath, but with him that stands here with us this day. There is also a man elsewhere I have revealed this dream to. He will reveal it to Jacob, but not before it is time. This garden you have seen is the Garden of Eden after the Fall, for my cherubim no longer need defend the gates of Eden. Satan desires to draw Jacob into wickedness, to prevent my holy calling upon his life. Tell my Israel he must wait upon my Spirit. Behold, I am the First and the Last. He who truly believes in me will receive eternal life and I will raise him up on the last day.' Then the Lord told me to relay the information to you. Jacob, Scripture teaches us that the testimony of two persons is required as confirmation of the truth. Have you ever experienced a dream such as I have described?"

"Yes. The vision you saw about the garden has plagued me since I was very young. It is just as you have seen it."

Pastor McGuire tore out the pages he had written in his notebook and handed them to me.

"Jacob, I want you to keep these notes as a reminder of this night."

He closed his notebook and put it back into his shirt pocket. Neither of us said anything for several minutes. We just stared upward at the night sky.

Finally, the pastor said, "As improbable as it may seem, the Lord is calling you to serve him in a very special way, Jacob. He has revealed your terrible dream to me to confirm his calling upon your life. You have been given supernatural gifts. I believe you are called to be an evangelist in the last days. In the prophecy of Hosea, Chapter Two, verse Twenty-three, Lord describes a period of great fruitfulness during the time of the end. I recall a similar description in Chapter Thirty-four of the Book of Ezekiel also. Jacob, the Lord has warned you not to misuse this gift. Satan, the enemy of our souls, always attempts to put barriers or great temptations in our way. It is as if God and Satan are playing a game of chess."

His comment reminded me of the parable Dr. Mike told me. It is clear I am in the front lines, a pawn in a chess match taking place in the invisible war of good versus evil.

"Pastor, I believe my pawn has reached the opposing king's row and I have been converted into a white knight."

The impact of the message and the words of Pastor McGuire are sinking into my consciousness like a knife. Pastor McGuire paused before saying anything more. "Jacob, it is clear to me God is going to teach you how to see into the spirit world. I know what you are feeling. It will seem strange to you at first, but do not be afraid. The Lord will enable you. Remember he has arranged for you to meet another person to help you at some time in the future?"

"Yes, I did hear him say I would meet someone, somewhere and at some future time."

"If you earnestly search for His truth, you will find it, and it will increasingly set you free. Do not forget. You must never misuse this gift. Practice complete obedience. Don't use it for your own purposes. Never depart from God's plan. If you remember, Aaron and Moses died in the wilderness because of their disobedience. They never entered the promised land. Know also gifts of prophecy and visions are under your control. Only use them as the Lord prompts you to use them. Is this clear?"

"Yes, I understand. I am relieved to know my dream is not just a quirk of my imagination. In the past, I've doubted about my sanity many times."

"Do not worry Jacob. The Almighty tonight has blocked the dark side's attempt to draw you into the occult. God truly has set you free from your dream. You will never experience it again."

"I want to thank you for sharing your gift with me. I am indebted to you for all time."

"Nonsense, Jacob. It would be an offense to the Lord if I failed to do so."

I am attempting to absorb all the pastor has said as we both quietly returned to my table. Terry was right. I certainly encountered the outer limits tonight. How else could this pastor possibly have known about my dream, except from the Lord. Yes, I am convinced this is real. I am being called by the Almighty to serve as a prophet and a seer into the spirit world. But why would God choose someone like me, a secular Jew? Why not choose some Orthodox guy? I need to find out more, a lot more about the Lord. I will need help from this pastor and from others as well.

"Pastor McGuire, I'm not sure at this point how to go forward with this. How am I to understand, much less put into practice, these two awesome gifts? I mean, you have to admit, this is pretty extreme stuff. How can I be certain I am taking the right path?"

"It will take some time, I'm sure. I suggest you begin by reading the entire Bible, both the new and old covenants. Pray constantly for direction from the Lord. He is faithful and will guide you. Let's keep in touch with one another. Here is my card. If you need help, don't hesitate to call or write to me. I promise to help you as much as I can. I'm here for you one hundred percent."

"Thanks, Pastor. I appreciate your offer, and I will keep the notes in a safe place. I really don't know how to thank you for ministering to me this evening. It has been awesome."

"Yes, it certainly has been. I will let you go now to check on how Hank and Emily are getting along."

As we arrived back to the table, I spotted Emily dancing with yet another student I did not know. Hank is looking a bit dejected back at our table. A steady parade of young graduates kept cutting in on Emily on the dance floor. At least there's safety in numbers. About eleven thirty, the party began to wind down and people started leaving. Everyone at our table took time to share cell phone numbers and addresses, promising to stay in touch. Allen gave me a big hug and thanked me for being his friend over the last four years. We promised to keep in touch with one another on a regular basis. He gave me his parent's address in Valdosta and I gave him the address of my parents in Omaha. Meanwhile, Emily is busy saying goodbye to her new-found friends. I took her by the hand and led her outside. She danced her way back to the car.

On the way back to the hotel, I asked, "Emily, I hope you didn't give any of those guys your cell number or address in Omaha?"

Emily retorted haughtily, "Well dear brother, I'm sorry but such information is not accessible for your pay grade."

I have no doubt Emily will enjoy pleasant dreams tonight while I struggle to understand this strange calling upon my life.

Covenant of Trust

It's now almost eight o'clock Monday morning. Last night's party wasn't over until late. It was 1:30 in the morning by the time I delivered Emily back to the hotel. I went straight to bed, and slept soundly all night. What a relief! The dream failed to show up. I'm hopeful it is truly a thing of the past. I now know this is a critical time in my life. Today I must decide the future course of my relationship with Rachel. After getting dressed I turned on the TV to catch the morning news. Fox is interviewing the Syrian delegate to the United Nations. He held little hope for an end to the violence and persecution in the region. Incredibly, our Government is providing military support for opposing sides in the area. As a Jew, my thoughts naturally run to the plight of modern-day Israel, surrounded on all sides by hostile neighbors and the rapidly growing Islamic State of Iraq and the Levant (ISIL). Already the fragile peace between two of Israel's neighbors, Jordan and Egypt appears to be broken. Iran has succeeded in trading the weak promises to forestall the bomb for the removal of sanctions, releasing millions of dollars to them. Those funds no doubt will end up financing the Shiite terrorist groups in the region. Meanwhile Russia has entered the chaotic region in support of the Baathist regime in Syria. They continue to make air strikes on the city of Aleppo. Damascus is rapidly turning into a heap of bombed-out buildings. This destruction is a clear fulfillment of Isaiah's prophecy in Chapter Seventeen. How many diplomatic failures will occur before Israel's neighbors unite to destroy Israel once and for all? I can't listen to any more of it. I turned off the TV, and went downstairs to breakfast.

It is after nine o'clock when I finished eating and returned to my room. I decided to call Emily to invite her to lunch at the DUC. A sleepy-sounding sister answers my call.

"Hello?"

"Good morning, Emily. It's your Bro. What do you have on your schedule today?"

"Hi Jake. I'm supposed to take a tour of the library this morning at ten o'clock. Why?"

"Well, I thought we might meet for lunch at the DUC, say about one o'clock. I'm buying."

"Oh, I'm sorry, Jake. I can't make it for lunch today. I already have another date."

"You have another date?"

"Yes. Hank invited me to have lunch with him at Bernie's."

"Well, perhaps we can take in a movie this afternoon?"

"I'm sorry Jake. I can't go to the movies with you either. I am scheduled to attend an orientation talk at Glenn Memorial Auditorium at 2:00. Also, I'm invited out to dinner this evening, so you're plain out of luck today."

"Emily, you just arrived in Atlanta three days ago. You need to slow down your social life. I don't like you dating guys you have just met and I don't know. Can't you wait until next semester when you get to know them better?"

"It's not what you're thinking, Jacob. Dad's friend Dr. Schroeder invited me to have dinner with his family at their home tonight."

"Oh, that's different. I'll call you tomorrow morning."

"Fine, Jacob. We can talk then. Bye."

"Goodbye, Emily."

After failing to schedule any time with Emily, I wasn't sure how to spend the rest of my day. I put on my sweats and took a morning run. I finished my run and grabbed a sandwich in the lunchroom. Someone left an old issue of Science on one of the tables. I picked it up and browsed through the pages. There's an article in the magazine about an archaeological dig in Israel. A private research team has begun a search for the location of the first temple of Solomon. Rachel is taking a double major of anthropology and archaeology in the graduate school. I thought, this article is right up Rachel's alley. I took the magazine back to my room with me. The more I know about her interests, the better. Anyway, it will give me something to read until it's time to go to dinner. I ended up reading the entire magazine, including the classified section in the back. A number of universities and research firms have placed ads offering jobs for geneticists and microbiologists. If my application to Johns Hopkins does not succeed, perhaps I can find a place at another school. I tore out the page and put it in the drawer of my end table. I spent the remaining time reading one of the books I borrowed from the library, the Book of Enoch. I keep losing my place in the book. My mind is pondering about what I will say to Rachel over dinner. The hours, minutes, and seconds pass slowly. Finally, it is time to leave for 1520 North Peachtree Battle Avenue.

Rachel is waiting on the front porch when I arrive. I quickly get out of the car and open the door for her. She gives me a broad smile as I help fasten her seat belt.

"I want to thank you for calling me this morning, Jake. I thought you might be having second thoughts about our dinner date."

"No way, girl. We are going to find a solution to this problem. Here, I came across a copy of Science Magazine. It has an article on a dig in Israel."

"Oh, thank you Jake. I will read it later at home. How did Emily like the party?"

"Emily had a blast. Hank has invited her to go to lunch with him today. Where do you want to eat? There's a nice French café on Wesley Road. How does a French meal sound to you?"

"A French restaurant is a wonderful idea, Jake."

The café is pretty much empty when we arrive. The room is stylishly decorated with French impressionist paintings on the walls. There are booths along each side of the room with tables in the center. The tables and booths are decorated with vases of freshly cut flowers and candles. We sat down in a booth near the back of the room and ordered coffee from Pierre, our waiter. Pierre brought us the menu. It is all in French. We sat quietly sipping our coffee for a few minutes, trying to decide what to order.

"Rachel, the only foreign language I know is Yiddish. Do you know any French?"

"I was about to ask you the same thing. I guess we are on our own."

Finally, I broke down and asked Pierre for his suggestions. He was only too eager to help.

"Oui Monsieur, I would recommend our special this evening,

the *chateaubriand beef filet, avec haricots verts, de pommes de terre au four.*

The waiter can see from my face, I am having trouble with the French.

"I see *Monsieur* you do not understand well the French language."

"Yes, but go on Sir."

"Let me continue in the English. The main course includes also the green beans and potato au gratin. To complement your dinner, I recommend a full-bodied Bordeaux. Perhaps you would like to begin your meal with our *soupe au pistou,* a vegetable soup with a sauce seasoned with garlic, basil, tomato and cheese. And *for une résistance finale, our Charlotte Malakoff aux Frances,* a cold Bavarian almond cream soufflé with fresh strawberries. While you decide, let me light your candle for you."

I looked across the table at Rachel. Her eyes met mine. They are soft and smiling. Her response clinched it. Never mind the cost. This is our special time.

Gathering my courage, I beaconed to our waiter. "Pierre, I am going to accept your suggestions."

Pierre is smiling as he retrieves our menus. "Monsieur, you will not be disappointed."

"Rachel, I need to share something with you before we discuss our engagement. It's about my dream. Do you remember me telling you I have experienced this dream almost my entire life?"

"Yes, I remember. I suggested you meet with Dr. Mike. Did you see him yesterday?"

"I did. He explained many things to me. I can't possibly share all of it with you now. But there's more."

"More? Like what else has your mind been up to? Let me guess. You danced with one of Alice's friends."

"No, I didn't dance with anyone. I might not have met Allen's pastor, Bill McGuire, had I done so. He attended the party in order to congratulate the grads. He dropped by our table to greet each one of us individually. Rachel, when he shook my hand, it was like his eyes were burning a hole right through me. He paused a minute, then he took a notepad out of his shirt pocket and wrote something down. Next thing I know he is asking me to take a walk outside with him. He said he needed to share something very important with me."

"Did you go outside with him?"

"At first, I tried to get out of it, but I couldn't come up with a good excuse."

"No doubt he wants to convert you to Christianity."

"I didn't know what to expect. I finally agreed to follow him outside. We stood there on the patio together for a few moments, just looking up at the stars. Then he drops a bombshell. He tells me Ha Shem has given him a message for me, what Christians call a word of knowledge. Ha Shem told McGuire I would meet someone in the future as a conformation of this prophecy."

"Sure, it happens every day. I would have immediately returned to my table Jake."

"No, it is for real, Rachel. While we are outside the Lord shows him a vision. It is a vision of my dream. Ha Shem tells him I will never experience the dream again."

"So, you slept one night without the dream and now you are certain it will never come back again? Really Jake, you are more gullible than I realized."

"I am completely certain my dream will never bother me again. Rachel. How could he have known the details of my dream? I have never spoken to another person about my dream except for you and Dr. Mike?"

"What else did he tell you? Are you destined to become a believer in Jesus?"

"As I say, the pastor wrote down everything the Lord told him in a notebook. He tore out the pages and gave them to me. Take a look for yourself."

I handed the notebook pages to her and waited while she slowly reads them.

She hands the notes back to me. She has an incredulous look on her face.

"This certainly is weird. The gist of the message seems to be you are a son of Aaron, and Ha Shem is calling you to become some kind of a modern-day prophet."

"Yes, it is hard to believe. Pastor McGuire says Adonoi clearly declared to him I am a prophet and a seer. After reading the message to me, the pastor said Ha Shem was going to verify the message is true and genuine. While we both were looking up at the stars, the pastor saw a vision. Here is the clincher, Rachel. The vision he described matches my dream to a tee, in every detail. He said my dream is actually from the dark side. Satan is trying to draw me into the occult. As long as I follow the right course in my life, Ha Shem declares I will never experience my weird dream again. I am free finally from this awful dream. It seems like I'm now on a kind of spiritual journey. Ha Shem told McGuire I would meet someone in the future as a confirmation to this prophecy. Where it will lead, I have not the vaguest idea."

"Jake, I'm changing my mind. I think you are a *meshugeneh*, a true crazy man."

"Really?"

"No, not really. Are you sure the pastor had no way of knowing about your dream beforehand? Perhaps Allen or another of your friends told the pastor about it?"

"No, Al could not know about my dream, Rachel. As I said earlier I've never told anyone about it except Dr. Mike and yourself. I haven't even told my family yet."

Rachel took back the note and stared at it for a long time. I can see she didn't quite know what to make of it. She is trying to find some explanation, some way to discredit the note. Finally, she gave up.

"Since you absolutely are sure this crazy dream is gone, perhaps future dreams will be more positive, such as how much you love me, Jake. I don't know what it means, but I can live with it. Perhaps in time you can forget you ever had such a dream. I do appreciate you telling me about your dream and your talk with the pastor. We must never keep secrets from one another."

"Rachel, I feel so much better after sharing this with you, getting it off my chest. I was worried. I thought you might decide to end our relationship. Are we ready now to discuss the most important thing in our lives, our engagement contract?"

Just then, the waiter arrived with the first course of our dinner. "Jake, let's eat our dinner first and discuss our engagement over dessert."

"Yeah, that's a good idea, Rachel. I'm hungry."

The food is great, but I'm still hungry, but this hunger is not for more food. Rachel is who I am yearning for. As she quietly eats her food, I watch her every movement. I can't keep my eyes off of her. I am looking forward to the day when she will be mine. We ate the soufflé dessert slowly. The last of the wine left us with a warm feeling. Now is the perfect time to address the reason for our meeting, my visit with Papa Kaufman.

"Rachel, I've thought a lot about our problem. I know you have as well. It occurs to me we have three options. The first one is to just go ahead and get married without your father's permission. As I said earlier, to elope would be an absolute disaster. Your family is very close. Marriage without his permission would drive a wedge between you and your father. I could never be accepted by him. This is not a viable option."

"I agree, Jake. I do not want to be estranged from my parents. It would be too painful for me as well as both of them."

"The second option is your father's suggestion. He is convinced we are not ready for marriage and should postpone our engagement until we resolve several issues. He recognizes this postponement will be very difficult. At this point he strongly feels we need to take a time out, to avoid contacting one another until such time we are able to resolve these issues."

"You mean he wants us to completely break off contact with one another? For how long?"

"He suggested a period of six months."

"Easy for him to say. Did you agree to that? It sounds like you went right along with everything he said. Really Jacob, I thought I was marrying a man who would stand up and fight for me. It sounds like you have acquiesced to all of his demands."

"No, I didn't agree to all of his suggestions, but I did hear your father out. Nothing is gained by arguing with him … starting a war. I need to keep the peace. Let's be clear. I did not promise anything. I simply told him I would consider his advice. That's all I said."

"Well, your approach may have been peaceful, but your results are not encouraging Jacob. Does Papa expect you to become an Orthodox believer?"

"Well, our different spiritual understanding is a major concern to him. His argument makes a lot of sense to me. He feels your intense faith, and my secular orientation will be a major problem for us down the road. He strongly believes the foundation of a successful marriage rests upon a common religious relationship, a marriage of three rather than two, one being Ha Shem. He used a Hebrew analogy to make this point with me."

"I know. I can hear him telling you the story of the Hebrew

spelling of the words for a man and a woman. He told you the fire of passion is not a good enough foundation for a stable marriage. I've heard it all before, Jake. According to him, forest fires in the wilderness often continue to burn for a long time, but ultimately they are extinguished by cold winter snows."

"Rachel, our fire is not going to go out any time soon, but let's be real. Up to this point my faith has been pretty much non-existent, whereas yours is strong. Your father is right. This can become a real problem for us at some point in our marriage. This problem doesn't have to exist forever and always. In fact, I've already begun to rethink my relationship to Adonoi. In view of this latest spiritual experience, my spiritual life is being greatly changed. I expect this faith issue soon will cease to be a problem."

"The bottom line is you are in agreement with Papa on the faith argument, correct?"

"Yes, I do think your father shared some pretty important Jewish wisdom with me."

"What other objections did he mention?"

"He also mentioned the challenge we would face, both in different grad schools and the separation it would entail. He asked what we would do if you became pregnant."

"Not much chance of getting pregnant with you in Baltimore and me down here, Jake."

"Rachel, your father does not feel we are ready to enter into marriage at this point in our lives. But it's not a complete downer. He said there is a real possibility he could agree to our marriage at some time in the future, providing we are able to resolve these issues. It won't be easy, I know. I can't imagine no communication between us for six months, but I'm willing to give it a go if you are."

"Jacob, I'm willing to try, but I have to tell you, I'm not as ready as you seem to be."

"So, this brings us to our third option. It is really an extension of the second one. We can agree to postpone our engagement and marriage for now, agreeing to wait for one another without a legal ketuba. I propose we go ahead and write up our own unofficial ketuba covenant to seal the bargain. Your father doesn't have to know we have entered into such a private engagement contract. This is what I believe we should do."

"So, you're proposing to me again."

"I'm proposing we enter into our own marriage contract now, but keep it under wraps. I'm willing to honor our private agreement until your father agrees to bless our engagement, even if it means we must wait until we both finish our schooling. I understand this will be more difficult for you. You will be postponing the time to begin our family. I understand how important family is to you. So, what do you think? Are you willing to make such a commitment?"

Rachel remained silent for several minutes as she considered my suggestion. It's evident she is considering the pain and sacrifice this decision will mean for her.

"Jake, first of all, your proposed ketuba is not a legal and binding contract without being signed by my father. Second, you are right when you say family is very important to me. I don't know whether I can wait until we both finish grad school. How much confidence can I have in you to keep your end of the bargain? You are a wonderful man. You are kind, compassionate, loving, and dependable. What's more, you are extremely intelligent, but Jake you have one serious weakness."

"What's that?"

"You still don't know do you? Do I need to tell you again? I have never met a man so absolutely clueless when it comes to understanding members of the opposite sex. You are completely naive when it comes to dealing with women. Apparently feminine wiles are beyond your ability to comprehend. You don't know how we think, how we feel, and most of all, how we can manipulate men."

"Oy! Name one guy who understands women?"

"You certainly do not. There is only one way I can agree to wait for you. First, you must sacredly vow before Ha Shem to faithfully keep this agreement. Second, you are going to have to agree to follow a plan designed to safeguard our covenant."

"You can depend on me, Rachel. I love you."

Jake, this will mean a new set of rules for you. But this time they are my rules, not from the rabbi. Think of them as an addendum to our ketuba."

"Sort of like the Bill of Rights to the US Constitution?"

"Yes, sort of like that, except it is 'Rachel's Bill of Rights.'"

I'm recalling the first set of rules. Now she is proposing another set. Her own personal rules for me to follow. I should have anticipated this. I'll never forget the day we played our second tennis set when this strong willed, no nonsense girl bluntly laid down the rules for courting. I wonder what new rules she has in mind?

"Rachel, I'm certainly ready to discuss this with you. What are your new rules?"

"First of all, you need to know these rules are not open to arbitration, Jacob. If you feel you cannot live up to them, our covenant is over, kaput. What's more, you are going to have to make my rules crystal clear to all of the young women you know, especially those you will be working with in the Genetics Department. Do you understand?"

"Yes, I understand. I intend to keep your rules just as I have kept the rabbis' rules. Do you have a list, or do we come up with a list together?

"I have the list in my head. I don't need to write them down. It's only five, not ten like the Torah. Like the Torah these five rules are written in stone."

I watch as Rachel lifts her hand and points at me with her index finger.

"First you must agree never to date another woman, even if it appears perfectly innocent to you."

I realize it would be unwise to challenge any of her rules at this point. I simply nodded my acceptance to rule number one. Now a second finger is pointing at me.

"Second, you must avoid being alone with another woman, even those you are working with in your department, single or married."

Oh boy, this one sounds hard, but I remain quiet and accepting it with a nod of my head. Now three fingers are pointing in my direction.

"Third, you must respect and protect our privacy. You must not share any intimate thoughts and concerns we have with any woman acquaintance, even in mixed company."

The words came out before I could stop them. "Rachel, what if I need to discuss my educational program with a woman advisor or someone on my committee?"

"Really Jake. It should not be too difficult for you to find a male advisor."

The four fingers are now being shaken up and down at me.

"Fourth, you must not give or accept even the smallest gift from another woman. I don't care if it is only a microscope slide or a test tube."

I am waiting for the thumb, but she now presents a clenched fist in my direction, ready to deliver a knock-out blow.

"Finally, Jake, you must agree not to travel in a car alone with a person of the opposite sex. You probably think these rules are too strict, and you are strong enough to resist the temptations coming your way, but I know better. You are not! The rules I am asking you to keep are for one purpose and one purpose only, to protect our relationship. If you vow before Ha Shem to keep these five rules, I will agree to wait for you as long as it takes. It will be our private ketuba, our covenant with one another. On the other hand, if you think these five rules are too hard or unreasonable, then we need to end our relationship right now, tonight. So, it's up to you, Jake. Do you think you can accept these rules?"

I didn't know quite what to say to her. This Rachel's torah seems pretty unreasonable to me. She is right on one point. I am naive when it comes to understanding women. She now is waiting breathlessly for my response. I need to answer her quickly.

"Rachel, I expected your rules to be tough, but honestly, this last one seems overly rigid. To require me not to work alongside a woman in the lab? After all, we are adults. I mean can't we just trust one another to honor our ketuba?"

My answer cut her like a knife. She says nothing. Tears are welling up in her eyes and streaming down her cheeks. She reaches for her napkin to stem the flood. I watch as she tries to gather enough composure to speak.

In a barely audible voice, speaking like she would to a poor helpless child, she tells me, "Jake, unfortunately, your response to the rules I have proposed falls short of the mark. I would like to accept your suggestion, but I cannot. I cannot change any one of these five rules and agree to marry you. I am so sorry, so very sorry. Our relationship must end tonight."

A flash of anger rises up from my stomach, taking me totally by surprise. Where did it come from? I try desperately to control my tongue, but I cannot. "Perhaps you will find Jeremiah Levitt more willing to accept your torah than I, Rachel."

"Who told you about Jeremiah Levitt, Papa? Jamie is a good friend, but he is not the man I love. Jake, you are the one I wanted to marry, not Jeremiah Levitt."

Now the tears really begin to flow. My angry outburst has placed a dagger in her heart. I think back to the first time I met Rachel. She so quickly intoxicated my soul. I was drunk with love for her then, and I still am. In spite of the straight jacket these rabbinical rules will place on my life, I cannot imagine living without this woman. My only solution is to agree to her terms, ipso facto.

"Rachel, it is evident to me your father has passed on his wisdom to you."

"No, not Papa, Jake. These rules came from the other side of the family They are from Hanna, my dear mother, and they are not impossible rules to keep. My father has kept them faithfully for many years."

"Rachel, I have no choice but to surrender unconditionally. I will keep your rules to the best of my ability. I love you with all of my heart and I can't imagine living my life without you at my side."

Reaching across the table, I carefully wipe the tears from her face with my handkerchief. It takes her several minutes to process my sudden reversal of field. She finally manages a small smile. "Are you sure, Jacob? I don't want to force you into something you will regret later."

"What's to regret? I'm just a poor shlemiel, hopelessly in love with a wonderful tsatskelah, a very beautiful and talented woman."

"I've gotten mascara all over your handkerchief. I'm sorry. I was so hoping and praying you would agree. I realize I am asking a lot from you. You men are so dumb about these things. I wish you were otherwise, but you aren't. That's why I needed you to make it official, to vow to keep this agreement before Adonoi. To be completely faithful to me, and to remain celibate until we marry."

After my earlier resistance, I knew she needed something more than a quick acquiescence to the rules. In my most official sounding tone of voice I agreed to her request.

"I hereby swear to keep this covenant, to live by these rules and any others my bride-to-be may ask of me, to remain faithful to her to the best of my ability, so help me Ha Shem."

I watch as my audible oath before Ha Shem transforms her sad face into an image of utter joy. She gives me a spectacular smile as she yields her heart to me again.

"Then it is settled. We will wait for one another until your Papa agrees to bless our marriage."

"I love you much more than I can express in words, Jacob. I promise before Ha Shem to be faithful and true to you always, to honor you and serve you all the days of my life."

"Rachel, I don't expect you to be perfect. You are a strong-willed person and I respect your spirit. We no doubt will have problems and disagreements from time to time, but with Ha Shem's help I'm sure we will overcome them."

"Do you have any problem with me preparing our ketuba for you?"

"No, I don't have a problem with you preparing our ketuba. I'm sure I will agree with whatever you propose. This doesn't mean you will be married to a mouse, Rachel. I will speak up when I see things we need to discuss."

"Yes, Jake. I do not want a mousey man for a husband. Now I have a little surprise for you. I really felt you would come to agree with me on these new rules. I wanted so much to make this a perfect evening. So, I have prepared a little something special in advance. I put the last touches on it while you were meeting with Papa. Jacob, this is our very own ketuba."

I watch as she pulls out a cardboard tube from her oversized purse. She carefully removes a parchment roll from the tube. I beckon to the waiter to clear the table. Rachel spreads the parchment out in front of me. What I see is a true work of art. Evidently skilled in calligraphy she has written the agreement with a quill pen and India ink. The text is in Hebrew and English. The margins are decorated with little blue birds interspersed with red pomegranates. A quote from the Bereshith, the first book of the Torah, is at the top in Hebrew and in English. It reads: "Therefore, a man shall leave his father and mother, and shall cleave to his wife, and they shall become one flesh."

"Rachel, this is beautiful. It must have taken you a very long time to prepare it."

"Do you like it?"

"Yes, except you have forgotten one thing."

"What is missing?"

"You have not added our Bill of Rights."

She smiles warmly and hands me a pen.

"Jake, I want you to read all of the provisions before you sign on the dotted line."

"I am reading it already. This is great, Rachel, but I see here we are supposed to give one another a gift of silver as a dowry. I don't have anything with me. I don't even own a silver dollar. What do we do?"

"Not a problem, Jake. Just sign it."

After signing the scroll, I hand it back to Rachel.

"Jake, do you remember Ha Shem rested on the seventh day, the day after he created man and woman?"

"I remember. Adam said 'Whoa man!' when he saw Eve."

"That may be. Do you have seven dollars in your wallet?"

"Yes, I think so. Why?"

"Just give me your seven dollars please."

Knowing this girl, I did not question her any further.

"Jake, I have been saving a silver necklace for just this occasion. I love it, but I'm willing to sell it to you for the bargain price of just seven dollars."

"Really? Do I have to wear it in public? People can get the wrong impression."

"You don't have to wear it, Jake. You can give the necklace back to me as my dowry, my *zekukim*."

"Awesome, you certainly have planned everything. I just wish I knew more about this engagement tradition. I could have given you something more elegant."

"You already have, Jacob. You have given me your heart. And now it's time to give me your zekukim."

I watch as she reaches back in her purse and hands me a small jewelry box. I open the lid. Inside is a beautiful silver amulet embossed with the chai, the Hebrew symbol for life. "My goodness, Rachel, this is so thoughtful of you."

"Turn it over, Jake."

I turn the amulet over and read aloud the inscription on the back. "Miss Rachel Kaufman has consented to become the wife of Mr. Jacob Cahn. She has brought to him from her father's house her whole heart and life. Mr. Jacob Cahn has confirmed his love for her as evidenced by the gift of this silver amulet."

Today's date is inscribed at the bottom of the silver amulet. It is impossible for me to contain my feelings. I carefully wipe the tears from my eyes before signing our ketuba. I handed the scroll back to

Rachel and she signed her name.

"Rachel, I want you to keep our ketuba out of sight for now."

"Yes, I will keep it out of sight in my hope chest. It's time to take me home now, Jacob."

I beckoned to Pierre the waiter to come to the table. I paid the bill and then asked Pierre if it was possible for us to speak with the proprietor of the restaurant? I explained to him I wanted to ask the proprietor if he was willing to do a small favor for us, to witness our engagement agreement.

"Oui Monsieur. I am sure the proprietor, Jacques Bideau, will be pleased to do so."

Monsieur Bideau, the proprietor, came to our table and agreed to witness our agreement. We thanked him several times. Shortly after he left our table, Pierre returned with an envelope. Enclosed was a gift certificate for a meal for two at his restaurant.

Rachel and I walked hand in hand back to the car. Once inside the car we sealed our agreement with a long tender embrace followed by our first kiss.

A vision suddenly flashed brilliantly before my eyes. The now familiar Hebrew letters, the yod and the hei, like a burning bush, they are blazing in front of my eyes. I carefully and slowly forced myself to draw back from Rachel and started the car. We both were lost in thoughts and said very little on our way back to her home. Back in my apartment it took me most of the night to manage finally to turn off my brain and fall into the state of hynagogia.

6

Satan's Emissaries

"No weapon that is formed against thee shall prosper; and every tongue that shall rise against thee in judgment thou shalt condemn. This is the heritage of the servants of the Lord, and their righteousness is of me, saith the Lord.
(Isaiah 54:17)

Dark Enemies

It is still dark when I awoke and opened my eyes. I stared at the clock beside my bed, trying to focus on the numbers. It is a quarter of three. Somehow, I knew something strange and sinister is about to happen. There is a rank odor in my room, like the smell coming off of a garbage dump. I didn't have to wait long. Suddenly I see a bright silvery object appear not ten feet from my open window. I try to get up from my bed, but I am unable to move even a finger. I don't like what I'm seeing. Four strange looking creatures are now in my room. They have large heads, oval black eyes, and spindly arms and legs. Plainly these creatures are not human. I recall the TV program I watched just last week about Unidentified Flying Objects, UFOs. They interviewed several people who had been abducted, taken aboard UFOs and mistreated by creatures like these now gathered around my bed. They are called Greys. Although I have never seen a UFO until now. I have always believed they existed. I tried again to move, but it is like my body is paralyzed. Waves of terror now are coursing through my body and my stomach is tied up in knots. I begin to hyperventilate. This is not good. Suddenly I begin to slowly levitate above my bed. The Greys are intent on moving me toward the open window. Surely this cannot be real. I must be dreaming. No, this is real, pure undiluted evil. In this chess game black has moved first. I must escape before these creatures take me with them. I try to scream, but my voice fails me. There's only one thing I can think to do. I prayed out loud.

"Adonoi, I need your help right now."

Suddenly my body drops to the bed and the Greys flee back to the silver ship. I sit straight up in bed. The rank odor of evil in the atmosphere has been replaced by the beautiful smell of incense and I feel like I am being bathed in warm oil. I sense I am not alone. I cannot see anyone, but I am conscious someone else is here with me in the room. Yes, I know Ha Shem is here. I sense his power, his holiness, and his love. It is an awesome experience. The depth of my soul is blessed. Words now are coursing through the air above my head in gentle, undulating waves. It is hard to explain what I am seeing. As the words float over me, I hear them spoken, but not audibly. The words in my head are telling me, "Soon you will see, soon you will understand and fully know."

Frightened by the awesome power of the words, I bury my head beneath my pillow and wait. It must have been about ten minutes later when I slowly took the pillow off of my head. I am back in the real world again. Framed by the small window in my bedroom I recognize the constellation of Leo, the Lion of Judah, in the night sky. I am reminded of Dr. Mike's words. Yes, there is more, much more to this thing called "The Way." I continued to gaze at the heavenly host, trying to make sense out of this latest surreal experience. My mind turns back to a book I once read about UFO abductions. Some of these people who are abducted later reported finding implants in various parts of their bodies. These implants have been surgically removed and examined. They appear to be transmitting high frequency radio signals It is believed these implants may be changing the DNA of the victims. The parallels to my experience are too close for comfort. What is the Dark Side's intent? Are they attempting to change my DNA or possibly blocking my ability to serve Ha Shem? One thing is perfectly clear to me. I will not be flying solo. I will not face these battles alone. With this comforting thought, I drifted back to sleep.

The sun is streaming through my open bedroom window when I awoke on Tuesday morning. Last night's incident is still vivid in my mind. Never have I felt such a sense of intimacy with my maker. He is real and I now know I can depend upon him, and know him intimately. I wait impatiently for the world to wake up, anxious to share this latest experience with someone. Dr. Mike said he would be attending a conference and unavailable today. Looking at my watch I realize it's too early to call Rachel. Perhaps Al and some of the guys are having breakfast in the café. I quickly got dressed and headed downstairs. No such luck. The café is deserted. I will have to catch up to them later. After returning to my apartment I picked up one of the books from the Library and began to read. I can't seem to concentrate. My mind keeps returning to last night, replaying my awesome rescue from the dark side. I finally put the book down at half past eight and walked down to the café again. This time I see Al sitting at his usual table in the far corner of the room, eating another full course breakfast. I filled my plate with scrambled eggs, fruit, and a short stack and poured myself a cup of coffee. He is immersed in thought as I sat down next to Allen, no doubt uptight about Thursday's speech.

"Good morning Al. Thursday is game time for you."

"Yes, it all starts on Thursday Jake. The weatherman said we are in for some heavy thunderstorms. They are left-overs from the hurricane hitting the Yucatan. I hope we don't get drenched outside on the Quad."

"I'm not concerned about the weather Allen. My angst is how well my family will get along with Rachel's family. I'm trying to get them together at Saturday's Legacy Brunch."

"I wouldn't get worked up about it. Both of the families have alums. I'm sure they will find plenty to talk about, old times and such. Did you pick up your robe and mortar board?"

"Yeah, I checked them out last week, plus I bought a new suit. I want to look sharp when the families get together."

"And how are you going to fit your flat mortar board over your pointy beanie cap?"

"I will use my superior Jewish ingenuity and put my *kippah* in my pocket. I am ready for the big event, but I don't have an umbrella. You don't own an extra one, do you?"

"No, I don't. You better pick one up today. I believe you are going to need one."

Although I'm anxious to tell Al about last night's experience, I realize my story would invariably end up revealing my dream to him. I just patiently listened to Al's summary of our graduation schedule. After breakfast, I returned to my room and called Rachel. She answered my call on the first ring.

"Good morning Jake. Do you still love me?"

"Good morning. Rachel. You don't have to ask me whether I still love you. I will never stop loving you, and you can depend on it."

"Do you think you can learn to humble your male ego long enough to tell me so? "

"Rachel, I love you. You are the light of my life."

"That's better. So, how did you sleep last night?"

I do not want to share last night's experience with her, but I don't want to lie either. "I had a hard time falling to sleep."

"I'm glad you finally got to sleep. I was thinking, if you need to learn about Orthodox Jewry, I can loan you some books from Papa's library. Just let me know."

"I'm not sure your dad would approve of you checking his books in and out to me. You need to ask him if he wants you to do so. By the way, Allen tells me they are predicting some heavy rain for graduation week. Be sure to carry an umbrella with you. Will I see you at the Class Day reception and banquet Thursday evening? You do know our class has chosen Allen to give the graduate student address, right? The reception begins at six, and the banquet follows from seven until nine."

"I'm sorry, Jake, but I can't go. Hillel has scheduled a rehearsal for Friday's Shabbat service at the Marcus Hillel Center. Papa is one of the speakers. The rehearsal begins at 5:45 and they are serving a dinner afterward. Our family will be seated at the head table at the dinner. I'm so sorry. I really wanted to hear Allen's message."

"I'm disappointed as well. I definitely will miss having you with me. The Commencement Legacy Brunch on Saturday would be a great opportunity for our two families to get acquainted with one another. Is your family planning to attend?"

"Yes, we are planning to attend the Brunch. I will ask Papa if our two families can sit together."

"Great! I will arrive early and save some seats for us. "

"No, you don't have to reserve a table Jake. Papa will take care of it. Will there be the four of you?"

"Yes, Mom, Dad, Emily and myself. Something unusual happened to me again early this morning. I need to talk to you about it in person. Can we meet for breakfast tomorrow at the DUC, say about seven o'clock?"

"You want me to meet you at the DUC at seven in the morning? Why so early?"

"I have an appointment with Dr. Mike at nine."

"Okay, I can meet you there at seven. You might have to see a face without makeup."

"Thanks Rachel. I truly love you. I'll see you tomorrow morning dear."

After hanging up with Rachel, I dialed the hotel to check on Emily.

"Hello Emily, it's your Bro."

"Good morning Jake. Are you ready for Class Day?"

"I'm as ready as I ever will be."

"I am so looking forward to your graduation. You must be very excited."

"I'll be glad when all of this pomp and circumstance is over with. Most universities have one graduation ceremony and it's over, but not Emory. Tomorrow's reception is for graduates and their families only. Are you going?"

"No, Jake. I have a date."

"Hank Bowman again?"

"No, Hank is going to your reception. I met this nice guy here at the hotel. He a freshman. His name is Ted Greenwald. Do I have your permission to go out with him? I'll be eternally grateful."

"You know Dad has placed me in charge of you this week. I need to think this over Emily. I'll get back to you on it."

"I can have Ted talk to you. Please, Jake? I really like this guy."

"We will see. As I say, I will get back to you on this. I really don't think it is a good idea at this time, but I will think about it."

"Okay. I'll wait for your call. Bye."

"Goodbye Sis."

As I hung up the phone, this conversation leaves me feeling a bit uneasy. I never know what Emily is going to do next. I should have just given her an outright no answer. I'll be relieved when Dad resumes his control of this little klip. I need to talk to someone about last night's experience. I took Pastor McGuire's business card out of my wallet and dialed the number at the church. A woman answered the phone. She tells me the Pastor is not in his office. He is working at the Atlanta Mission Shelter downtown and suggests I contact him there. I wrote the number down she gave me and thanked her. When I dialed the Shelter, Pastor McGuire answered my call.

"Atlanta Mission, Pastor McGuire. How can I help you?"

"Hello, Pastor, this is Jacob, Jacob Cahn."

"Hello Jacob. Good to hear from you. What's up?"

"Pastor, I wonder if you have time to meet briefly with me sometime today? I realize this is short notice, but I experienced something rather unusual last night and I would like to get your take on it. Is there any chance we can meet for lunch someplace?"

"It so happens I am free today for lunch, Jacob. Why don't we meet here at the shelter? The Atlanta Mission Shelter is located downtown on Howell Mill Road northwest, between eighth and tenth streets. Do you think you can find us?"

"Yes, of course. Shall I meet you there at noon?"

"Noon is fine. You can park in back of the building. There are several reserved parking spaces that you are free to use. I'll meet you there in the parking lot at twelve."

"Thank you, Pastor. I really appreciate it. I'll see you at noon."

At 11:30 I left my apartment. I decide to take my backpack with the books and the return envelopes. Perhaps the pastor will agree to write a recommendation for me. I took the elevator to the parking garage. I fired up Old Blue and headed for downtown Atlanta. Pastor McGuire is waiting in the parking lot for me when I arrive.

"Hello, Jacob. You are right on time."

"Pastor, I really appreciate you taking time out from your busy schedule to meet with me. Where is a good place for us to have lunch?"

"We will have lunch right here at the shelter, Jacob? Just follow me."

"Pastor, this is a pretty old-looking building. It looks like it once may have housed a manufacturing plant of some kind. Is there some history to the facility?"

"Yes, the building originally was a textile mill during the early 1930's. All of those plants are long gone now. This is the oldest shelter in Atlanta. The Atlanta Mission opened in 1938 to feed the hungry during the depression years."

As we entered the building I expected to find a drab interior. Instead I found a hallway with walls brightly painted and decorated with scenes of Atlanta city life. I noticed a number of doors into offices located along the main hallway. At the end of the long hallway are doors opening to a large cafeteria. The cafeteria is filled with people eating their lunch at nicely decorated tables.

"I must say I am impressed. Pastor. The shelter is really awesome. Do you serve lunch every day?"

"Yes, we serve three meals a day. We make very good use of this old building. We have invested a lot of energy and funds to develop the facility, as well as our other satellite shelters. We are still a little short on room and funds. The demand for services has skyrocketed recently. Jacob, I am going to treat you to a meal from our kitchen. I believe the menu today is French onion soup, salad, and hush puppies, plus the beverage of your choice, coffee, tea, or soda."

"You are serving hush puppies? I'm looking forward to sampling some of those."

We waited patiently for our turn to be served. Pastor Bill chose a table near the back of the room where we could have some privacy.

After we finished our soup and hush puppies, I related my latest experience. I describe the silvery disk parked in mid-air outside of my window early in the morning, and the terror I endured as the Grays levitated my body above my bed. Then I told him of my wonderful rescue in response to my prayer for help. I tried to describe into words my sense of the presence of Ha Shem in the room with me, but I couldn't adequately explain what I experienced.

"Pastor, do you have any insights about this latest experience of mine?"

"Well, Jacob, I think I do. Before I give you my take on it, I want to hear what you think it means?"

"One thing I do understand. My prayer was answered quickly. Adonoi definitely came to my rescue. He actually spoke to me, at least sort of spoke to me."

"Sort of? What do you mean by sort of?"

"Well, I didn't hear an audible voice exactly. I saw these words wafting in midair over my head. As these words passed overhead, the sound of them resonated in my mind. It is like I heard the Lord speaking to me, but only directly in my mind."

"I find this very interesting, Jacob. Perhaps he is showing you something about the capabilities of a human mind and brain. What words did you see?"

"I saw and heard just a short sentence. He told me the following: 'Soon you will see, soon you will understand, soon you will fully know.' As I said, I saw the words with my eyes and heard the sound of the words in my mind, but not audibly out loud. I became aware of the power of those words. The image and sound of the words was followed by something even more profound. I became aware of a holy presence right there in my bedroom. I'm certain it was Ha Shem. I have no doubt he is real and doing something very special in my life. At the same time, I'm feeling very vulnerable. Frankly I'm afraid to get too close to him. Why does he want to use a secular Jew like me instead of a thoroughly trained and skilled pastor like you? I don't know whether to welcome these strange events or to try to avoid them."

"I certainly understand what you are saying. I can tell you the fear you experienced in response to God's presence is completely normal. Your awareness of the presence of evil also is valid as well. Can I ask you a few more questions?"

"Yes, of course."

"Have you done something out of the ordinary just before you experienced this latest episode?"

"No, I don't think I have done anything unusual. Yesterday I visited the MARBL collection at the Woodruff Library and met with Doctor Michael Stein. I spent a few hours with him while he reviewed new books. After we finished with the books, I had the opportunity to share my recurrent dream with him. Dr. Mike arranged for me to borrow a number of books on dreams from the library. I took four of them back to my room, plus an extensive list of references on the subject of dreams."

"Did you read any of those books yesterday?"

"Well, I did browse a few."

"I assume you still have the notes I gave you at the grad party last week?"

"Yes, I keep them in a drawer next to my bed. I read them every now and then."

"Satan always attempts to bind us up, to restrain us from doing anything other than living a meaningless life in this world. I believe you have him worried, Jacob. Do you recall the warning the Lord gave you?"

"I will have to review your notes again."

"I am referring to the warning the Lord gave you about misusing your gift. The Lord warned you to not walk in the ways of Balaam. The Devil desires to tempt you into misusing your anointing. Do you know what the way of Balaam means?"

"I'm not sure. I assume it is to help me avoid misusing my gift by trying to make money from it."

"Making money from the gift of prophecy has occurred many times in the past, but the warning he has given to you goes beyond the making of money. He is warning you to avoid being drawn into the mystery religions, the occult, and sorcery. God hates these sins because they supplant Himself and place Satan on the throne. God declares them abominations. The Lord is cautioning you to remain faithful to your calling and to avoid opening a portal to the dark side."

"Pastor, I really feel overwhelmed by all of this. How do I deal with spiritual darkness? What does Ha Shem mean by saying I will know soon?"

"The words over your head reveal you have not yet arrived at the point where you completely understand your destiny. Jacob, I am convinced the Lord is preparing you for something very awesome, something special. As far as what soon means to God I can only say you will not be late in knowing, nor early. You will come to know just at the right time. Let me say a few words about this thing we call prophecy. First of all, prophecy is not necessarily something you see in the future. Prophecy is basically a communication between the natural mind of man and the supernatural realm. It can be about past, present, or future events. There are two parts to prophecy. The first part is the reception of the message. It is a God-given gift. But communication is not the only thing involved. The prophet must correctly interpret the message. The correct interpretation depends upon another spiritual gift, the gift of wisdom. Right now, you only have experienced the first part of the prophecy. If you had received the second gift, the gift of wisdom, we would not be talking about this today. You have still much to learn. Tell me again what the Lord told you in those words floating over your head, Jacob."

"He said, I soon will see, understand, and fully know."

"This is step one of his revelation for you, Jacob. You should be encouraged because God is not through with you yet, not by a long shot. I wish I could help you more, but I have no idea where this is all going to lead. I must admit I have never encountered anything like these experiences before. Have you still not made the decision to accepted Yeshua as your Lord and Savior?"

"Well, I pray a lot more now and I am aware of my faults, my sins of the past. Allen Schaeffer has explained the Christian faith several times to me, but I haven't been ready to declare Jesus as my Lord and Savior. More than one God is a problem for us Jews you know."

"Yes, polytheism is difficult to understand. Considering these supernatural events in your life, it really is shocking to me you have not yet placed your trust in Jesus. No question God is dealing with you on this amazing supernatural level. I must admit I don't understand it."

"I don't have a problem with others believing in Jesus as the Messiah. As for myself, I have continued to follow the advice of my father like a good Jewish *mensch*. My belief is purely monotheistic. I have been taught we serve one God. The Shema describes the Almighty as one, not three, or three in one. I interpret this passage as ruling out the Christian world view? My father has very little regard for Jesus, often referring to him as 'that man.' Let me tell you, those words are not expressing a favorable opinion about Jesus."

"I would like to straighten you out on this issue of the trinity, Jacob. I realize the mention of the name of God violates the teachings of the rabbis, but I need to use the word in order to explain the Christian doctrine of the trinity. Can you excuse me if I use the holy name?"

"I really don't have a problem with it."

"Let's proceed. The correct translation of Genesis Chapter One and verse Twenty-six says, 'Let us make man in our image after our likeness.' The Hebrew word for God used here is the word Elohim. The word is a plural noun. In English, we require the plural subject 'us.' The insistence by the Jewish rabbis on there being one God is based on the Shema found in Deuteronomy Chapter Six and verse

Four. Here the word for God is Yahweh, translated as Jehovah in the Protestant Bible. It is in the singular case. The Lord uses this word when speaking to the Nation of Israel. Here is the reason I believe God uses a different word here. How could Israel have maintained their monotheistic belief in a single God, or even understand the trinity when their neighbors worshiped a plethora of other gods? I believe the cultural and historical context of the time explains why the Lord used the singular form rather than the three distinct forms used in the New Covenant. Once Yeshua and the Holy Spirit are fully revealed to mankind, the time is right for the God to reveal the trinity, God's three-part nature, Father, Son, and Holy Spirit. Do you understand what I am saying?"

"Yes, I understand your argument."

"Let me give you a quick analogy. Water exists in three forms, liquid, ice, and vapor, but it is all water isn't it."

"Yes, your analogy does hold water. Recently I became aware of the fact some Jews have accepted Yeshua as their Messiah while still retaining their Jewish traditions. They usually refer to themselves as 'completed Jews.' But what's to complete? Aren't we all born Jews, sons of Abraham?"

"Jacob, I don't have time today to sufficiently explore the evidence for you to support the claims of Jesus as the Messiah. So, I'm going to give you some homework. I want you to read several scriptures for me. Read them with an open mind and no preconceptions. Pretend you are a Hebrew scholar, a rabbi. Will you do this for me?"

"Yes, I will try. I already have several books in my room to digest."

"This assignment will not take a lot of your time. First, I want you to read a few scriptures from the Old Testament, the Tanakh. I want you to take some notes on what is occurring. Then I want you to read some passages from the New Testament and compare those readings to the events in your notes. Do you have a pen and something to write on?"

"I have a pen and I'll use the back of this paper place mat."

"Here is the list. In the Tanakh I want you to read: Psalm Twenty-two, in Isaiah Chapters Fifty-two and Fifty-three, in Zechariah Chapter Eleven, verses Twelve and Thirteen. Then in the New Testament, your assignment is to read Matthew Chapters Twenty-six, Twenty-seven, and Twenty-eight. In Mark, Chapters Fifteen and Sixteen, in Luke Chapters Twenty-two, Twenty-three, and Twenty-four, and in John Chapters Eighteen through Twenty. Have you taken a statistics course?"

"Yes, I have taken statistics as well as logic."

I want you to using statistics and logic to compare the Tanakh readings with the New Testament readings."

"I can do it."

"Good! I want you calculate the probability of obtaining matching events between prophecies given hundreds of years before the birth of Jesus with the events recorded in the New Testament? Let me know what you find. Do you have access to a Christian Bible?"

"Yes, there's a Bible in the nightstand next to my bed. I believe the school puts one in every room."

"No doubt the Gideon Bible in your room uses the King James translation. The translation is fine if you have studied Shakespeare, but you might want to borrow or purchase a Bible with a more modern translation."

"I will read the scriptures you suggest, but Thursday is Class Day. I'm not sure when I can get my comments back to you."

"I understand. Can you agree to pursue this project after you arrive in Baltimore?"

"I should have some time on my hands before classes start. I will give this Biblical comparison thing a try. Do I get extra credit for this assignment?"

"If you follow through on this, you will certainly earn extra credit. You might even get rid of all of your debts."

Pastor Bill's face is smiling. I can tell he is pleased I have agreed to do the study of the biblical passages.

"I do appreciate your advice, Pastor Bill. There is one other matter you might consider helping me with."

"Just name it?"

"I need to obtain three letters of reference for the Institute at Johns Hopkins. I know we have just met, but you already know a lot about me. Would you consider writing a recommendation and forwarding it to the Institute?"

"Sure, Jacob. I am more than willing to write a recommendation for you. Do you have the address with you?"

"I have something better. I have a self-addressed envelope here in my backpack.”

I managed to locate the envelope and handed it to Pastor Bill. “I hate to rush away, but I have to get back to the campus. I need to check on my sister Emily. She is staying at the Omni downtown. I promised Dad I would keep tabs on her until they arrive Friday morning.”

“I remember meeting Emily and I can understand your concern. She certainly is a live wire. You need to chaperon her carefully, Jacob.”

“I plan to watch her like a hawk. I’ll be on my way then. You have been a great help Pastor, and I very much appreciate it.”

“God bless, Jacob. I believe you have my card with my phone number and email. Let me hear from you soon.”

“I will let you know how things are going. Thanks for the lunch. I really enjoyed the hush puppies.”

“God speed Jacob.”

I left the homeless shelter and arrived back to the campus about half past three. Emily called on my cell phone just as I pulled into the parking lot. “Hello, Jake.”

“Hello Emily. Is everything okay? “

“Yes, it’s all good. I’m calling you from the hotel lobby. Do you remember me telling you Ted Greenwald asked me to go out with him? He wants to take me to see this play *Peachtree Battle* at the Ansley Park Playhouse Thursday night. Ted tells me it’s the longest running play in Atlanta’s history. I really want to see this play Bro. Can I go?”

“Emily, I told you I needed to think about it and would get back to you.”

“Well Ted is here with me right now, Jake.”

“That’s all well and good Emily, but—”

“Won’t you just talk to him now Jake?”

“Emily, I don’t need any new problems. Dad expects me to keep watch on you. I would need to chaperon you two, and I can’t go on Thursday. As you know Thursday is Class Day. Why don’t you two go to the play with me tonight instead? I will pay for my ticket.”

“No can do, Jake. Ted has two tickets for Thursday night. They are sold out every night this week.”

“I haven’t met this Ted before. How am I supposed to know if he’s kosher?”

"With a name like Greenwald you are asking me if he is kosher? C'mon Jake, he's a nice guy. I know you will like him. Anyway, you will be off to Baltimore next week. So, what's the difference?"

"I plan on leaving next Tuesday. You said Ted is there with you now?"

"Yes, he is right here. Will you just talk to him please?"

"I suppose so. Put him on the line."

I hear muffled talking. Evidently Emily has covered the phone with her hand. No doubt she is cluing Ted in on what to expect from her older brother. I hear a new voice on the line.

"Hello Jacob. This is Ted. Emily says you want to have a few words with me."

"Yes, Ted I do. Emily tells me you also are enrolling at Emory this next semester."

"Yes, I am. I hope to major in business."

"Ted, you need to understand our father has given me strict orders to supervise my sister Emily while she is here in Atlanta. My father insists I accompany her on any first date. Unfortunately, I cannot chaperon you two Thursday evening as it is Class Day. Since I do not know you I am going to have to ask you to postpone your date until another time."

"Yes, but I have two tickets for this Thursday night."

"Since the play is so popular, I expect you can still get a refund."

"You don't have to worry about Emily. I will take good care of her and get her back to the hotel on time."

"My father's rules are not guidelines Ted. Since you are Jewish, think of them as an addendum to the Law given to Moses. I am sorry, but I am not willing to approve your request. Perhaps you can just go down to the playhouse and sell your tickets to another couple."

"Is there some way I can change your mind, Jacob?"

"No Ted, there's no way I can allow this to happen. Please put Emily back on the line."

"Okay."

"Hi Bro. I heard everything you said. You don't have to worry. Bye."

"Not so fast Emily. I want to hear you say you understand and you are not going out with Ted. Why am I feeling you are going to ignoring my rules? Are you planning to go out with him anyway? I hope not, because I will be checking on you. Don't make me go looking for you in Atlanta in the middle of the night. I will if I have to and I will thoroughly embarrass you by dragging you out of that play. Do you hear me?"

"I hear you Jake. You sound more like a communist enemy instead of a brother. Goodbye."

I hear a click on the other end of the line. I'm feeling very uncomfortable with her response. Did I hear her say she is not going out with this guy? I just have to hope Ted has taken my decision to heart.

I took time to call the playhouse to check their schedule. They said Thursday's play will end about eleven thirty.

I have some time on my hands. I remember Mike saying the Society for Neuroscience conference is being held at the Omni all this week. I thought of returning for another visit, but decided to stay out of the rain in my dry apartment About nine thirty I fixed myself a peanut butter sandwich and put the tea kettle on the stove for a cup of tea. Waiting for the tea kettle to boil I began to consider my assignment from Pastor McGuire? Am I bound too tightly by my particular world view to undertake such an unbiased study? What would Rachel think if I was to declare I believed in Jesus as the Messiah? I wondered also if I would ever discover my particular life's calling and destiny? It is ten o'clock when I decided to go to bed. I set my alarm clock for six. I can't afford to miss breakfast with Rachel at seven. I wonder what she looks like early in the morning? It wasn't long before I drifted off into never-never land.

Giants in the Land

I am sleeping soundly when my alarm clock rudely wakes me up at six. I am pretty sure it is Wednesday. I turn off the alarm and manage to struggle into the shower. The thought of sharing breakfast with Rachel is intoxicating. My eyes fall on the stack of books on my kitchen table. I put the books I needed to return in my backpack along with the recommendation envelopes. I dressed in my best jogging sweats, grabbed my backpack, and lit out on a dead run for the DUC. I arrived at seven on the nose. Rachel is already seated at a table near the window.

"Good morning, Rachel. You look wonderful. I like your hair hanging down like that."

"Good morning Jake. I have an excuse for my straggly hair. I hope you realize I do not customarily get up this early in the morning. Why didn't you pick a spot closer to my house?"

"You are right, Rachel. I should have suggested we meet someplace closer to you. Will you forgive me?"

"Yes, of course. So, what is it you need to tell me?"

"Shall we eat first?"

"No, we can order breakfast after you tell me what is bothering you."

"Okay, but my story may spoil your appetite."

"I'll take the chance. What is so important you need to see me this early in the morning?"

"Rachel, I experienced another strange thing. I knew something weird was going to happen that night. When I woke up at midnight, I knew something evil was in my room. The room was simply reeking with a terrible smell. It smelled like a garbage dump. I tried to move my arm, but I could not even move one finger. I looked outside my open window and saw this silver UFO craft. Then four Greys with spindly legs and arms and black oval eyes began moving me from my bed, attempting to take me to the UFO. All I could think to do was pray. Immediately when I did so, the creatures dropped me back onto my bed and scrambled back to the UFO. Afterward, I became aware of another presence in my room. The vile odor is replaced by a sweet smell incense."

"Are you serious, Jake?"

"Yes, it is pretty wild. The next thing I know I see words wafting through the air above my head. I hear the sounds of the words in my head."

"What words?"

"The words said, 'Soon you will see, soon you will understand. Soon you will fully know.' There is no sound, yet these words are echoing in my brain. Then it was over."

"Jake, I must say you are stretching my ability to handle these weird experiences of yours."

"Yes, I know. But I still need to tell you these things. You need to know all about me. I don't want to hold anything back from you. If you remember, we promised we would be completely honest with one another."

"Yes, I remember. I will add this last experience to my data bank."

"I appreciate your trust in me, Rachel. Really I'm not a complete fachadick."

"No, *Bubele,* you are wonderful and I love you. We will deal with this thing."

"I love you Rachel. Thanks for hearing me out about this latest excursion of mine. Let's order some breakfast."

We spent another few minutes talking with one another over breakfast. To spend time alone with Rachel is wonderful. It is the best medicine a doctor can prescribe. I feel like a new person. Rachel left at 8:30.

I called Dr. Stein on my cell phone to check on our appointment. He said he still is available to meet with me at nine o'clock. As I am leaving I spotted Professor Adams, my Biology instructor, sitting at a table near the doorway. I decide to give him an update on my career decision.

"Hello Dr. Adams. Can I join you?"

"Hello Jacob. Yes, have a seat. Have you had your breakfast?"

"Yes, I have. I can't visit with you very long as I have a nine o'clock appointment. I just wanted to let you know about my future plans. I found your comments about the emerging fields in genetics very convincing. I am taking your advice and changing my major."

"Really! I thought you were planning to enter the medical school here at Emory."

"Yes, medical school was the plan. After a considerable amount of time arguing with myself, I finally decided to go for broke and apply to the McKusick-Nathans Institute of Genetic Medicine at Johns Hopkins. They have agreed to accept my application. Of course, I realize the application is only the first step."

"Yes, they carefully screen their applicants to the program. Johns Hopkins is the most respected genetics program in the Nation. I certainly hope you succeed in being accepted. Go for it, Jacob. You are a very capable young man."

"Dr. Adams, would you be willing to refer me to them? They require three recommendations. I am sure your recommendation would be looked upon with favor as you are a well-respected biologist."

"Well thank you, Jacob. I would be more than willing to recommend you. By chance do you have an address of the Institute with you?"

"As a matter of fact, I just happen to have an envelope in my backpack."

I lost no time in opening my backpack and handing him one of the envelopes.

"Very good, Jacob. When will you be visiting the Institute?"

"I plan on leaving right after the commencement activities are over."

"I will get right on this for you."

"Thank you, Doctor."

"Keep me posted, Jacob."

"Yes, I will. My appointment is across the campus, so I better leave."

"Okay, Jacob. Nice of you to share your plans with me. Have a great day."

I quickly made my way across the campus to the Woodruff Library. It is five minutes after nine when I arrive. Dr. Mike is waiting for me. I opened my backpack and placed the books on his desk. He checked them back into the library.

"I see you have finished quite a few books over the last three days."

"Well, I skimmed each of them to see if they were of any help."

"Did some of them extend too far beyond your sense of reality, Jacob?"

"To be honest, Dr. Mike, they did. In fact, I found some of them to be, in a spiritual sense, quiet dark and foreboding. It may be because of the bad experience I had early this morning."

"Do you want to tell me about this dark experience?"

"I'd rather not. It was very upsetting to me. "I can if you insist."

"Why don't you just give me a quick summary of it and leave out the scary part."

"Yes, I can do that. Yesterday morning I had an encounter with a UFO. When I awoke yesterday there was an odor in my room like the smell of a garbage dump. This space ship was parked right outside of my open window. There were these three Greys in my room trying to take me into their craft. I was frozen and unable to move. I asked Ha Shem to come to my aid, and he did so. They immediately fled back into the UFO and sped off to who knows where. Here is the good news. The rank odor was suddenly gone and my room was anointed with the smell of a sweet aroma. Then I see these words wafting in the air over my head. They are not audible, but somehow I hear them in my brain."

"What were the words, Jacob?"

"The words are saying, 'Soon you will see, soon you will understand, soon you will fully know.' Then it is all over. I told Pastor McGuire about it at lunch yesterday, and Rachel about it just this morning."

"I see. Well, that is quite a story Jacob. These creatures do exist. They are demonic. The fallen angels have given demons these weird bodies. There are holy angels and fallen angels. Both hold watch over the Earth and do battle in the Second Heaven. The fallen angels mated with human women. Out of those unions came the Rephaim, the giants. This is mentioned in the Torah as well as in extra-biblical writings such as the Book of Enoch."

"Do you think I was in any danger during this episode?"

"Yes, Jacob, I believe you were in a lot of danger. Do you know what was going on there, Jacob?"

"I have no idea."

"You most likely are being targeted by these satanic beings for manipulation."

"What kind of manipulation?"

"They usually are attempting to produce hybrid super humans as in the past. However, in your case I believe they were attempting to prevent you from carrying out your divine purpose."

"Done in the past?"

"Yes, we have many stories and legends about such hybrids and embodied demons. We even have physical specimens, but very highly placed persons in government have largely succeeded in hiding most of this evidence from the public. All through Scripture we read about these creatures called nephilim. Did you ever wonder why the Lord instructed Joshua to totally destroy the Canaanites? The DNA of these people groups was corrupted. The Lord destroyed these groups to protect Israel's genetic seed., but they continue to reappear. Isn't it interesting how your visions touch upon your future vocation? Perhaps Adonoi desires to show you some of Satan's nasty handiwork."

Dr. Mike paused to glance at the clock on the office wall.

"Jacob, it is lunch time and I have a class at one o'clock. We will have to postpone this exploration for another time. I'm very sorry. Can I give you a rain check?"

"Yes, of course. I do have one final question for you. Would you consider writing a reference to Johns Hopkins for me?"

"Certainly, I would be happy to do it for you son."

I reached into my backpack and withdrew the last of my three envelopes and handed it to Dr. Mike.

"Thank you, Dr. Mike, for everything. I appreciate you putting up with my bag of big troubles, my *tsuris*. I'm sure a recommendation from you will help my chances for acceptance. I will send you my address as soon as I find a place to live."

"I will look forward to hearing from you, son."

"I hope you will not be offended if I ask one more question? It is of a personal nature."

"We have shared personal thoughts before. What is your question?"

"Twice now you referred to me as son and have used the word on two other occasions."

"So, what is your bottom line, Jacob?"

"I am wondering if you might be considering me as your adopted son?"

"You have seen through me, Jacob. I have a special affection for you. In my future written communications, I will use the word with a capital letter."

"These meetings have been a very great blessing to me. I feel a deep sense of admiration and affection for you also, Dr. Mike. May Adonoi bless and keep you."

"So, my spiritual son has given me his blessing. I receive it Jacob. You now are truly a member of my *mishpachas*, my spiritual family."

We shook hands and then Dr. Mike gave me a bear hug. I felt a bit blue as I left him. I will not see this man's face for quite a while. I will miss him.

A few drops of rain began to fall as I made the journey back to my apartment. By the time I arrive, the rain is coming down in sheets. Allen's weather report was quite accurate.

I decide to take a quick shower, just to warm up. After my shower, I dressed in some dry clothes, and had lunch. Then I settled down to read one of the books Dr. Mike had selected. Before I knew it, the sun was setting. I caught the evening news at six. After the headlines, I turned the TV off, ate some Dinty More stew for dinner, and went to bed early. Tomorrow is Class Day.

7

The Overcomer

"Have not I commanded thee? Be strong and of a good courage;
be not afraid, neither be thou dismayed; for the Lord thy God
is with thee whithersoever thou goest."
(Joshua 1:9)

Class Day

I am up early on Thursday morning, not sure how to start my day. The Honors Program does not start until 5:30 tonight. The reception for honors students and their family members starts at eight o'clock and is supposed to be over by nine. Rachel and her family cannot attend the Class Day celebration. Her father is one of the scheduled speakers at a meeting of Atlanta Jewish Leaders. Even though she is graduating with honors, her father expects her to attend his meeting. At least they will be attending the Legacy Luncheon on Saturday. I went down to the lunchroom for a snack. The place is deserted. I grabbed a cinnamon roll and a cup of coffee and took them up to my room. I want to spend some time revisiting the Book of Enoch. The beginning of the book is quite incredible. Enoch declares he has written this book to the elect and righteous living during the End Times described in the Book of Revelation when Ha Shem shall destroy all the wicked and godless people in the world. I believe this day is fast approaching. The Book of Enoch describes the descent of the two hundred Fallen Angels on Mount Herman located in Southern Syria. These angels who chose to mate with human women, corrupting mankind's DNA. The result of this unholy union produced the Nephilim, the giants of old. It seems there are certain locations where connections are made between our physical Earth and the Second and Third heavens. There is the bible passage where Jacob, my namesake, sees a ladder with angels passing up and down between earth and heaven. I recalled the mountain where God met with Moses. I turned on my laptop and began to search for other examples of mysterious connections between heaven and earth in the Bible. The tower of Babel, located in Mesopotamia came up with the story of Nimrod. Nimrod apparently was a giant, a possible hybrid Nephilim.

Chapter Six in the Book of Genesis describes the Fallen Angels who chose to cohabitate with human women, producing hybrid offspring, the giants. The Book of Enoch gives much greater detail about these two hundred Fallen Angels who descended upon Mount Herman in Southern Syria. In Matthew Chapter Sixteen Jesus takes his disciples to this same area and declares the powers of Hell will not prevail over his church, effectively reclaiming this portal from Satan. There is a confirming passage in Chapter Twelve of the Book of Joshua where it states King Og, the last of the giants, ruled over this region. Are there giants on the earth? It appears they may still be here.

My search led me to an example very close to home. There is a controversy whether such a portal to the heavens exists on Mount Graham, in Arizona. The local Apache tribe opposed the installation of a large telescope on the mountain because it was used by their shamans to open a gateway to the heavens, seeking spiritual guidance. The tribe took legal action. They lost the case. If the tribe had used the mountain as a burial ground, their claim most likely would have prevailed. This outcome suggests there was strong political pressure influencing the court's decision in favor of NASA, the University of Arizona, and the Vatican, the joint users of the facility. Among several astronomical facilities built on the mountain is a very large binary telescope used to study the sky in the infra-red spectrum. According to information on the internet, the Vatican astronomers began observing UFOs, in the early 1960's. The Vatican still remains closed-mouth about what they are observing.

The Vatican has said only something is coming toward the earth. What really caught my attention is the acronym for this telescope. The official name is the "Large Binocular Telescope Near-infrared Utility with Camera and Integral Field Unit for Extragalactic Research." Yes, the acronym for this telescope is LUCIFER. They named this telescope after the Angel of Light, the Devil. The Vatican has signed on to the idea aliens from other planets and galaxies are visiting our globe, intent on taking up residence here on our planet. Several recent news reports caught my interest. For some mysterious reason, prominent political and religious leaders recently are taking trips to Antarctica. There are rumors of an alien presence there, a group they refer to as The Nordics. At the end of the World War II, the Nazis apparently established a submarine base in Antarctica. Admiral Byrd launched an invasion of this base, but was repelled by a fleet of UFOs. I found a very interesting post by a Messianic rabbi on YouTube who found several intriguing passages in the Book of Enoch, perhaps relating to Antarctica. In Chapters Eight and Nine of Book One, Enoch describes a chain of seven mountains aligned roughly north-south, the center peak being the highest. This mountain range Enoch describes as being located in a wasteland to the South. According to the passages in Chapters Ten, Eighteen and Nineteen of the Book of Enoch, seven stars (the Fallen Angels) are taken to this mountain range and buried beneath the rocks by the Angel Raphael. This imprisonment of the Fallen Ones was done because of the sin of cohabitation with women on the earth. They were to remain imprisoned until Judgement Day. Enoch describes this place, located in the South, as a barren land where the sun never sets. Interestingly, there is a chain of seven peaks in Antarctica. The central peak, Mount Vinson, is the highest peak on the continent, rising 16,067 feet above sea level. Some speculate these Fallen Angels are still there, imprisoned beneath the earth for thousands of years. Although the Book of Enoch is not included in the canon of Scripture, I find it very interesting. I am left wondering why there is so much interest in Antarctica now?

I recalled the opinion Allen shared with me when I first met him. He took time to explaining the differences among various Christian denominations. I remember him saying the Vatican, under the new Jesuit pope, is departing from the true doctrine of the Scriptures. He said Catholic theologians are in the process of developing an entirely different bible, supposedly including text recognizing contact with aliens. Allen felt this contact began long ago, before the time of Noah, and involves sharing of higher technology, just as described in Enoch. He said this process has been accelerating in recent years. It seemed preposterous to me at the time, just too radical a story. I decide to humored him. I don't want to start any argument with my new friend. After studying this Book of Enoch and reading to some of the postings on the internet, I'm not as skeptical.

I finished my sleuthing on portals and gateways to the heavens at half-past four. I needed to get out of my apartment for a while. I'm anxious for this whole commencement week to be over. Since I already have obtained the three recommendations necessary for entrance to the Institute, there is no reason to schedule any meetings today.

I put my new suit on for the first time. I could not locate my old raincoat. I put my overcoat on, grabbed my umbrella and left for the Glenn Memorial Auditorium. The rain is still falling. I drove my car to the main campus and parked it at the Fishburne Parking Deck. I am a half hour early when I enter the Auditorium. I locate the two reserved seats near the front of the room, one for Alice and one for myself. Since Allen is the keynote speaker, he will be seated on the podium with the other dignitaries.

The Glenn Auditorium is used jointly by Emory University and the Glenn Memorial United Methodist Church. Emory uses it as an auditorium and the Church uses it as their main sanctuary. The building is named to celebrate the life of Reverend Wilbur Fisk Glenn, an Emory alumnus who for fifty years served as the Methodist minister and leader of the denomination in the South. The Glenn is used primary for music rehearsals and performances.

Allen and Alice arrived shortly after I did. By six o'clock the room is packed full with a mob of very excited graduates. The Methodist Pastor of the Glenn gave the invocation. Then the Dean of Students took the podium and began by reviewing the commencement schedule, emphasizing once again our obligation to attend the Legacy Honor's luncheon and the Torch and Trumpet Soirée on Saturday, the Baccalaureate Service followed by brunch at the Dobbs Center on Sunday, and the Commencement Service on Monday. He then made a few remarks congratulating the graduates on their achievements.

The Dean gave a very nice introduction for Allen, listing his many achievements in the school and his community involvement. Everyone present took note of the Dean's glowing admiration for his achievements and his calling to the ministry. He is greatly respected by the students for his Christian dedication and commitment. Alice is excited as Al comes to the podium. I am confident my friend will give a great speech. He certainly did so. I'm sure all of the graduates, including myself, will remember the challenging message he gave. The title of his talk, To Reach for the Stars and Beyond, was very well received. Unlike the fiasco experienced during the rehearsal, he kept the crowd's attention through the entire speech. As Allen ended his talk, the room burst forth with applause and bravados. Alice is wildly clapping her hands. That's when I first saw she is wearing a beautiful diamond ring. I have to smile. Al is not going to let this girl get away from him. The Dean made a few closing remarks and then released us to attend the Senior Class Reception being held at the Emory Conference Center Hotel in town. I volunteered to drive Allen and Alice to the hotel rather than try to negotiate the trip using the shuttle service. I felt rather out of place without Rachel at my side. Allen and Alice sat together in the backseat. After a great meal, we returned to the apartment complex at 9:45. I went straight to bed. Tomorrow is the Legacy Brunch. I again will miss having Rachel at my side. She is attending the Hillel Shabbat Service. I finally was able to turn off my mind and get some sleep.

My sleep didn't last long. I am jolted awake by the sound of my cell phone. Half asleep, I struggled to locate it in the dark.

"Hello, who is this?"

"It's your sister."

"What time does your watch say?"

"My watch doesn't say anything Bro. I'm just calling to let you know I am safely back at the hotel. I'll see you tomorrow. Goodnight.

"Not so fast young lady. I distinctly told you not to go to the play. Did you decide to go out with this guy anyway?"

"The play was great. Afterward, Ted simply insisted taking me to this favorite place of his. There is a little bar and grille right downtown near the playhouse. They have a great band and a dance floor. Ted's a great dancer."

"Emily. This is tsuris. You are in big trouble. I am going to have to tell Dad about your failure to keep the rules we agreed on?"

"Sometimes you are a real schmuck Jake. I hope you realize I didn't have to call you tonight."

"No, you assumed I called last night to see if you are at the hotel. I planned to call, but I fell asleep instead."

"Why don't you calm down, Jake. Have a cup of coffee. Everything is great. Ted and I had a wonderful time."

"Emily, I am having a hard time controlling my temper. You are busted. I'm going to tell Dad about this tomorrow."

"Go ahead. Be a tattler. See if I care. Goodnight!"

I am now talking to a dial tone. At least Dad and Mom will be arriving tomorrow and I can relinquish this task of playing nurse maid to Emily. I wonder who Dad is going to blame when I tell him about this? Will he come down hard on her, or will I end up getting the brunt of his wrath for failing to rein her in? This is not going to be a favorable time to discuss my decision to become a geneticist. I should have checked up on her right after the Class Day event was over. There's no sense in crying about it now. What's done is done. I turned over and went back to sleep.

Clash of Wills

It is Friday morning. My expectation of a good night's sleep did not happen. I couldn't shut my mind down and slept fitfully through the night. At 6:30 my feet hit the floor. I wake up thinking about my next encounter with Rachel's father. Rachel said their family will not be attending the Honors Reception or the Torch and Trumpet Soirée on Shabbat. She did say they will be going to the Legacy Luncheon Saturday morning and the Baccalaureate Service on Sunday. The Legacy brunch is my first opportunity to get our two families together.

I thought a short morning jog might help me clear my head. I quickly dressed in my sweats and headed for the main campus, stopping long enough at the DUC to consume a sweet roll and coffee. I arrived back in my room by nine o'clock. My conversation with Emily last night is still bothering me. I called the hotel to check on her. A sleepy voice answers the phone.

"Hello, Ted?"

"No Emily, this is your brother. Do you remember me?"

"Oh. It's you. I thought Ted might be calling to invite me to breakfast this morning. What time is it anyway?”

“It’s a couple of minutes after nine. Emily, the reason I called is to let you know I am paying all of you a visit this morning. I need to talk to Dad about my plans next year. Unfortunately, I did forget to check on you at 11:30 last night. What time did you say you arrived back in the hotel?”

“I don’t remember.”

“Let me jog your poor memory. Do you remember we agreed you would return to the hotel immediately after the play?”

“Yes. I remember.”

“I’m going to have to tell Dad about this revolt of yours.”

“Like I told you last night Jacob, I don’t care what you say. You are so weird. Don’t you ever get tired of reading those books of yours? Really, you need to get a life. When will you stop buddying up with your male friends and take someone of the opposite sex out on a date? Pretty soon people are going to think you’re weird.”

“For your information Emily, I have a girlfriend. Her name is Rachel and she is Jewish. In fact, she is Orthodox. You will get to meet her tomorrow at the Legacy Luncheon.”

“Wow! What a surprise. I can’t wait to check her out. Is she a bookworm like you?”

“Yes, and she is very smart, not a smart-aleck like you. I will be arriving in about an hour, so be prepared.”

“You might see Dad, but you won’t see me. If Ted doesn’t take me to lunch, Mom and I are going shopping together. I need a new dress for the Honors Reception and the Soirée.”

“Okay then. I’ll see you when I see you. Be prepared. Dad will have fire in his eyes. As a graduating Legacy student, I will be receiving a Legacy Medallion to wear at the Commencement. It will be held at Glenn Memorial Auditorium at nine o’clock sharp. Don’t be late.”

“Do I have to dress up for this Legacy thing Jake?”

“Yes of course. It is to honor the previous graduates of Emory, including Dad.”

“I hope they have some light food. I’m on a diet.”

“I have no idea what they will be serving. Just be ready to go tomorrow.”

“What’s a Legacy student anyway?”

“These are the grad students whose relatives are members of

the Emory Alumni Association. That's me! There will be an address by the Vice President of the University. You don't want to miss his talk."

"Why should I need to hear what he has to say? I'll be more interested in meeting the graduates wearing their medallions."

"You said you wanted to meet my girlfriend. This is your opportunity. Rachel and her family will be sitting with us on Saturday."

"I'm puzzled. You said Rachel's family is Orthodox. How can they attend this brunch on Shabbat?"

"Just because you are Orthodox doesn't mean you can't attend a special function like this, especially when your daughter is being honored."

"Okay. What's next after that?"

"After the Legacy Brunch, I hope to take our two families out to dinner. Do you want to go with us to dinner?"

"Yes, of course."

"Now let me remind you about the Baccalaureate Service. It will begin at nine o'clock on Sunday at the Glenn. I know it is hard for you, but you have to be on time for all of these activities. Do you understand?"

"Yes, Bro, I'll be on time. Will I like the Baccalaureate Service?"

"Yes, but I can't take time to discuss it with you now."

"Jake, someone's knocking on my door. It's probably Mom. See you later."

"Okay, I'll see you when I see you."

I took time to take a quick shower and dress in my new suit before leaving for the hotel. I arrived twenty minutes late because of the heavy traffic in the CBD. I parked in the hotel garage and took the elevator to the third floor. Dad's and Mom's room number is 314. I knocked and Dad opened the door.

"Good morning, Jacob."

"Good morning Dad. How was the flight?"

"We experienced some turbulence. Your Mom kept looking out the window at the wings flapping up and down. She was very glad to get her feet on solid ground again."

"I can imagine you were pretty glad when the plane landed."

"Yes, for sure."

"Dad, I need to talk to you about a decision I have made. After several weeks, I have decided not to enroll in the medical school program here. Instead, I have sent an application to Johns Hopkins in Baltimore."

"I know the school has an excellent medical program, but I'm disappointed. You know our roots are pretty strong here at Emory."

"True. However, I do not want to pursue the general practice major."

"What?"

"Yes, I have applied for admittance to the McKusick-Nathans Institute of Genetic Medicine in Baltimore."

"*Zer schlect!* What you're telling me cannot be. We have made plans for you to take over my practice in Omaha. How on Earth did you come up with this idea?"

"Dad, it is because I want to enter the field of human genetics. I love it and I believe I can do well as a geneticist. You will be proud of me one day."

"And just how are you going to find the *gelt* to pay for a graduate program at Johns Hopkins? Don't look to me to foot the bill. The cost will be twice the cost at Emory. No, I won't permit you to do this."

"As far as the money goes, I will be in line for grant money, enough money to pay for my tuition, books, and even my living expenses. I also plan to find some work at the University."

"Like doing what, a dishwasher in a cafeteria? No, I'm not allowing you to do this Son."

"Dad, I'm now twenty-one years old and fully capable to chart my own course in life. Please try to understand. This field is exploding. New knowledge is pouring out like water down the Big Muddy in Springtime. These emerging fields may free mankind from the curse of diseases which have plagued mankind for thousands of years. I am totally committed to doing this Dad."

"I believe Emory also has a genetics program."

"No, the genetics here is limited to post-doctoral research only."

"Have you applied to the medical school here at Emory? You know there is a deadline on applications."

"No, Dad. I have not applied."

"I don't want to discuss this with you any more today. Your mother is going to be very disappointed in you, Son."

"Perhaps she will. I don't want to spoil this weekend for her. Why don't we put this topic on hold and revisit it after commencement?"

"For your mother's sake, I'll agree to set it aside for now, but I want you to submit your application to the Medical School right away before they close applications."

"I'll look into it. I assume Emily and Mom have gone shopping."

"Yes, they left shortly before you arrived."

"I would like to take all of us to lunch when they get back. I can show you around the city. There has been a lot changes since you were here last."

"I'm not in a mood for a tour of Atlanta. We can take a look at the city some other time."

"Whatever you say Dad."

"Jacob, has Emily been behaving here? Has she followed your instructions?"

I don't want to totally disrupt our family's peace twice in one day. I decide to give Emily a break.

"Well, not perfectly. Let's just say I am quite willing to hand her back to you Dad. She has been running on high test fuel here."

"I guess I understand what you are saying."

"Dad. I can pick you up for the Alumni Reception and then take all of us to the Soirée? Do you have your tickets?"

"Yes, I have them."

"Do you still have use of the rental car?"

"Yes, I rented it for a week."

"You won't need the car. I can meet you at the Legacy Brunch tomorrow at eight o'clock."

"No, I want to pick you up at your dorm."

"Okay, I'll wait for you there."

I left the hotel feeling very insecure. I failed to win Dad over. I have no other plan in mind to bring closure to my future plans. Wild thoughts are winging through my mind. Why not just steal Rachel, and flee to Paris? We can go to the Riviera. Such a plan is pure fantasy. No, I will struggle on, looking for a more reasonable solution. When I arrived back at the dorm.

I felt the need for some moral support. I went to the game room. Allen, Terry, Hank, and a couple other guys are there.

I asked, "Hey, anyone up for a game of pool?"

I got no response. All of them are involved in a discussion about their future plans.

Hank asked me, "Jacob, we want to hear what your future plans are."

"Hank, I have been given a spiritual calling by Ha Shem."

Allen's ears perked up. "You are hearing from the Lord, Jake?"

"Yes. I have been having some unusual dreams and visions at various times during my time at Emory."

Allen asks, "Can you tell us about these dreams of yours?"

Hank adds, "Jake, I want to hear about these dreams. We all are very interested. Can you go into greater detail?"

"I'm sorry guys, but I don't have the time now to say anything more."

Allen asks, "Can we convince you to meet with us again after commencement?"

"Al, I doubt if we can get a group together. The guys will be too busy preparing to leave."

I suppose you are right. I'll check with everyone after graduation to see who can make it."

I returned to my room and dressed for the Honor's Reception and the Torch and Trumpet Soirée. Then I went downstairs to the lobby, found an easy chair, and sat down. I began thinking about all the myriad things going on in my life. At half-past seven, Dad pulled up in front of the dorm. We left for the Miller-Ward Alumni House and the Torch and Trumpet Soirée downtown. At the entrance to the Miller-Ward Alumni House we were assigned to a table and given name tags. The large room is decorated with colorful banners and flags. There are round tables decorated with flower center pieces. Each table has eight place settings. An usher escorted us to Table 34. A family of four already is seated at the table. The family introduces themselves and I introduce them to my family. Seated on stage in the front of the room are all of the department chairmen. A band of trumpeters dressed in Scottish kilts is standing at their right side. Suddenly the trumpeters blow a loud trumpet fanfare. The Dean of Students comes to the microphone and invites everyone to join him in the Pledge of Allegiance. The Dean then asks Rabbi Ibrahim Ben Nun to open in prayer. The Rabbi gives the Priestly Blessing over the honor students and their families. Then the Dean welcomes everyone and introduces the University's department heads. Each department head comes to the podium and reads off the names of their honor students. The room is filled with the noise of shouts and loud clapping. The Dean then invites a young woman graduate to the podium. She sings the Emory University song, a capella. Next, we are served our meals by the catering staff. It is a meal of southern fried chicken, shrimp, crab cakes, oysters, and clams, topped off with hush puppies and wild rice. Dad gives us permission to eat the crabs and shellfish. The waiters fill our glasses with sparkling cider. As we eat our meals, our two families are sharing easy conversations. After the dessert of pecan pie and ice cream we said goodbye to our table guests and made our way back to the car. Then it is on to the Soirée at the Emory Conference Center Hotel.

Our family arrives at the Emory Conference Center Hotel a little after nine o'clock. Dad pays for valet parking. I have no trouble finding Allen and Alice. They are sitting at a table in the vestibule. Just then Terry and Hank show up. I introduced everyone to Mom and Dad.

When we entered the ballroom, a live band is playing Cuban music. The guys took turns dancing with Rose, and with Emily of course. With her new yellow dress, Emily is quickly in demand by a host of male graduates. We all joined in on the electric slide. The evening went by quickly. It was half-past one in the morning before I got to bed. I went to sleep looking forward to Saturday's Legacy Honor's Program when my family will meet the Kaufman family.

8

The Graduate

"Then the king spake unto Ashpenaz, the master of his eunuchs,
that he should bring certain of the children of Israel,
and of the king's seed, and of the princes;
Children in whom was no blemish, but well favored,
and skillful in all wisdom, and cunning in knowledge,
and understanding in science and such as had
ability to stand in the king's palace."
(Daniel 1:3–4)

The Legacy

I awoke on Saturday morning with great expectations in my heart. I quickly dressed in my new suit and ate a breakfast of cold cereal and coffee. Then I took time to shine my shoes before leaving for the Glenn Memorial Auditorium. I have every hope this day will turn out well.

I arrived in front of the auditorium ten minutes early. I waited there for the two families to arrive. The Kaufman Family arrived five minutes later. The sight of Rachel takes my breath away. She is wearing a trim blue suit, and her dark brown hair is glistening in the bright morning sun.

"Good morning Mr. and Mrs. Kaufman. Good morning Rachel.

Isn't this a magnificent morning?"

Hanna smiles as she adds, "Yes, it is just a perfect morning, Jacob. I just love the smell of the air after a summer rain, don't you?"

"Yes. Everything smells so fresh and clean."

I walked over and took Rachel's hand. "You look ravishing in your blue suit, Rachel."

"Thank you, Jacob. You look sharp in your suit as well, except your tie is not under your collar. Let me fix it for you."

As Rachel is tucking the tie under my collar I said, "Quite a change from the attire I wore when we first met, wouldn't you say?"

"Yes, I must say this is a major improvement."

Mr. Kaufman is touched by the morning as well. Looking over the campus he remarks, "This campus is so beautiful. We are so blessed to live in this city. I love this Emory University. Don't you agree, Jacob?"

"I certainly do, Sir. The Administration certainly knows how to celebrate a graduation in fine style. I believe Rachel told me you have arranged for our two families to sit together this morning."

"Yes, Papa has reserved a table for us very close to the front of the Auditorium Jake."

I kept looking expectantly back toward the parking lot, wondering if Emily is keeping her agreement to be on time.

"My family should be arriving any time now."

That is when I spotted Emily's bright yellow dress in the crowd. "Oh, I see them now."

I waved to Emily and she returned my wave. I introduced my family to them when they reached us.

"Mr. and Mrs. Kaufman, these are my parents, Joel and Rose Cahn, and this young lady is my sister Emily. Dad and Mom, I want to introduce you to Rachel's parents, Samuel and Hanna Kaufman. I guess I don't need to tell you, this girl standing beside me is Rachel."

After greeting one another we made our way inside. Mr. Kaufman stopped briefly at the reception table to pick up our tickets and find the number of our table. The table is located front row center. Every table is decorated with a center piece of beautiful flower arrangements. There is an Order of Service at each place setting, listing the participating dignitaries and the names of the Legacy Graduates. An envelope also has been placed inviting those present to contribute to the Alumni Association. After everyone is seated at our table, Rachel and I walked to the side of the auditorium to take our designated places in line with the other graduate students.

The ceremony began with an entry procession called "The Feast of Reason and the Flow of Soul." The Atlanta Scottish Pipe Band entered, followed by University dignitaries, faculty, and finally the Legacy graduates. We stood at the rear of the platform as the Vice President of the University welcomed the gathering. The Invocation was given by the Bishop of the Atlanta Episcopal Diocese. After a short speech by the Vice President, he read the names of the graduates and we walked up to the podium to receive our Legacy Medallion. After receiving the Medallion, the graduates returned to their tables. The brunch is buffet style, and each table is released in turn. Mr. Kaufman and Dad soon are discussing politics over our meal. Meanwhile Mom and Mrs. Kaufman are involved with a lively discussion about their favorite recipes. Emily began asking Rachel a million questions about campus life. She went on non-stop. I soon realized I'm the odd man out. I smiled a lot and tried to appear interested. There is no opportunity for me to discuss our relationship with Mr. Kaufman. At least I did have the satisfaction of watching our two families freely interacting with one another during the brunch. All too soon, we said our goodbyes and everyone left to go their own separate ways.

Dad turned to me and asked, "Jacob, do you want to go back to the hotel with us?"

"Dad, I'd rather not if you don't mind."

"Okay. If you change your mind, give me a call on my cell phone."

I said goodbye to them and walked slowly back to my apartment. What a bummer! To say I am disappointed is a major understatement. I have only three chances to gain Mr. Kaufman's ear, and I already have blown one of them. Only two opportunities remain, tomorrow's Baccalaureate and Monday's Commencement service. I needed some time to just cool out. I will call Rachel later. In the meantime, I want to work on the task Pastor McGuire assigned to me. I spent the entire afternoon working on it. I soon discovered the prophecies of the Tanakh matched the New Covenant accounts very closely. According to the Pastor's plan. I applied the tools of statistical analysis to the problem. I carefully rejected paired cases where a dependency might exist, where the actual event could have been manipulated after the fact. I succeeded in identifying a valid number of discrete events, adequate to subject the data to a valid test.

The result of the test is quite amazing. I obtain a result of 0.001 at a confidence level of 95 percent. In other words, the probability is very low of that the matched events mentioned in the data could have occurred by chance. Even with the results I remained stubbornly resistant to the acceptance of Jesus as the Messiah. I cannot truthfully dismiss the position of the new covenant scriptures. This Jacob will be wrestling with this issue for many days in the future.

It is after six when I finish my assignment. I cooked up some spaghetti and made a pot of coffee. After finishing dinner, I decide to turn on the TV and catch a ball game. Fox is broadcasting the Georgia Tech North Carolina game in Greensboro. The seventh inning stretch gave me the opportunity to call Rachel.

"Hello Rachel, this is your lover boy calling."

"Where are you, at a ball game?"

"No, I'm back in my cave. I'm watching the ACC playoff game in Greensboro. What are you up to?"

"Not much. I'm trying to decide what courses to take this Fall. I think I'm going to take a language course, perhaps Arabic."

"Arabic will come in handy if you plan to work in the Mideast."

"Have your plans changed since we talked earlier today?"

"No. I'm still planning to go to Johns Hopkins to pursue a career in genetics, Rachel. I'm very serious about this."

"You realize we will be separated by many miles Jake."

"Yes, I am aware of it."

"Jacob, I need to know where you stand. Are you committed to our relationship or not?"

"Rachel, you know I am totally committed to you."

"So, why didn't you tell my Father about our commitment at this morning's Legacy Brunch?"

"I am as disappointed as you are Rachel. Emily dominated your attention, and our parents paired off along gender lines. I was the odd man out. I never did have an opportunity to discuss our relationship with Mr. Kaufman. Anyway, if I did bring up our engagement and your father restated his objections, I believe Dad would have come unglued. The last thing we need is an argument between two patriarchs. I just smiled and tried to look interested in their conversation. What do you think I should I have done?"

"I have no idea. I'm sorry things didn't turn out better."

"I'm glad you aren't upset with me Rachel. I had three opportunities to present our case to your father. I blew one and I'm now down to just two more, tomorrow's Baccalaureate and Monday's Commencement service. I assume your parents will be attending the service tomorrow?"

"Yes, of course. They wouldn't dream of missing it."

"Can we meet on the Quad afterward?"

"Yes. Perhaps you will have a better chance to talk to Papa then. Jake, don't forget to wear your cap and gown."

"You are sounding like a wife already, Rachel. I hung it on my closet door so I wouldn't forget it. I'll see you tomorrow morning."

I returned to the ball game. The Bulldogs fell short. The final score was six to five. I turned off the TV after the game and called the hotel. There is no answer. Dad is probably taking the family on a tour of Atlanta. No doubt they are having dinner some place other than the Omni. I read for an hour or so and then went to bed.

The Baccalaureate

I awoke on Sunday morning at 5:30 before the alarm went off. I put on my sweats and took a quick jog to clear my head. It is another beautiful morning. I arrived back in my apartment at the Clairmont Residential Center at 6:15. I bought a roll and a cup of coffee from the machine downstairs and took them up to my room. I showered and dressed for the Baccalaureate. Then I called the hotel to check on the family. They are all up and ready to make the drive to the campus. I told Dad to look for the signs pointing to the Fishburne Parking Deck. It was eight o'clock before I left for the Quad. I arrived just before the 8:30 deadline. As I arrive, the marshals on the Quad are attempting with limited success to line everyone up for the processional. Rachel found me shortly after I arrived.

"Hi Jake. Have you found your family yet?"

"Not so far. When I called them this morning to tell them where to park they were about ready to leave. Are your folks here?"

"Yes, Mom was up early and very excited. Jake, if it was up to Mom we would be married already."

"Yes, I believe she is on our side, on our same wave length."

There still wasn't any sign of my family. The Glenn only holds twelve hundred people and there is a line of people waiting to enter.

"Rachel, I believe my family will end up sitting in the tented area. We will have to locate them after the ceremony."

"Yes, most of the seats are already taken."

By nine o'clock the grads are assembled and the bagpipers led us into the Glenn Auditorium. As we entered, the University Candler Choir, accompanied by the pipe organ, begins singing A Mighty Fortress Is Our God. The dignitaries are seated on the dais. We, in our caps and gowns, file in and stand at the front of the room. The Bishop from the Atlanta Episcopal Diocese asked everyone to stand for the invocation. After the prayer, we are seated. The Senior Rabbi from the Atlanta Temple then took the podium and gave a charge to the graduates. I was impressed by the sincerity of his message. His word, based on Chapter Twelve in the Book of Proverbs, was certainly inspired. It was "out of the box." He gave a clear warning against allowing the pride of intellectual achievement to blind us from our divine calling to minister to the needs of humanity. In a firm voice, the rabbi declared, "It is Ha Shem who will judge the acts of every person. Nothing hidden in the world is unknown to our Creator."

He admonished us to serve mankind and Adonoi with fear and trembling. After the rabbi's charge to the graduates, the Candler Singers sang a medley of church music from around the globe, featuring classical, contemporary, folk, and spiritual genres. The University President then took the podium and delivered the keynote address. His message paralleled the advice of the rabbi, to serve our families, communities, governments, and the world at large. The program then closed with the recessional. The dignitaries exited down the central aisle followed by the graduates to the sound of the organ playing a Bach fugue.

Once outside, Rachel and I located my family. Dad suggested we avoid the brunch crowd. He offered to take us to lunch in town. I recommended the French café on Wesley Road where Rachel and I ate last Monday. Everyone agreed to my suggestion and we walked to the Fishburne Parking Deck. Emily asks me if Ted Greenwald can come to lunch with us. Before I have a chance to say no, Dad tells her he can come. Emily gets Ted on her cell phone and tells him he is welcome to join us at the restaurant. Emily hands the phone to me and I give Ted the directions to the restaurant.

When we arrive, Ted is waiting for us in the parking lot. He's dressed in a suit and tie. I introduce him to everyone. He seems like a nice enough guy. I decide to forget his previous indiscretion. The restaurant is just opening. I introduce everyone to Monsieur Bideau, the Proprietor. Pierre, our waiter, escorts us to a private room in the rear of the restaurant. Emily tells me she is really impressed. Before we sit down, I slip Dad the gift certificate I had received from our previous visit.

"Son, I can cover this. You can use this later."

"Dad, there is no sense me taking it to Baltimore."

My argument prevailed and he kept the gift certificate. Everyone at our table is staring at the French menu. I decide to help move things along. "Pierre, can you suggest a light lunch for us."

"Oui, Monsieur Cahn, *Je recommande une Potage Parmentie. Une Salade Nicoise, una Oeufs a la Fondue de Fromage, et Tarte Aux Pommes.* Oh, you must try our onion potato soup, a Mediterranean salad, and our special apple tart for dessert. It is very elegant."

Emily quickly answered, *"Votre suggestion est très élégant."*

I am impressed. "Emily, where did you learn to speak French so fluently?"

"Jake, don't you remember? I took four semesters of French in high school?"

"I honestly don't remember, but I'm glad you came along today Sis."

Everyone enjoyed our lunch together, especially Rachel. She described our previous visit to the restaurant. Rachel's Mom wanted to know more about our relationship. Rachel nodded her head toward me. I decide to give her a general answer, carefully avoiding the issue of our engagement.

"Rachel and I have enjoyed our friendship this past semester, but we haven't made any definite plans for the future."

It was a boldfaced lie, and I think Hanna knew it. Sam Kaufman remained as quiet as a tomb. A few minutes past two o'clock we all decided it was time to leave the restaurant. My second opportunity passed without any breakthrough with Papa Kaufman. Rachel thanked my parents several times for the lunch, saying how much she enjoyed getting to know them. Mr. Kaufman also graciously thanked Dad for the lunch.

Then Mr. Kaufman tells me, "Jacob, I will save you the trouble of taking Rachel home. She will go home with us."

It is more of a command than a kind offer.

Dad asks, "Son, do you want to go back to the hotel with us?"

"Thanks for the offer Dad, but no. I have some things to catch up on before tomorrow's ceremony."

As we arrive back at the apartment complex, Dad reaches into his coat pocket and gives me an envelope. We said our goodbyes and I watched them drive away. Suddenly I feel very alone. It is as if I am standing on the shores of a desert island and watching the last ship sail away into the sunset. I returned to my room and changed clothes. I didn't feel like being alone. I went downstairs to the recreation area. As I enter the room, Terry Johnson is playing a game of pool with another guy. I didn't know him. Terry introduced me.

"Jake, this is my friend Phil O'Reilly."

I sat down to watch their game. Phil soon made short work of Terry. I offered to play the winner. I am hot. After winning the lag, I make two balls on the break and then proceed to clean the table.

Phil is impressed. "Jake, I was going to suggest we put some money down on our next game, but I've changed my mind."

Allen walks in just as I am explaining to Phil my lack of expertise in pool. "Phil, I certainly do not clean the table every time I play pool."

Allen pipes up and says, "That's right, Phil. He only does it when there's some money riding on the game."

Winning the game was good medicine for me. It felt good to just relax and not have to ponder future issues in my life. We men enjoy play, and the competition is a large part of it. The four of us ended up playing pool until dinner time at six. Then we went to dinner together at the café. After dinner, we hung out in the lounge and shared our thoughts with one another. Phil is an interesting guy. His major is archaeology. I mentioned my girlfriend Rachel Kaufman is also an archaeology major. He said he knows Rachel and has taken several classes with her. Terry then shares his thoughts on his future vocation. He wants to become a large animal veterinarian. I shared my plans as well. Allen is quiet, content to just listen. Of course, we all know his future plans. About nine o'clock we parted company and I returned to my cave.

I am thinking about all the separate plans we shared with one another. Soon we will be leaving, going our separate ways. I will miss them, especially Allen. I thought also about the rabbi's warning to avoid being caught up in the success of our vocations, losing sight of the needs of our fellow human beings. I wonder if I can make any meaningful contribution to mankind in this field of genetics? Only time will tell. I took a shower, shaved, and turned in.

Commencement

Monday morning is another beautiful cloudless day in Atlanta. I once more put on my new suit and donned my cap and gown. I hung the Legacy medallion around my neck and left for the Quad. I arrived on the Quad a little after seven. Rachel is waiting there for me in the northeast corner of the Quad. The faculty marshals took almost half an hour to get the candidates lined up for the processions. They formed us into four lines, one at each of the four corners of the Quad. At exactly 8:00, the pipe and drum corps led the faculty, board of trustees, the deans, vice presidents, and all of the award recipients to the front of the Quadrangle. Next the faculty marshals, marching behind banners, led the graduates from the four corners into the Quad. We are followed into the Quad by the elderly alumni. After being seated, Emory's president took the podium to preside over the exercises. He invited the commencement speaker, Surgeon Raymond J. Johnson., a man with gifted hands, to come to the podium. The talk by this black man is very inspiring. He came from very humble beginnings. His mother encouraged him to overcome many obstacles: financial, cultural, and spiritual. I felt his presence here today is a powerful witness of the South's success in overcoming the dreadful racial injustice of the past. Following Dr. Johnson's commencement address, the president takes the podium and confers honorary degrees and awards. Each honored graduate comes forward to the podium and receives his or her award. Allen Schaefer, our student-selected Class Day Speaker, is given special mention. The academic dean of each school then asked each of the student marshals to come forward as representatives for the graduates of each of the schools. The graduates from the schools are asked to stand to receive their degrees en masse. At the conclusion of the ceremony the degree candidates are marched from the Quad to their particular schools where they will receive their diplomas and attend smaller receptions. The graduates of the College of Arts and Sciences are told to remain on the Quadrangle for their diploma ceremony. I tried to spot my family and Mr. and Mrs. Kaufman in the crowd, but I could not find them. I know they are there somewhere in the sea of faces. As each graduate's name is called, the graduate walks to the podium to receive their diploma from the Dean of Students. This is taking a very long time. My name is called before Rachel's. I eagerly go forward to shake the Dean's hand and receive my diploma. I returned to my seat and carefully removed the ribbon. I open the roll just far enough to read the name on the diploma. I'm smiling. The diploma reads "Jacob Cahn" in fancy script. Yes, they finally have succeeded in spelling my last name correctly. It's official! My undergraduate days are over.

By 4:30 the last of the graduates received their certificates. The Dean of Students dismissed us. Rachel first spotted her parents in the crowd. Taking her by the hand, we worked our way through the crowd. They are all smiles as we reach them.

Mr. Kaufman is saying, "*Mazel tov tokhter*, mazel tov!"

He gives Rachel a warm hug and adds, "I am so pleased with you Rachel. This day gives me so much joy, so much *mekheiyeh*. I'm looking forward to your continuing studies in the graduate school. This is a great day for our family, baruch Ha Shem."

Mrs. Kaufman is very excited as well. She turns to me and says, "Jacob, this is the first college graduation I have ever attended. It is so wonderful."

I replied, "Mine too, Mrs. Kaufman, this is my first graduation."

My comment slowly sinks in. "My goodness, this must be your first college graduation."

We all have a brief laugh. Hanna continues to rave about how much she is enjoying the graduation ceremony.

Mr. Kaufman shakes my hand and repeats Mazel tov over and over again.

"Congratulations, Jacob. I know you have a great future in front of you."

My family now joins us. Dad, Mom, and Emily also are all smiles. Everyone shakes hands. The ladies are soon hugging on one another as if they are old friends. They are sharing their thoughts about the future of their two graduates.

　　　　　　　　　　　　　　　　　　　　　　　　

The conversation is interrupted by Dad's cell phone. He excuses himself and steps aside to take the call. A concerned look has crossed his face. He is nodding his head as he listens to the person on the other end of the line. I walk a bit closer to hear what is going on.

"Yes, I understand. Are you able to locate Dr. Milton at the Hospital? … I see. Is the patient now stable? I am down here in Atlanta. … Yes, I understand. … No, it won't be necessary. I can take the red eye and be there late tonight. I will call you from the airport as soon as I arrive. Continue the life support. Make every effort to contact the next of kin or someone from the family. Find out if she has a rabbi or pastor. I will call you just as soon as I arrive."

I am not surprised by the call. I have heard many such calls before.

Dad hangs up his phone and rejoins us. He explains "I'm sorry folks. I must apologize. I have an urgent situation, a matter of life and death. I must return to Omaha as soon as possible. I will be assisting in a heart transplant operation tomorrow morning. I am so very sorry to interrupt this celebration with these two wonderful young people, but I must return to Omaha immediately. Jacob, I will take your Mom and Emily back to the hotel on my way to the airport. They can return tomorrow as originally scheduled. I need you to take them from the hotel to the airport tomorrow."

"Of course, Dad."

Mom is clearly upset. She has visited the woman several times in the hospital and they have become very attached to one another.

Mom is pulling on Dad's arm, asking. "Dear, can't I return with you today?"

Dad is firm. "No, I wouldn't be able to get you on a flight this soon. Jacob will take you to the airport tomorrow. Your plane leaves at 2 o'clock in the afternoon."

I try to convince her Dad's way is best. "Mom, this works for me. I need to be out of the apartment by 2:00. I will check out earlier. I will drop by the hotel later this evening and take you both to dinner. You can check out of the hotel tomorrow by noon. I will pick you up then."

Rachel turns to her father and asks, "Papa, I have some books in my locker I need to turn in. Can Jacob help me take them back to the Library?"

"Do they have to be turned in today?"

"No, but Jacob will be leaving tomorrow and they are pretty heavy."

"Very well, but I want to see you back home directly after turning them in. We have planned a little party in your honor at the house for you and some of our friends."

"Okay, Papa. I will come directly home afterward."

Rachel and I said goodbye to everyone and walked across the campus to her locker. She opened her locker. There are only three small books and her cap and gown, carefully folded in a plastic bag in the locker.

"Rachel, if I had known you had so many books, I would have borrowed a cart from the library."

She smiled. "Jake, can you take time to return my cap and gown to the rental store for me?

"Yes, of course."

I carried her books back to the library and dropped them into the return chute. Then we walked, hand in hand, to the parking lot at Clairmont Center. Old Blue is waiting.

I am trying to sort out the day's happenings. I cannot think of any words of hope for Rachel, or for myself either. It especially hurt when her father did not invite me to attend her celebration party. Also, Dad's sudden exit put the kibosh on my last opportunity to speak to her father about our ketuba.

As we reached the car I asked, "Can we just sit here a few minutes and talk about today? I am feeling like there's a nine-hundred-pound gorilla sitting on my chest, and it's too heavy a burden to carry alone."

"I feel the same way, Jake. It is certainly not the way we planned this day to end."

"Well, one thing is abundantly clear. Your Papa certainly did not want me to attend your party this evening. I guess we now know his true intentions."

"Yes, your perceptions of Papa's intentions are right on the money Jake. We both must find some way to get through the rest of this day. You are handling this better than I am, and I love you for it. Please don't forget to turn in my cap and gown. Don't take it to Baltimore with you. Perhaps I should write you a reminder note?"

"No, I'll be sure and drop it off before I get out of town."

I turned the key in the ignition and we began making our way back to North Peachtree Battle Avenue. I did not know what to say, so I turned on the radio. We listened to the music, saying nothing. There are several cars parked in front of the house when we arrived I noticed one of the cars has New York plates. I pulled into the driveway and turned off the engine.

"Rachel, I will be leaving for Baltimore on Wednesday morning. I will call you before I pull out of town."

"I'll will wait for your call Jake. Where will you sleep tonight?"

"I'm not sure. I probably will rent a motel room in town. I may be meeting some of the guys tomorrow. I'm not sure."

"Please remember to call me before you leave Atlanta."

"I won't forget to call, Rachel."

There is nothing else to say. I watch as Rachel walks to the front door. It is a lonely drive back to the campus. I changed out of my suit and then called the hotel. Emily answered the phone.

"Hi, Bro. Are you coming to take us to dinner?"

"Yes, I'll pick you up in front of the lobby at seven o'clock."

"Can I invite Ted to come along?"

"No, absolutely not. Mom will need to have some quiet time with just the two of us. She is very concerned about the lady having the operation."

"Please, Jake?"

"No, not tonight Emily. See you at seven."

I hung up before Emily could muster another reason to include Ted. Then I decided to take a short nap. My head is reeling with a "level 10" headache. I fell asleep almost immediately and didn't wake up until half past seven. I called the hotel.

"Mom, I overslept. I'm sorry."

"I understand, Jake. You don't have to take us to dinner. I really don't feel up to going out anyway. We will have room service bring us up our dinner."

"Are you sure?"

"Yes, I'm fine with it."

"Okay, I'll see you tomorrow at noon."

I hung up the phone. Only then did I realize I am hungry. I opened the refrigerator. It still contained three eggs, an open can of tuna fish, and a jar of mayonnaise. The only bread I have left are two hotdog buns. I warmed up some day-old coffee and sat down to my last dinner in the apartment. Then I turned in.

9

Parting of the Ways

"Then Jacob went on his journey,
and came unto the land of the east."
(Genesis 29:1)

Atlanta Exodus

I awoke early Tuesday morning with new conviction. It is time to close this chapter of my life and get on with becoming a geneticist. The lunchroom is probably not open. My pantry still contained a box of oatmeal, but no milk. Coffee creamer seemed like a viable substitute. I brewed a fresh pot of coffee. That was breakfast. Packing is next on my agenda. I rounded up all of my gear and carried it to my car. As required by the management, I cleaned the apartment and dumped my trash in the trash bin at the end of the hall. I placed the Gideon Bible back in the end table drawer. I finished all of this by eight o'clock. After checking to be sure I hadn't left anything behind, I walked to the admin office and turned in the keys to the apartment. I signed the release form for the lady at the desk. She tells me I am officially checked out. My four years at Emory now are history. Dad asked me to take Mom and Emily to the airport and see them safely off. I left for the Omni hotel. On the way, I returned the two caps and gowns. Mom and Emily are waiting in the lobby when I arrived. Looking at their pile of luggage I realize I don't have enough room in my car for their baggage plus my own. The only choice I have is to unload my stuff and store it at the hotel temporarily. I presented my problem to the doorman.

He asked, "Did your family stay here at the Omni?"

"Yes, they are just now checking out of the hotel."

"Let me guess. You have just graduated from Emory and have checked out this morning from your apartment. Your car is loaded with your stuff."

"You have described my situation exactly."

"It happens every year. You can use our luggage cart to bring your things into the lobby. There's a storage room behind the registration desk. Be sure to mark your things."

"Very cool. I really appreciate this." I wheeled the cart to my car and loaded it with all my gear. Then the doorman helped me unload the cart into the storage room. I loaded Mom's and Emily's luggage into my car and we left for the airport.

Mom was in a good mood. She had talked to Dad earlier in the morning. She gave me a rundown.

"Jacob your father caught a flight just leaving Atlanta and arrived at Omaha in time for the operation. The operation went well. Your dad sends you his love and is sorry he had to leave so soon. Will you be driving to Baltimore this evening?"

"No, I won't be leaving tonight. I want to get an early start. I'll check into a motel room tonight and leave early tomorrow morning. I'm supposed to meet with some of the guys at the Clairmont Complex this afternoon."

"I'm glad you won't be making the drive tonight."

"I will give you a call just as soon as I arrive and get settled in Baltimore, Mom."

I pulled into the departure lane in front of Delta and unloaded their luggage. A red cap checked in the luggage for me. I told Mom I would stay to watch their plane leave. Then I parked in the short-term parking lot and walked to the terminal to see them safely off. They departed on schedule. I called Dad and told him the flight departed at 2:00. Then I returned to the Omni and reloaded all of my gear. I pulled into a gas station, checked the oil, water, tire pressure, and filled the car with gas. The oil was a quart low. I purchased three quarts of oil, one to top off the car and two more for the trip. I asked the attendant if he knew where there was a Christian book store. He told me there is one across the street and down a block.

After paying the attendant in cash, I realize my wallet is feeling lighter. Then I remembered Dad's envelope in my jacket pocket. Dad had tucked ten crispy hundred-dollar bills inside the envelope.

This really made my day. Perhaps he is softening his position about my application to Johns Hopkins.

I drove to the Christian book store. There is an older lady at the register as I enter. Her nametag reads Mrs. Owens. "Good, morning Mrs. Owens. My name is Jacob Cahn. I need to select some Biblical reference books? Can you help me?"

"Good morning, Jacob. I'm sorry, but I can't leave the register. I'm the only one on duty today. Our reverence books and bibles are on the back wall of the store. Why don't you just select some books and bring them to the counter. I can help you decide the better ones for you."

"Fine, I'll take a look at them."

I took a shopping card and wheeled it to the back of the store. I was able to find a King James study Bible containing a Greek and Hebrew/Chaldean dictionary, a Strong's King James Concordance, an Interlinear Hebrew/Greek/English Bible, a book containing eight of the Apocrypha texts, and two complete commentaries, one by J. Vernon McGee and one by Charles Missler. I wheeled the cart back to the front of the store."

The lady manning the register is impressed. "My goodness, you are a very ambitious young man. You have made excellent selections."

"There was one Bible I did not find. I wonder if you also sell Gideon Bibles?"

"No, we don't sell them. The Gideons give their Bibles away free of charge. The translation is the same as your King James Study Bible, so you don't really need one."

"I wanted one because of the helps they include."

"Yes, those are very helpful. The owner of the store happens to be a Gideon. Let me take a look. He sometimes stores some of them here in the storage room. I will have to lock the register and the front door."

"I will stand guard for you."

The lady locks the register and front door, then goes back to the storage room. She soon returns with a brand-new Gideon Bible.

"I found only one. It must be meant for you, Jacob."

"Yes, I believe it is."

I paid for the books in cash and made my way back to Old Blue. I again counted the money in my wallet. I still have a little over seven hundred dollars left. I pulled into the Motel Six and registered for one night. I remembered the lady in the Christian book store's theft concern. I unload everything and took it inside. I did not want to run the risk of someone breaking into my car overnight. Once in my room, I called Allen on my cell. "Hello Allen."

"Hi Jake. What's up?"

"Al, I remember you and the guys wanted to meet with me after the Commencement. I know we are going to meet on Wednesday, but I'm going to leave tomorrow morning. Can you round up the guys and meet this afternoon instead?"

"I don't know? This is short notice. I doubt if we can round up everyone so soon. Can you just stay one more day?"

"Allen, I really want to leave tomorrow if at all possible."

"Well I can put out a text message and we will just see what happens. Where shall we meet, in the lunchroom at the Clairmont Complex?"

"Yes, Clairmont is fine."

"Okay, I'll just get the word out and we will see what happens. What time do you want to meet, Al?"

"We could meet at five and then find someplace to have dinner. There's a Red Lobster here near my motel on Central Avenue."

"Well, we might want to just eat here at the Student Union and save some money."

"Okay. Go ahead and text the guys. I will see you at five in the Clairmont lunchroom."

"Fine, Al. I'll see you there at five."

I headed to the University and pulled into the parking area at the Clairmont Complex. It is only a quarter past three. I am one hour and forty-five minutes early. I took another jog across the main campus and back. I set a steady relaxed pace. I took a few minutes to sit on a bench in front of the Woodruff Library. The sun is very bright. It is casting long shadows of the campus buildings on to the campus lawns. The now quiet campus put me in a nostalgic mood. I realize this day marks the end of a very special time in my life.

Memories of the last four years are flooding back. Truly the students are at the mercy of their instructors who inflict intellectual pain upon their lives. The struggles I experienced at Emory are now over. The struggles were worth it. I sat under great professors and lab assistants. Many took me under their wing and helped me through the rough spots. I thought of all those special friends who unselfishly shared their lives with me. I will treasure these special friends of mine. I bonded very closely with, Allen, Hank, Terry, Pastor McGuire, and Dr. Mike. Foremost in my thoughts are those precious times I spent with Rachel. Will my future days at Johns Hopkins ever approach the happiness I have experienced here? Probably not. I am feeling again very alone now. I retied my shoes and set a course back to the apartment complex. I decided to check out who might be in the lunch room.

The lunch room is practically empty when I entered. Allen is seated at a large table with Hank. No one else is present.

"Hi Guys. I didn't hear anything from you, Al. Are we waiting for some others?"

"No, Jake, this is it. I checked with Terry and several others and they aren't coming. They have too many other things going on."

"I see. What do you want to do guys?"

Al says, "I think we should go ahead and cancel our get-together, Jake."

Hank agrees. "Jake, we can organize a meeting on the internet."

I asked, "Do we want to eat something here at the cafe?"

Hank looks dejected. "Jake, I would rather not."

"Okay, I understand. Do you have any quick questions? If so, I will try to answer them."

There are no questions. It is plain they have been overwhelmed by the busy week.

"Since you have no questions, I propose we officially cancel our meeting. I'm hungry and I still have my meal ticket. I'm going to go dine at the elegant Student Union."

Hank says, "I lost mine, but I doubt the cashier will ask to see them."

We all agreed to my suggestion. Allen suggested we race one another to the Union. Terry ran like a deer. I came in second. The special on the menu was pot roast. Hank anointed it with a lot of ketchup. As we ate, Al began to reminisce about our undergrad years.

"Jake, do you remember when Hank was twirling that blond girl at the Soirée and let go of her? She ended up in the trumpet section of the bandstand?"

"I don't remember seeing anything like that Al."

Terry says, "Al, when Hank let her go, Jake was outside talking to Pastor McGuire."

Hank's face is turning red. "Please guys, erase that memory from your memory banks."

We spent two hours sharing our experiences.

Allen lapsed into deep thought. Then he began to share his heart with us.

"Gentlemen, I have enjoyed sharing my life with you here at Emory. As you know, I will continue on to become a medical doctor and will be leaving the safety and security of this wonderful Nation of ours, the United States of America. The Lord has been telling me the time of his return is very imminent. As I go to the underdeveloped world, I do not know if I will ever return here again. There are very bad things going on overseas, especially to Christians. Alice and I will always treasure this time spent here with all of you. Please pray for us."

I was touched. "Yes, Al, I certainly will keep you in my prayers. We are brothers are we not?"

"Yes, Jake, I feel very close to all of you. I will leave it at that. I love all of you as Jonathan loved David."

I watched as Hank dug a handkerchief out of his pocket, trying to keep the tears from flowing. I did the same. We stood up from the table and hugged one another. It was time for us to depart. I arrived at the motel on Center Street at nine o'clock. I parked and locked the car. A tall burly man in his late sixties was at the registration desk. He had long hair, and a full beard. His bare arms were covered with tattoos. The name Roger was his shirt.

"Hello, Roger, I'm Jacob Cahn, room six."

"My name ain't Roger. It's Don. A week ago, someone left this shirt in the closet."

"Okay. Do you serve breakfast here?"

"Yeah. Breakfast hours are from six to nine-thirty. They serve cold cereal, milk, toast, scrambled eggs, orange juice, fruit, yogurt, and muffins. There's coffee, but I need to warn ya. There's chicory in it and it's strong! You don't need to use a cup. Never learned to like chicory coffee myself."

"Thanks Don. I'll use a lot of sugar."

I returned to my room and called Rachel. I so wanted to see her one more time. I do realize we need some time to heal, but it is just too hard to accept.

"Hello, Rachel. It's me."

"Hello Jacob. I'm so glad you called. Are you on the road or still in town?"

"I'm at the Motel 66 on Center Street. I thought perhaps we can see one another one more time before I leave."

"I wish we could, Jake, but Papa has insisted we make a clean break of things. A time of healing he calls it."

"He said the very same thing to me when I met with him in his study. I promised him to eliminate all communication with you for six months. I guess the answer to a meeting has to be no."

"Yes, the answer is no. Six months is a long time, and your graduate study in Baltimore will be even longer, four or five years."

"Yes. Even with this accelerated schedule, it will take me at least four years. It could take six."

"This sounds like the beginning of a very long separation."

"Yes, I'm afraid it is. To encourage you, the cost of my training will be covered by generous stipends. Even before I receive the degree they will be launching me into basic research. I should be making good money from those research grants. I'm looking at no longer than three years before we can be married. We don't have to wait for me to finish the degree."

"Thank you for encouraging me, Jake. I suppose you need to get your rest. We can't meet, but if you happened on your way out of town to drive slowly by the house, and happened to glance up at my window on the second floor about seven o'clock tomorrow, you might see me there."

"If I knew how to play a guitar I would stop under your window and play you a love song."

"I'm engaged to a modern-day troubadour no less. I love you Jacob, and I always will. May Ha Shem keep you safely under his wings."

"Goodbye Rachel. I will miss you."

"Yes, I will miss you also lover boy."

I closed my cell phone, turned down the covers and tried to go to sleep. Sleep didn't come until one in the morning. I woke up at six. I quickly dressed. The free breakfast was not yet ready. I quickly checked out of the motel and drove through McDonalds. Breakfast was an Egg McMuffin and a cup of coffee. I arrived in front of her home at seven. As I drove slowly past her house she is standing in front of her window as she promised, throwing me kisses over and over again. I know I will cherish those kisses for a long time to come. My emotions are welling up inside of me. Handkerchief in hand, I pressed down on the accelerator and drove away.

Reaching the onramp for Interstate 285, I turned south. At the I-20 interchange I headed east toward Columbia, South Carolina. I reached Columbia by one o'clock. Stopping at a shell station I topped off my tank and checked the oil. The engine already is a quart low. I bought a tuna sandwich, a cup of coffee, and two quarts of 40 weight oil. I added one quart of the forty-weight and was on my way again. An hour and a half later I reached I-95 just outside of Florence. Turning north, another hour brought me just south of Fayetteville. I stopped at a gas station on Highway 87 to refill my tank and check the oil. It was only a pint low. Old Blue is doing better now. I topped it off with the heavier oil and then looked for someplace close by to eat lunch. The station attendant told me about a Pizza Hut located just a few miles ahead on Cedar Creek Road. I pulled into the Pizza Hut a little past two o'clock. The place is crowded with a group of high school kids. From the bedrolls on the roof of the bus I guessed they are headed for summer camp. As I entered the room there was one open chair at a table. Two young boys are seated at the table with an older man. No doubt he is a chaperone or leader of the group. I invited myself to their table.

"Hi guys. Mind if I share your table?"

Their leader says, "No problem. Take a seat."

"Where's your group headed?"

"We're on our way to the Smokey Mountains. Where are you going?"

"I'm on my way to Baltimore. I will be entering grad school there. My name is Jacob."

The older man says, "I'm John. This guy on my right is Henry, and this other ugly dude here is Carl."

Their pizzas are just coming out of the oven when I ordered mine.

"So, where are you headed in the Smokies, John?"

"We're headed for our church camp near Spirit Mountain."

"I've never heard of Spirit Mountain. Where is it?"

"It's about 30 miles west of Ashland."

"The name sounds like a good place for a church camp."

"Yeah, these kids are pretty stoked about it. Carl here is Jewish. He recently accepted Jesus as his Savior."

"So, you are a completed Jew Carl."

Carl swallows his pizza before he answers, "Yes, you can say I am."

"To date Carl, I have met only one other Jew who believes Yeshua was the Messiah."

John is quick to correct me. "Jacob, you should say Jesus is, not was the Messiah. He is alive."

"I stand corrected. What convinced you to believe Jesus is the promised Messiah, Carl?"

"I know he is the Messiah because I met him."

"You met him? What do you mean?"

"Just what I said. I was praying after my dad died of a heart attack and Jesus came into my room. He told me dad was in heaven with him."

"Was this a dream, Carl?"

"No, I actually saw and talked to Jesus. It was awesome. He told me I am going to serve him as a missionary. Now do you understand why I am on this trip to camp, Jacob?"

"Yes, I understand. Quite a story Carl. I wish you well."

John asks, "You seem to know a lot about completed Jews, Jacob. Are you one also?"

"Well, I know about Jesus, but I can't say I know him personally like Carl does. I do believe Jesus was a real person, not simply the figment of someone's imagination. Our religion is monotheistic. According to our Torah there is only one God. We do not believe in polytheism."

"Christianity actually shares your view of one God, Jacob. His divinity is expressed as three persons, God the Father, the Son, and the Holy Spirit."

"Yes, I know this is the Christian belief. My good friend from Emory is a born-again Christian and intends to become a missionary doctor. His wife is a completed Jew. I recognize Judaism and

Christianity are closely related. Given time, I intend to search this issue out more deeply for myself."

At this point, Henry decides to enter the conversation. "Jacob, I can tell you are a very sharp individual, but you will never convince yourself Jesus is the Messiah with your mind. You will only find Jesus through your heart."

"Why so, Henry?"

"This is because a person's world view most always overrules factual information. It trumps a person's reasoning every time. Only a heart change will move you beyond your current view of reality."

"You are pretty sharp yourself, Henry. So how does a person get a heart change, from a Christian cardiologist?"

"Of course not. This only happens through the third person of the Godhead, the Holy Spirit. Human reasoning cannot provide you with faith."

Just then the bus driver made an announcement, "Lunch is over everyone. We are running late. Please put your trash in the trash bins and re-board the bus."

These three young lions took time to pray with me. I wished them well.

I set off once more on my journey, feeling a growing sense of anticipation. My goal today is to reach Richmond, Virginia, some two hundred miles ahead. As I leave Cedar Creek Road behind and merge onto I-95 I encountered a long string of army vehicles heading north in the slow lane. They no doubt are coming from Fort Bragg and heading for their summer training camp. They will be training for a far different kind of job than my three young friends I met in the pizza place. I took time to pray for their safety.

I've always had a patriotic streak in me. At one point in my junior year in Omaha I considered postponing my undergraduate program to serve in the Military. I was offered the opportunity to apply for admittance to West Point. Had I accepted their offer, I would have saved my dad a lot of gelt. I liked the idea, but my family was dead set against it. I remember Dad asking, "Jacob, do you know what they train you to do in the Army? They train you to kill people, people you don't even know."

He convinced me to forego the idea. His suggestion was for me to apply for admission to Emory, his alma mater instead. I still feel a twinge of guilt. These young recruits are standing in the gap for people like me, putting their lives on the line.

Just outside of Rocky Mount I pulled off to get something to eat. I found a Hardees just west of town. I took the drive-through lane and ordered a hamburger, fries, and a chocolate shake. When I pulled back onto I-95 it was five o'clock. I set my course for Richmond. At five minutes to six I crossed the Roanoke River. Ten minutes later I am in Virginia. A sign shortly after crossing the state line says "Virginia no longer collects Interstate tolls." No need for coins. The sun is setting when I reach the outskirts of Richmond and the Petersburg I-295 bypass. It has been a very long day. I took Exit 45 and pulled into the La Quinta. Judging from the signs posted along the highway, La Quinta has the lowest rates of the better motels. It is after seven by the time I get checked in. The room has a microwave, a refrigerator, and a coffee maker. Also, they serve a free breakfast. This works for me since I did not bring any food along. Tomorrow is the big day when I will arrive in Baltimore. I went straight to bed. I was asleep before my head hit the pillow.

The sound of Richmond's morning rush hour traffic woke me up at seven. It is now Thursday. I am anxious to get back on the road. After a shower and a cup of instant coffee in the room, I walked down to the breakfast area. I ate a quick meal of scrambled eggs, fried potatoes, toast, more coffee, and orange juice. Then I checked out. I'm *On the Road Again*. Yes, I like country music. I decide to stay right on I-95 through Richmond. It is slow going. Once I pass the beltway and I-64, the traffic begins to thin out. Three hours later I reach the Washington DC Beltway. I decide to stay on I-95 through the middle of the city. Perhaps I can get a quick look at the Capital. Now I know why people complain about the way Washington DC streets are laid out. I became totally confused after crossing the Potomac and ended up going south on I-295. I took the first exit and managed to find the onramp going north again. A half hour later I arrived on the outskirts of Baltimore. The Interstate dumped me right downtown onto Russell Street. I slowly drove past Camden Yards, the Orioles ballpark. I pulled over to take a look at the Baltimore map Terry gave me. I plotted a path to the inner harbor. I will eat a late lunch at the harbor and then find some place to stay for the night.

Even on a Thursday night the place is alive with people. I turned east onto Pratt Street and found a parking place. There is a restaurant close to the parking lot, Phillips Seafood House. It is in a restored building named the Power Plant. As I walked into the restaurant I am surprised to see stained glass light fixtures hanging from the high ceiling and antiqued mirrors covering the walls. Large windows on one side furnish a view of the harbor. The open kitchen allows you to watch the chefs tend their pots of steaming crabs and grills filled with treasures from the sea. I decide to treat myself to a combination seafood dinner, including crab cakes, oysters, mussels on the half shell, scallops, and rockfish served in a butter sauce. I could get spoiled by this city's non-kosher food.

As I walked back to my car, I took Terry Johnson's phone number out my wallet. Why not try to contact him? If he's home, perhaps he will offer to put me up for the night at their farm in Timonium. I called him on my cell. A man answered.

"Hello, Is this Mr. Johnson?"

"Yes, and who am I speaking to?"

"Sir, my name is Jacob Cahn. I'm a friend of Terry's. Is Terry perhaps at home?"

"Yes, but he's working somewhere out in the barns. Can I help you with something?"

"Sir, I just arrived in Baltimore this afternoon. I am applying for admission to the graduate school here at Johns Hopkins this Fall."

"Oh, yes. Now I remember. Terry mentioned you. You are the boy who wants to become a geneticist. You say you just now arrived in Baltimore?"

"Yes, I arrived just an hour or so ago."

"Do you have a place to stay in town?"

"No, I don't. I wondered if you would mind putting me up for the night until I can find a place to stay?"

"Yes, of course. I'm sure Terry will be very pleased to see you again. Are you familiar with the Baltimore area?"

"Not really. Terry did supply me with a map. Is your farm easy to find in the town of Timonium?"

"I can give you enough directions to get you to the downtown area and Terry will meet up with you there."

"That sounds great Mr. Johnson. I really appreciate your offer."

"Not a problem. Do you know where East Pratt Street is?"

"Yes, I'm on it right now. I just finished eating a great seafood dinner here at Phillips Seafood House."

"Do you have a pencil and paper?"

"Just a minute. I'll just trace the route on my map. … Okay I'm ready."

"Take Pratt Street east to South President Street and turn north to the onramp for I-83. Take I-83 north about thirteen miles to Exit 16A. This will drop you onto West Timonium Road. Go east until you get to West Aylesbury Road. Take a right turn and look for the SuperFresh market on the right side of the road. Terry will meet up with you there. He will be driving a red Ram pickup truck with the name 'Johnson Thoroughbred Farms' on the door. Do you have the route marked?"

"Yes, I have it marked. I will be leaving right away, so I should arrive there within a half hour or so."

"Fine, Jacob. Terry will meet you there."

I cranked up Old Blue and headed for Timonium. By now the sun is setting on the horizon, turning the clouds a brilliant orange. It was an easy trip to the SuperFresh. Terry was waiting for me when I arrived. We shook hands and then he had me follow him. It only took another ten minutes to reach the farm. The farm is simply magnificent. We pulled into the farm on a wide driveway flanked with white oak trees. The farm buildings are painted white and the grounds are manicured as neat as a pin. It reminded me of a landscape painting you might find in some art gallery. The farm is fenced and cross-fenced with white rail fencing. There are perhaps thirty or so thoroughbred horses grazing in the lush pastures and a few more in the paddock area. The entrance road ends in a wide circular drive. A large US flag is flapping from the flagpole located in the center of the circle. I parked Old Blue in front of the beautiful Victorian style home. Terry showed me into the house. On the entry way walls are pictures of horses, trainers, and jockeys. There are pictures of the family gathered next to horses in the winner's circle. As we entered the great room, Terry introduces me to his parents, Mr. and Mrs. Johnson.

"Jacob, this is my mother Eldora and my dad Howard."

I shook hands with Mr. Johnson. He then invited Terry and I to follow him to the game room. Several large glass cabinets in the room contain ribbons, awards, and trophies. It is evident the Johnsons run a very successful thoroughbred operation. Terry and I sat down in comfortable leather chairs with his dad and talked awhile about school. It has been a long day and I caught myself several times nodding off.

Mr. Johnson is quick to understand my situation. "I see you have had quite a day, Jacob. We can talk some more in the morning. Right now, you need to get some shuteye. I need to warn you. We get up early here. Breakfast is at five."

"I have to admit I am pretty tired, Mr. Johnson."

"Let's dispense with the formalities, Jacob. My first name is Howard, but everyone except my banker calls me Howie. Can I call you Jake?"

"I'm fine with Jake. Let tell you say I certainly appreciate your hospitality. Since I'm not good company, I believe I will take you up on your offer."

Terry got up and said, "Let me help you take your gear upstairs, Jake."

We unloaded my suitcase and a few books from my car and then climbed the stairs to the second floor. There is a spare bedroom at the end of the hall. Terry points out the bathroom across from the bedroom.

Terry explained, "This is my older brother Ted's room. He is now in California with a few of our horses."

"Are they stabled at Santa Anita, Terry?"

"I wish! No, they're at the Fairplex track in Pomona. Our string is a cut under the horses racing at Santa Anita."

"When I was a kid I used to go to Aksarben, the track in Omaha. The name of the track is Nebraska spelled backwards. They closed the track a few years ago."

"Yes, I know. We raced there a few times."

"Will Ted be coming home after the Fairplex meet?"

"No. Ted is following the California fair circuit. He will come home in January to help during the breeding season. I know you will eventually be renting an apartment close to the campus, but anytime you want to get away for a while, I'm sure Dad won't have a problem with you using Ted's room here. If he does come home, we have a couple of bunkhouses as well. If they don't suit you, there are several tack rooms used by our grooms."

"I want to pay you a visit from time to time, Terry."

Terry opened the door to Ted's room. It is a large room, more like a combination of bedroom and office. A roll top desk and book cases are along one wall. Some thirty or so trophies are displayed on top of the bookcases. On the opposite wall are pictures of past winning races and ribbons. Terry showed me the walk-in closet and said I am free to store my gear there. He opened the drapes in front of a large bay window. The window looks out toward the east side of the farm. I see a small stream winding its way through the undulating lush green pastures. There are scattered groves of white oak trees beyond the pastures. The scene is illuminated by the light of a full moon just rising above the trees. The view is simply breathtaking.

"Terry, you have a little paradise on earth here."

"Yes, this paradise is the result of over seventy years of hard work, Jake."

"Did your dad inherit the property?"

"No, Dad purchased this property in 1945 after he returned from the war. Dad came from very humble beginnings. Sometime I will have to tell you the story of how the Lord made it possible for him to purchase what was once just overworked tobacco fields."

"I'll look forward to hearing his story. I think I better turn in now. I'm yawning. The long day's drive has left me feeling a bit dizzy. Thanks for everything, Terry. I really appreciate this."

"You are very welcome. Jake. Get some sleep. We can talk at breakfast in the morning."

Terry closed the door behind him and I stored my gear in the closet. As I roll into bed, I am thinking about my agenda for Friday. I will call the Institute and let them know I'm in town.

The Promised Land

I awoke early Friday morning to the sound of crowing roosters. I had a good night's sleep. I felt great, even though I had gotten to bed late. After breakfast, I returned briefly to my room to make my bed and straighten things up a bit. It is still way too early to call the University. Over breakfast Terry tells me he is going to give one of their horses a morning work. I decide to walk out to the training track to watch. When I arrive Terry already has completed working the mare and is busy hosing her down, brushing, and cleaning her hoofs.

"How did the workout go, Terry?"

"Not good, Jake. I'm not thrilled with this girl. She just doesn't seem to be interested in running a lick.”

“How old is the mare?”

“She’s a three-year-old. We took her to the track earlier this year, but she had the same problem. Some horses just love to run and some don’t. She’s a loafer. She is content just to follow behind the other horses around the oval.”

“She certainly is a good-looking horse.”

“Yeah, she has good conformation, but unfortunately you can’t see what God puts into their thick heads when you breed them. Perhaps you will discover some way for us to figure it out before we spend so much time and money training them.”

“Perhaps so Terry, but I wouldn’t hold your breath. What are you going to do next after you finish washing her down?”

“I’m going to feed the yearlings in the barn and then put them out to pasture. Do you want to help?”

“Sure, but you’ll have to show me what to do.”

“No problem. Let’s put this gal on the walker and then we’ll go feed.”

It took us until half-past nine to finish feeding the yearlings, and turn them out to pasture. Then it was time to muck out their stalls and put new bedding down. After finishing the bedding, I returned to my room. I’m anxious to make my telephone call to the Institute. The procedure followed by the Institute is a little different. Since they accept so few applicants, new students are carefully pre-evaluated. Although I have been encouraged to apply, my admittance to the program is not granted until prospective students meet with the Director and staff. I will be interviewed by selected professors, both in the Institute as well as from other schools. I’m still not sure whether I can move things along right away or whether I will have to wait until sometime in September. The first thing I need to do is to let the Program Administer, a Ms. Roberta Capoletti, know I’m in town. I took a deep breath and dialed her office number.

“Good morning. You have reached the Administrative Office of the McKusick-Nathans Institute of Genetic Medicine. This is Ms. Roberta Capoletti. How may I assist you?”

“Good morning Ms. Capoletti. My name is Jacob Cahn. I recently submitted my application to your Institute for admittance to the human genetics program for this Fall semester. As I understand it, my first step is to contact you to set up a series of interview appointments. I realize this is a bit early to be contacting you. Is it possible to schedule my appointments any time soon? I am staying with a friend in nearby Timonium."

"Welcome to Baltimore, Mr. Cahn. You are not too early. We like to begin our interviews for the Fall semester as soon as possible. I believe you are coming to us from Emory University?"

"Yes, I arrived from Atlanta just yesterday."

"First off, we will need a telephone number so we can contact you."

"My cell phone is 402-661-2336. Let me also give you the Johnson's home phone number here in Timonium. I am staying here with a friend. Their number is 410 663-5153."

"Jacob, I have arranged for the Department to take two other applicants to dinner this Monday evening. Are you free then?"

"Yes, my schedule is open."

"Let me give you a bit of background information. It is our policy to assign one of our students to act as a guide for new applicants. Student guides assist our applicants in getting oriented to JH. We will select a student guide for you. Your guide will be taking you to your various interviews and introducing you to the members of the faculty."

"I see."

"I have a young lady in mind who has volunteered to serve as a guide. It is short notice, but she may be able to attend the dinner as your guest. Would you rather come alone, with someone else, or with this young lady if she is able to attend the dinner, Jacob?"

"I don't suppose you have a male guide available?"

"No, not at such short notice."

"I understand. Is it possible for her to meet me at the restaurant? Otherwise, I think I need to just come on my own."

"She can meet you at the restaurant. Her name is Ramona Bergman. I will have her call you to let you know if she can attend. Otherwise I will get back to you. I believe I already mentioned our staff, including Director Hobbs and two other applicants will be attending the dinner. Dress is informal. Slacks and a sport shirt are fine. Do you have a pen and paper ready?"

"Yes, just one moment … I'm ready."

"We have made reservations at the Oceanaire Restaurant located at 801 Aliceanna Street. We will be meeting in the Mahi-Mahi

Dining Room at seven o'clock. We will look for you there. Do you have any questions?"

"I understand the school usually schedules the interviews to follow directly after the dinner engagement. Will I begin interviewing on Tuesday?"

"Yes, we usually schedule interviews for the following day, unless the applicant feels he or she needs more time to prepare. Are you comfortable with holding your first interviews on Tuesday, Jacob?"

"Yes, I am as ready as I ever will be. I've looked forward to this for a long time."

"That's the spirit. You will be meeting with professors from our staff as well as from other departments at various locations. Let me give you a few more pointers, Jacob. Try to stay relaxed. You wouldn't be here if we weren't interested in you. Be prepared to ask questions and respond to their questions as fully as possible. The more you talk, the less they have to. Also, be ready to discuss your fields of interest. Don't limit your field to a single concentration. Give them a feel for the type of research you are interested in. Finally, be prepared to address your overall view of human genetics and how it fits with your personal philosophy. Do you have any other questions I can help you with?"

"I do have one or two questions. Otherwise I think you have covered everything for me. Will you be getting back to me with the interview schedule?"

"You will receive the incoming student brochure and the interview schedule soon, perhaps at the dinner."

"I wonder, can you tell from my file if you have received my three recommendations?"

"Let me take a look, Jacob. Yes, we have received them. One last thing. I suggest you wear comfortable shoes on Tuesday. You will be doing a lot of walking."

"I will be prepared to walk, or run if need be. Thanks for your helpful information. I greatly appreciate your help Ms. Capoletti. Have a nice day."

"The same to you Jacob. I wish you every success on your interviews and we will see you on Monday evening."

I hung up the phone and went downstairs. Terry is in the game room, sitting at a desk watching the replay of a race on his computer.

"Hi Terry. Are you watching one of your horses running at Fairplex?"

"Yeah, this horse sometimes forgets to switch leads on the turn. I'm trying to figure out what his problem is."

"Terry, I called the Institute and talked to a Ms. Capoletti. She is scheduling my interviews with the Director and his staff. They will be taking myself and two other incoming students to dinner Monday evening at a place in town called the Oceanaire Restaurant located on Aliceanna Street."

"Wow, they're spoiling you, big time."

"Is it a special place?"

"You bet it is! It is a very high-class joint."

"Well I hope they are picking up the tab."

"I'm sure they will be, Jake."

"Terry, I think I'll just hang out here and read a book if you don't mind."

"Sure Jake. I'm about ready for a cup of coffee. How does a cup of joe sound to you?"

"That would be great."

"Can I also get you something to eat? Mom baked an apple pie this morning."

"I've never refused a cut of apple pie in my whole life."

Terry disappeared into the kitchen and returned shortly with two cups of hot coffee and two pieces of apple pie a la mode.

"Terry, this is just what this prospective geneticist needs at this point."

"So, what is the book you are reading about, Jake?"

"The book is discussing some of the latest information on the cell's design. I need to be able to say something intelligent when I meet with the staff at the Institute. The book describes these new research tools: the latest electron microscopes, spectrometers, chromatographs, ultracentrifuges, and many other modern techniques such as the CRISPR/Cas System. These new devices and techniques are just awesome. Geneticists are truly pioneers. It is like embarking on one of Columbus's voyages, discovering distant new lands."

"It sounds very complicated Jake."

"Yes, it is, but the molecular world within the cell is absolutely fascinating. I am keenly looking forward to next year's graduate program. Of course, I still need to get accepted. I understand they put you through a tough selection process."

"I'm sure you are going to make the grade, Jake."

"Terry, as I remember, you are entering veterinarian school here at JH."

"Yes, my time at Emory is over. I guess you know why."

"I don't remember you telling me anything about why you left Emory. What's the story?"

"Jake, all through high school my girlfriend Cindy Suzany and I were inseparable. I loved her dearly."

"She has a nice name, Terry."

"Yes, and she is sharp to. We planned to get married. She wanted to become a doctor. She decided to take her pre-med at Emory. At first, I didn't really believe in the old adage 'absence makes the heart grow fonder.' Then her letters became fewer and fewer. So, I made the decision to follow her to Emory and enter veterinary school there. My plan was a day late and a dollar short. After my first year at Emory she dumped me for a guy from California. I did manage to complete one year of the veterinary degree at Emory. I transferred my school records back here to JH. So here I am, back in Timonium, trying to forget her beautiful name."

"I'm so sorry, Terry. Did you know I'm facing a similar problem? Perhaps we can cry on one another's shoulder."

"Perhaps we can, and you can help me with my microbiology class."

"There's no reason we can't study together. I believe I will be taking biology courses on the Homewood Campus."

"Yes, we will have to do this. By the way, Homewood has some great facilities. The smart classrooms allow you to access audio as well as video media from the lectures any time you want to. Emory didn't have this level of instructional support."

"I haven't been on the campus yet, but I did take an internet tour of Homewood. I was impressed. The facilities seem to be the state of the art."

"They definitely are."

"I may just take a visit to the Homewood Campus on Saturday. Would you want to come along as my guide?"

"Sure. I don't have anything planned except feeding in the morning. If you pitch in to help we can leave by about ten o'clock."

"Okay, let's do it."

We finished our pie a la modes and I settled back to continue reading my book on cell design. Terry replayed the race video a few more times, and then turned off the computer.

"I'm going to do some work in the barns Jake. I'll see you at lunch."

"Okay, Terry. You know where to find me."

I lost track of time reading the book. The grandfather clock in the great room just finished striking eleven when the phone rang. I answered it for them.

"Hello?"

"Good morning. Is this the Johnson farm?"

"Yes, it is."

"Is a Jacob Cahn available?"

"You are speaking to him."

"Jacob, my name is Ms. Ramona Bergman. I am currently in the genetics program at JH. Ms. Capoletti contacted me and asked if I was available to take on a new applicant under my wing. I told her I am quite available. She assigned me to you as your personal student guide. I will be helping you find your way around JH. Also, I am invited to attend the applicant dinner with you on Monday evening. I thought I should give you a call and let you know you have a blind date, Jacob."

"It is nice to hear from you Ms. Bergman. I certainly appreciate your help in getting me acclimated here."

"Please don't feel you have to be so formal with me, Jacob. I'm Ramona to you. I understand from Ms. Capoletti you are staying with a friend in Timonium?"

"Yes, I am. My friend Terry Johnson is finishing up his veterinary program this Fall."

"I see. Let me give you my cell number. Do you have a pencil and paper?"

"Yes. Go ahead."

"The area code is 443 and my number is 855-2565. I want to see you get off on the right foot. Please feel free to call me any time. Can I have your cell number, Jacob?"

"Yes of course. It is an Omaha number. I will be getting a new number here in Baltimore as soon as I find a place close to the campus."

"Fine. Let me have your old number. You can update me with the new one later."

"Okay. My cell phone is area code 402, 661-2336. Thanks for volunteering to show me around and I look forward to meeting you Monday evening."

"I look forward to meeting you as well, Jacob. Have a great day."

Terry is listening to my call. "Well Jacob, you already have a blind date for Monday's dinner. Pretty fast on the uptake I would say."

"Yes, it seems I do have a date."

"Did I hear her say she is quite available?"

"She did sort of place emphasis on her availability, didn't she?"

"She may be very hungry Monday night, if you know what I mean. You better be careful."

"Terry, she's probably just doing what they have asked her to do. So, what buildings do you suggest we take a look at on Saturday?"

"We will park in the visitor parking lot near Mason Hall and walk across the Decker Quad to Hackerman Hall. Hackerman contains what they call a "living style classroom" on the ground floor. Upstairs we can take a look at the Divinci Machine developed at JH. It is capable of performing robotic operations, plus the area is a fully functional operating room. It's pretty interesting."

"What is a living style classroom?"

"This type of classroom is hands-on. In Hackerman, engineering students work together to actually build robotic devices, and other cutting age technologies. At the east end of Decker Quad is Garland Hall. This is where the University President has his office, as well as the academic advisors. To the north of Garland is Hodson Hall. We can visit one of the smart classrooms there."

"Are these smart classrooms the totally automated study facilities you mentioned earlier?"

"That's right. East of Hodson Hall is Levering Hall. It contains a food court, a lounge, and the Student Union. Beyond Levering is the Glass Pavilion Auditorium. It is a small auditorium where plays are performed. I could go on and on. We won't be able to take in all of Homewood on one trip. You will have to explore most of it on your own, unless you want to ask Ms. Ramona Bergman, your personal student guide, to show you around."

"I will have to weigh how often I ask this young lady to help me. Rachel and I have made a number of rules about my relationships with the feminine gender."

"Good thinking. You will have to keep a wary eye out for future storms on the horizon. As far as I am concerned, I have no hesitation whatsoever. I think it is about time for lunch. Let's see what we can whip up in the kitchen."

Terry and I sat down to a lunch of homemade chili and crackers. Afterward, Terry went out to train a two-year old from the starting gate. I again settled back to reading my book. Mrs. Johnson poked her head in to let me know dinner would be ready at six o'clock.

"Jacob, I'm wondering, do you have any special diet requirements, such as eating only vegetables?"

"Not really. I can't eat ham, otherwise I'm quite omnivorous."

She thanked me and then left the room pronouncing the word omnivorous over and over. I wondered if she is headed for a dictionary? Just as I was finishing my book, Mr. Johnson and Terry came back into the house early. Mr. Johnson didn't say a word, but went straight upstairs.

"Terry, your dad seems really upset about something?"

"He is. We had a problem in the pasture. One of our best two-year-old colts got tangled in his lead rope and the fence. He went into panic mode and ended up bowing a tendon."

"Is it a serious injury?"

"It will heal, but not soon enough to start this year. I had to give him a shot of tranquilizer while Dad held him down. Then we coated his leg with salve and wrapped him. We had a hard time convincing him to get up and walk back to his stall."

"Is your dad going to be able to handle this? he seems very angry."

"He'll be okay by dinner time. When something like this happens, he goes upstairs and asks the Lord to forgive him for losing his cool. You'll see. In a few minutes, he'll come back downstairs a changed man."

True to Terry's word, Howie Johnson came down the stairs ten minutes later completely changed. Smiling, he hugged his wife and said, "Hon, let's eat."

Apparently, all is right again in his world. Howie prays over the meal and we sit down to a dinner of pot roast and dumplings. After dinner, we men retreated to the game room to watch the submarine movie *Run Silent, Run Deep*. After the movie, I excused myself and returned to my room. I set the alarm for five o'clock. As I lay in bed, I pondered Mr. Johnson's behavior. The incident revealed an inner strength I do not have or understand.

The alarm rouses me at five Saturday morning. Downstairs the table already is set. We sat down to a breakfast of scrambled eggs, hot cakes, and hot black coffee. Terry detailed our plan with his dad to visit the Homewood campus after feeding time. I helped Eldora clear the table, then went outside to help Terry feed.

After feeding and cleaning up, we decide to take Terry's truck to the campus. After touring the buildings on Terry's list, we ate lunch at the Student Union.

After lunch I asked, "Say Terry, I'd like to take a look at the tennis courts?"

"Sure, Jake. They are right next to the Ralph O'Connor Recreation Center. Everyone refers to it simply as the Rec Center."

We walked down to the Rec Center. There are six courts located adjacent to the Rec Center. Terry showed me the recently remodeled lacrosse field. I wondered if my tennis skills would enable me to play lacrosse. I just might give it a try. Next, Terry took me along St. Paul Street to show the off-campus apartment complexes and the book store located in the Charles Common Building on East 33rd St. We didn't return to the farm until dinner time. I excused myself after dinner, and returned to my room to read some human genetics research articles.

I fell asleep in the easy chair and didn't wake up until three in the morning. I stretched out on the bed and was soon back to sleep. The alarm woke me at five. I rushed to get dressed. When I arrived downstairs, I am puzzled to find a quiet kitchen. Then I realize it is Sunday. Terry told me yesterday the family sleeps in on Sunday. Breakfast is at eight o'clock. I decide to take a morning jog on the training track. As I am completing my fourth mile, I see Terry loading hay from a haystack in the pasture onto the quad and heading for the barns. I decide to help him.

"Terry, I forgot to sleep in this morning."

"I thought I heard your door close about six. I see you have been taking some laps on our training track. Did you try out the starting gate?"

"No. Besides running four miles, I took some time to enjoy this beautiful morning."

"Jake, after we finish feeding the stock, I will be leaving with the family for church. Do you want to come along?"

"Well, I don't know. I will have to think about it over breakfast."

"Okay. Feel free to do whatever you want to do. Once a month Dad, Mom, and I play with the worship band, so we will be leaving at a quarter to nine for practice."

"I didn't know you played. What instrument do you play?"

"I play the guitar, Dad plays the drums, and Mom is on the keyboard."

"So, you are a musical family."

"Yeah, you could say so. Mom is really the one with the talent. Thanks for your help with the feeding, Jake. I didn't get started on time this morning."

Terry and I finished feeding and returned to the house. After breakfast, the family dressed for church. I told them I preferred not to go today, but promised to attend their church sometime soon. It was fine with them.

Howie said, "just enjoy yourself, Jacob. We will be back about one this afternoon."

I asked, "Would you mind if I use your internet connection, Howie?"

"No Jake, go right ahead. The password is written on the bottom of my computer."

The family left for church and I retrieved my computer from the back of the car. I spent the whole morning reviewing the latest research papers posted on the American Society of Human Genetics (ASHG) site. They returned at one-thirty. After a late lunch with the family I returned to my room to review some more of the papers from the ASHG site. By evening I felt ready for Monday's meeting. After dinner, Terry and I played chess. It was a close contest. We both won two games.

After the fourth game Terry asked, "Jake, I want to know why you didn't take my pawn with your bishop?"

"Gee, Terry, I didn't want to jeopardize my bishop by moving him to your side of the board."

"Not so! You just didn't want to take my pawn. I think you're hornswoggling me."

"I couldn't be, because I've never heard such a word before. Perhaps I'm fachadick, you know, just confused."

"I doubt you are confused, Jake. Can we agree to play fair and square?"

"Sure, if you insist."

I kept to our agreement and won the next two games. Terry looked dejected and defeated.

"Jake, it must be hard for you to remain so humble?"

"Who's humble? I just don't know how to lose, but thanks for showing me how."

We both enjoyed a great evening together. I now know this future veterinarian a lot better. At eleven o'clock we called it a day. I turned in and promptly fell asleep, thinking about tomorrow's dinner at the Oceanaire Restaurant.

10

The Challenge

"The Lord shall cause thine enemies that rise up against thee to be smitten before thy face. They shall come out against thee one way, and flee before thee seven ways."
(Deuteronomy 28:7)

The Battle is Joined

The alarm woke me on Monday morning at five. I have been looking forward to this day for a very long time. Tonight, I will meet Dr. Jonathan Hobbs and many of my future professors. They will administer the painful rod of learning to this young *hamisch,* the complex path to becoming a geneticist. After breakfast, I asked Eldora if she would mind pressing my dress shirt and slacks. Then Terry and I went out to feed the stock. It is another beautiful Maryland morning. We just finished feeding when a semi with a set of doubles pulls into the farm. Terry tells me they are sending a string of horses to his brother in California. Terry took the forklift and began loading the two trailers with enough clover hay, oats, and water for the long trip. The driver tells me it will take him over a week to deliver the horses. I watched as they loaded the young three-year olds into the trailer. Terry is bandaging their legs to protect them during the journey. Then he put blindfolds on the skittish horses. Each animal has its own individual stall in the trailers. I watch as Terry securely ties each horse to a railing to prevent any mishap. Then he gives each horse a tranquilizer shot before sending them on their way.

We returned to the house for a late lunch. I am feeling a bit nervous about this evening's event. I returned to my room to try to relax. Eldora had neatly hung my slacks and shirt on the closet door. I sat down in the easy chair and drifted off to sleep. It was four o'clock before I woke up. I rushed to get dressed in my suit. When I went downstairs, Terry was in the game room.

"Terry, I'm not sure just when to leave for the restaurant. I'm supposed to be there by seven."

"It won't take you longer than thirty or forty minutes to arrive to the restaurant and get parked. It is only 15 miles from here to the harbor. You just need to stay on I-83 south. The Interstate ends at South Presidents Street, right in front of the restaurant."

"I don't want to take any chance on being late, so I'm going to leave by six."

"Okay, but it will put you there early."

Terry is correct. I arrived at 6:35, way too early. All I can do is park in the restaurant parking lot and wait. For the next twenty minutes, I spent my time watching the sailboats pass in and out of the harbor. At five minutes to seven I entered the restaurant. The maître d' met me at the door.

"Are you here with another group, Sir?"

"Yes, I believe our group is meeting in the Mahi-Mahi Dining Room."

"Can I have your name, Sir?"

"My name is Jacob Cahn."

I watched as the maître d' crosses my name off his list.

"Am I the first to arrive?"

"No Mr. Cahn, you are actually the last person to arrive. Right this way please."

His comment sends a chill up my spine. Was I late? I glance at my watch. No, it is still a minute or two before seven. Relieved, I followed him to the rear of the restaurant. He ushers me into an elegant room. There are large glass windows overlooking the harbor. A young woman is standing at the doorway.

"Good evening. I'm Ms. Ramona Bergman. You must be Jacob Cahn."

"Yes, I'm Mr. Cahn. Good evening Ms. Bergman."

She gives me a wide smile and extends her hand. "I believe I gave you permission on the phone yesterday to call me by my first name. You are my summer project Jacob."

"Yes, I believe you did say to call you by your first name, but I'm not entirely sure just what summer project you are referring to."

Ms. Bergman responds to my words with an easy-going laugh, and pins a nametag on to my coat. She is a strikingly beautiful young woman. She is wearing a stylish black dress. It looks good on her. Her long black hair reaches down to her waist, complementing her attire perfectly. Coupled with her broad smile and warm greeting, I realize Terry's warning is right on the money. My ketuba agreement will be severely tested this evening. She ushers me to the end of the table and introduces me to Dr. Hobbs, the Director of the Institute. This intellectual giant of a man, the Director of the most famous genetics school in the world, is nothing like I envisioned him to be. He is a short, rather portly man with bright blue eyes peering beneath bushy eyebrows. His warm smile clearly reveals a jovial personality. Right away I know we will get along well with one another.

"Good evening Jacob. It's so nice to meet you. I hope you are enjoying our fair city of Baltimore?"

"Yes, I certainly am Sir."

"What have you had time to see since your arrival?"

"Actually, I have been introduced to the hinterlands of Baltimore. I'm staying with a friend from Emory, Terry Johnson. His family has a thoroughbred farm in Timonium."

"I see. I imagine they have given you a warm Maryland welcome."

"Yes, they have. Their son Terry is just now finishing his degree in the veterinarian school. He took me on a tour of the Homewood Campus yesterday."

"I see. You certainly are getting off to a quick start, like one of your friend's sprinters no doubt."

"I must admit I am eager to begin."

"Ramona, why don't you introduce Jacob to the rest of our staff and to the other two applicants and their student guides."

Ramona quickly introduces me to everyone around the room. There is a name card and packet of materials at each place setting. Dr. Hobbs moves to the end of the table and stands a minute, waiting for everyone to gather at their places. Ramona is seated next to me. Dr.

Hobbs then asks everyone to be seated and warmly welcomes us to the Institute. He quickly outlines tomorrow's schedule for meetings. After this brief orientation, Dr. Hobbs holds up his menu and says, "Ladies and gentlemen, I want you to take advantage of this evening's choice of entrées. As for you prospective graduate students, please feel free to order whatever your heart desires tonight. Soon you will be eating food from cafeterias and vending machines."

The waiters begin to take our orders from the menu, and what a menu it is. I decide to choose the chef's suggestion. It consists of four courses, beginning with a bowl of New England clam chowder and a Caesar salad. For my antipasto, I selected a helping of seared Faroes Island salmon served with an Asian crab crust and something called a dynamite sauce. The chef did not exaggerate when he described the sauce. It made my eyes water. For dessert, I chose cheesecake covered with a blueberry sauce. The table conversations dwindled as the dinners are served. As we finish our meals, the conversation begins to pick up again. The other two applicants at the table are males also. Joe Harmon is from Baltimore and received his BA from JH. The second fellow, Craig Simpkin, hails from Los Angeles and received his BS in Biology from the University of Arizona at Tucson. Both are goys. I could tell they both are very bright and intensely interested in the field of genetics.

Dr. Hobbs and the staff are engaging us in friendly conversations, asking us about our families and undergrad experiences. Director Hobbs turns to Craig Simpkin and begins asking him about his life in California. This is how the inquisition began.

"Mr. Simpkin, as a Californian, do you view human genetics as a radical profession?"

"No Sir, I do not. I have avoided the excesses of lifestyles followed by many of my California undergrads. My interest in human genetics began when I worked at the Cedars of Lebanon Hospital in Los Angeles. I was offered a summer internship following my junior year in high school. I worked in the pediatric unit. The doctors and nurses are dealing with major problems stemming from the effect of drug use by mothers of these children. They suffer from serious damage to their metabolic processes. I realized any meaningful solution will have to come from research at the cellular level."

"I see. So, your desire is to work in the area of fetal development?"

"Yes Sir, I hope to, but I am still considering other research possibilities."

"Is there any overarching philosophy directing your interest in genetics, Craig?"

"No Sir, I am willing to allow the pursuit of knowledge to take me where it will."

"Ah, we do have a careful man in our midst. You are playing it safe tonight Mr. Simpkin. I probably would do the same if I were in your shoes."

"Mr. Harmon, would you care to share your hopes and vision with us?"

"Yes, I would like to say a few words. I grew up right here in Baltimore. I attended JH after high school. I received my BS in pre-med at JH. My father is a doctor here at the hospital. His specialty is cardiology. He is one of the first to use robotics to perform heart surgery. I believe you are familiar with the Divinci Machine he helped to developed here."

"Yes, we are all familiar with your fathers work here at the hospital, Joe. What I would like to hear is how your contacts with JH prompted you to become a geneticist. Can you speak to us about your particular interest area?"

"Because of my father's career here at JH, I was introduced to the potential benefits to the medical profession from this exploding area of genetic science. My particular interest lies in the study of the immune system and its response to disease. I feel confident human genetic research soon will be able to improve the body's ability to ward off harmful organism, especially the intrusive retro-virus diseases."

"Solving the retro-virus problem is a very worthwhile goal indeed, Mr. Harmon. I'm sure you are already aware our Institute is working very hard on solving the puzzle of the immune system. Currently our research into the cell's reaction to broken protein fragments is looking very promising. Dr. Willard, would you like to make any further comments on this?"

"Yes of course, Joe. We must all realize we are now on the cutting edge of understanding these responses to infection and how the cell has fortuitously preserved remnant proteins to erect barriers to disease. I look forward to your contribution toward achieving the goals you have mentioned."

"Mr. Cahn, how did you first decide to pursue a career in human genetics?"

"Sir, I had a great professor at Emory, Dr. John Thomas. He took me under his wing and really was instrumental in my decision to abandon my earlier plan to follow in my father's footsteps to become a general practitioner."

"In what way did this instructor convince you, Jacob?"

"There were several ways. One of the first things he did was to suggest I become a student member of the American Society of Human Genetics. The information I found from the Society played an important role in convincing me to change my major."

"Are you interested in research?"

"Yes, I am very interested in researching the human genome."

"What concentration do you see as most compelling?"

"If I were to select one particular area, I would say it is research into fetal development of neurons of the cerebellum."

"Why do you desire to pursue fetal development, Jacob?"

"I believe there are many mysteries to be discovered, especially related to the actions of precursor neurons. Our brain's capabilities are utterly beyond other mammals. Another related interest of mine is human memory. We read in the literature of early man's ability to memorize huge amounts of information, yet we are not even sure today where memory is even stored? I do not believe it is stored in the neurons. They seem to function entirely as electrical circuits. I don't believe the neurons are the source of our reasoning abilities, and our creativity."

"Would you say we might discover how such a brain evolved from lower species?"

"No Sir. I would be very surprised to discover such an evolutionary pathway from lower mammals."

"Do you have reservations about the tree of life Jacob?"

"Yes, sir, I do not believe we are the product of blind naturalistic evolution."

"Not everyone here this evening has been given the gift of chutzpah you are bringing to table Mr. Cahn. Would you care to describe how you view the origins of this complex human brain for us?"

"I'm not sure it would be wise for me to address this topic further tonight. I don't wish to be labelled as a radical student activist."

"On the contrary, Jacob, I often ask my students to present their philosophical positions. You need not fear reprisals. Let me say as a way of encouragement, I also find many problems with our vaunted position of natural selection and progressive mutation. No doubt you have read Professor Michael Behe's book Darwin's Black Box."

"Yes Sir, I have. I also have read his second book, The Edge of Evolution where he demonstrates the fallacy of consecutive favorable mutations."

"Can we say you are an adherent to the Behe position called intelligent design then?"

"No Sir. Not in the sense it is presented by Dr. Behe and others who agree with him."

"What is your problem with his position, Jacob?"

"The case Behe makes for intelligent design rises and falls on the validity of the concept of irreducible complexity. Behe has been criticized for his conclusion such complexity is sufficient proof of the actions of a Creator. I believe his critics rightly point out the error of his position. The basis of his argument really rests on the philosophy of a "God of the Gaps." This philosophy is not a valid argument. It does not prove gaps in the archaeological record are necessarily filled by a Creator God. They could simply be missing. Therefore, it cannot be considered a non-falsifiable negative argument. His position does not furnish adequate proof of a Creator God."

"You are correct to recognize this criticism, Jacob. So how do we get beyond this and still accept a supernatural Creator? Are we condemned to accept a wait and see position, expecting to discover at some time in the future whether natural processes are valid after all?"

"No, Sir. It is my considered opinion the indicators of design in the cell's chemistry amount to a strong positive case for the active involvement of the Creator. Instead of arguing for a creative origin based upon a negative assumption, namely the inability of natural processes to generate life's systems based upon our admitted lack of knowledge, I would appeal for support for an active creative origin based upon a positive hypothesis."

"Can you suggest an example of such a positive hypothesis, Jacob?"

"I believe there are several. I can highlight one example among others. Consider the language of life itself, the language of the double helix. It is a language based upon the mathematical base of four. Such a language system points to an intelligent source beyond naturalism.

It is as if we are to put a chimp in front of a computer and expect him to compose the novel War and Peace. This language of the cell is a very compelling argument for me. Its simplicity of design is simply breathtaking. In my opinion, the beauty of the language of the cell far surpasses the greatest art masterpieces known to man."

As I concluded my response, I noticed Professor Frederick Altschul from the Institute is becoming very agitated. I realize my comments have surfaced a very aggressive opponent. Yes, there are enemies of Ha Shem in this bastion of modern science.

Dr. Hobbs then asks for responses from his staff.

"Would one of my colleague's care to comment on Mr. Cahn's position?"

It did not take long for Professor Altschul to respond.

"I would like to ask Mr. Cahn whether God has recently provided funds for him to carry out his research. Perhaps he has had a conversation with God at the burning bush?"

Altschul's cynical reference to the burning bush drew a few chuckles from those gathered around the table. My intuition is telling me to keep my trap shut at this point, but I ignored it. "Sir, I don't expect the Creator of the Universe to speak to me from a burning bush, but I don't rule out the possibility he might be interested in my research into his creation. Thoughts of a more inductive flavor might reveal hidden aspects of this very complex field. I believe Einstein referred to such a source as responsible for his development of the theory of general relativity."

What came next from Professor Altschul was unexpected and very nasty.

"I should expect such a response from someone with your cultural background Mr. Cahn. Many in this field will not be inclined to help foster research under the auspices of the Almighty. Have you considered the ramifications of your position? For example, you may experience difficulty finding members of the Institute willing to serve on your graduate committee."

Wow! I have uncovered a hidden time bomb in this supposedly egalitarian institution. What followed was a period of awkward silence. Professor Hobbs took up the task of restoring peace to our gathering.

"Thank you for sharing your position, Professor Altschul. I'm sure we will have ample opportunity to revisit this topic at a later time. I wonder, would another candidate care to comment on Mr. Cahn's philosophical premise? Perhaps one of the student guides would like to comment?"

Dead silence followed. Dr. Hobbs finally broke the chill. "I seem to be lacking volunteers. We apparently do not have another young lion in our presence this evening."

Dr. Hobbs took a look at his watch and called an end to the disrupted dinner meeting.

"Well, ladies and gentlemen. I believe it is time to end this very interesting time together. Tomorrow will come early. We will meet in my office in the Broadway Research Building at eight o'clock sharp. Ms. Capoletti will furnish you with your interview schedules. You will be meeting with our staff as well as professors from other schools on BRB, East, and Homewood campuses. If you come by car you can park in the hospital parking garage off of Broadway. We will validate your ticket. If you park at the Homewood campus, I recommend you take advantage of the visitor parking areas and then take a shuttle to our building. Your student guides will meet you in my office and help you navigate your way to the interviews. Thank you all for coming. We are looking forward to having many more conversations like this in the near future. I will see you tomorrow morning."

I gathered up the student guide materials and shook hands with Dr. Hobbs and the other professors. In an attempt at reconciliation, I made a point to shake hands with my protagonist, Professor Altschul. He shook my hand and smiled, but his body language gave away his underlying bitter resentment of this Jew boy from Omaha.

Miss Bergman met me at the door as I am leaving.

"Jacob, I wonder if you would mind dropping me by my apartment. I was here at the hospital and I just took a cab to the restaurant. It's rather late and I'm not sure how soon I can get a cab."

"I suppose I can give you a lift. Where's your apartment located?"

"I'm renting a brownstone just south of the hospital. It won't be very far out of your way."

"You will have to be my navigator. I'm not familiar with the brownstone area of Baltimore. My friend Terry did describe it. He said it is the party zone of Baltimore and I should avoid it."

"Yes, it's known as a party area, but rents are a lot cheaper there. Also, it is not far from the hospital, and close to the Inner Harbor."

"Are you an Orioles fan?"

"Of course. I enjoy taking in a game when I have the time. Are you inviting me to a game?"

"No, I just wondered if you like the Orioles team."

"Yeah, I like the team. You know the seafood at the Inner Harbor is great and there are all kinds of other attractions. Can I show you around sometime soon?"

"Sure, sometime we can take a tour of the harbor. By the way, Ramona, I need to warn you. My car is still full of all of my stuff. I haven't taken time to unload everything yet. The trunk is full and I loaded my small refrigerator into the front passenger seat. I can squeeze you into the backseat if you don't mind."

"No problem Jacob. I'm perfectly willing for you to squeeze me in the backseat of your car any time."

There is no mistaking this underlying message. I let it pass without a response. I showed her to my car and opened the back door for her. She quickly hopped in and we headed out.

As we turned onto Eastern Avenue she began to comment on the dinner meeting. "Well you certainly stirred the pot this evening, Jacob."

"I know I came off very blunt with the Professor. Do you think I hurt my chances?"

"I'm not sure. One thing I do know, you definitely made a lasting impression on Dr. Frederick Altschul. He had it coming. No, I don't think you harmed your chances for acceptance to the program, not at all. No doubt you picked up the fact he is a strong atheist. Professor Altschul is known for paying homage to the great god of natural selection. He is not an admirer of our Nation's Judeo-Christian culture. He probably would not have reacted to a Jew playing it safe, but to a Jew espousing an omnipotent Creator? This is another matter altogether. You pulled his chain Jacob, and there was no way he is going to let it pass without comment. Don't expect any support from his direction."

"I understand. At least I found out the way the wind blows in the Institute. I thought it was cool when Dr. Hobbs asked for comments; it was followed by dead silence, even from the staff. Did you notice everyone, including the professors, were looking down at their plates? They were searching for a place to hide from the shrapnel, like under the table. I suspect Altschul must have most of them under his thumb."

"It is good to get things out in the open early on, Jacob. Turn right at the next corner. My apartment is the third house from the corner. Have you found a place in town yet?"

"No, Ramona, I haven't. I will probably begin to look around after the interviews are over. No use renting a place when I might not be selected."

"Let me tell you something Jacob. In my humble opinion, your discussion tonight was a slam dunk."

"Do you really think so?"

"I don't just think so, I know so. You dominated the discussion period. You revealed exactly what they are looking for. They don't want a follow the leader type. They want an innovator, someone with new ideas. Remember, you are invited here because they are interested in you, not just because you simply filled out an application form and sent it to the Institute. They know more about you than you realize … Here we are. Why don't you park the car and come in for a few minutes? I'd like to show you my apartment."

"Ramona, thank you for the invitation, but it's not a good idea. Since we will be working together, I need to let you in on something."

"Sure, go ahead, tell me you are married."

"No, I'm not married, but I am sort of engaged."

"Sort of engaged? What are you saying. Are you engaged or not?"

"I have a girlfriend in Atlanta I dearly love."

"So, why aren't the two of you married?"

"She comes from an Orthodox family and her papa doesn't think I would make a very good son-in-law. So far, he has refused to approve our engagement."

"Jacob who are you marrying, papa or his daughter?"

"It's complicated. Fathers of Orthodox girls have the last word on engagements. There is a required contract called a ketuba. The father has to sign it. Otherwise, no engagement, and no marriage."

"Jacob, as a single girl with the last name of Bergman I am quite familiar with a ketuba.

"As I say, it is complicated. You see, the two of us have an unofficial ketuba. When her father refused to consider a real one we went ahead and made one up on our own. We had the proprietor of a French restaurant we were attending sign it. Also, I need to tell you something about Rachel. She is a no-nonsense kind of girl. She gave me a list of 'thou shalt nots.' I promised to abide by them. One of those rules I'm violating with you right now. I promised never to be alone with a young lady, and here I am taking you home."

"Well I will let you off my hook this evening Jacob, but don't expect me to cross you off my list. Thanks for sharing. This information will be very helpful in planning my next strategy."

I quickly opened my door and walked her up the steps to her apartment. "I'm also glad we have had this conversation, Ramona. It will make things a lot easier for both of us."

"True, but I need to warn you. Not only am I not taking you off my list of eligible bachelors, I am moving you up to the top of the list."

"Surely you are joking. I am very much in love with Rachel. We both hope to find a way to marry someday soon."

"Let me give you a geography lesson, Jacob. Atlanta is a long way from Baltimore, and I've never put much credence in the saying 'absence makes the heart grow fonder.' You can do worse. In my opinion, a Ramona in the hand is worth more than a girlfriend in Atlanta with a stubborn papa. Goodnight Jacob."

"Goodnight Ramona, and thanks for the warning."

I fired up Old Blue and headed back to Timonium, feeling pretty discouraged. I've been three days in Baltimore and I've already broken our covenant. I must admit, Ramona is easy on the eyes and she definitely does not wear an Orthodox neckless around her neck. A guy could do worse.

Allen often said, "The spirit is willing, but the flesh is weak." I'm beginning to understand exactly what he meant. How can I even explain this transgression to Rachel? I arrived back at the farm at half past eleven. The household is asleep. I quietly made my way up the stairs to my room. I set the alarm for five and went to bed.

A new dream came shortly after I fell asleep. I felt no fear with this one. It began with a declaration from Ha Shem:

I have created man in my image. Rejoice in his frame. It is an abomination to change what I have fashioned. The seed is holy. Thou shalt not mix my kinds. It is a sin unto death. My Spirit will take no pleasure in those who do so. They shall not enter into my rest. Turn away from those who seek to alter the fruit of my hands.

The next thing I saw was a vision of a great number of huge giants. Suddenly the earth opened its mouth and swallowed them. Screaming, they were cast down into the bowels of the earth. The vision and message quickly ended. It left me wondering. Is the vision possibly a warning related to my encounter with Dr. Altschul, or does it have a more futuristic meaning? I filed the experience in my memory bank and soon fell back to sleep.

Courage under Fire

On Tuesday morning, I wanted to remove any possibility of repeating my experience with Ramona the night before. I parked my car in the visitor parking lot on the Homewood Campus and took the shuttle to the hospital. I arrived at the BRB at half past seven and took the elevator to the fourth floor. Ms. Capoletti is standing at the elevator. She escorts me to Dr. Hobbs' office. Ramona already is in the room when we enter.

Dr. Hobbs stands up and welcomes me, shaking my hand. "Good morning Jacob. I hope you had a good night's sleep. We will be putting you through your paces today. I see you have taken Robbie's advice and are wearing some walking shoes. Please, take a seat."

I sat down next to Ramona and Dr. Hobbs took a seat at the end of the table.

Ms. Capoletti asked, "What do you prefer Jacob, coffee or tea?"

"Coffee with a bit of sugar for me Ms. Capoletti."

"Okay, just one lump. Jacob, the staff around here uses my nickname when we are meeting informally. Everyone refers to me as Robbie."

"I heard Dr. Hobbs use the name. I assume your first name must be Roberta?"

"Yes, you are correct Jacob. I believe Dr. Hobbs brought some donuts in this morning. I'll be right back with the donuts, tea, and coffee."

Robbie left the room. As we waited for her to return, I began to glance around the office. In addition to the large conference table, there is a large mahogany desk located in front of a bay window. The window looks out over the busy streets of the city. The walls of the office are decorated with several oil paintings of the human cell. One large oil painting at the end of the room is awesome. It depicts the DNA double helix in brilliant colors of green, red, and blue with a background of what appears to be the cell wall. On the wall behind his desk is another large oil painting of the Institute's founder, Victor A. McKusick, the "Father of Genetic Medicine." Dr. McKusick is the principle author of the book Mendelian Inheritance in Man. There are autographed photos of James Watson and Francis Crick who first reported the double helix structure in April of 1953. Nine years later, in 1962, they shared the Nobel Prize in Physiology with Maurice Wilkins for solving the mystery hidden in the cell nucleus. Wilkin's picture hangs near the window. There also are photos of several other prominent figures in the field, such as Francis Collins of the National Human Genome Research Institute and Craig Venter from Celera Genomics who together first sequenced human DNA in 2007. There are signed photos on the wall of foreign researchers, such as Walter Fiers from the University of Ghent, John Mattick, Executive Director and Marcel E. Dinger of the Garvan Institute of Medical Research in Australia. The displays in the room are overwhelming. I realize I am sitting in the very heart of the institution responsible for the greatest scientific discoveries of the 21st century. I have entered the "promised land." There are giants here.

Shortly before eight o'clock the other two applicants, Craig and Joe, arrive with their student guides. After Dr. Hobbs welcomes the two applicants, Robbie explains the procedure we will all follow and hands out our individual interview schedules.

Dr. Hobbs then explains the day's schedule to us. "Gentlemen, your student guides will escort you to each of the interviews. You will meet with three faculty members this morning and two others in the afternoon. This morning's interviews will all be located either in this building, in the hospital, or at other facilities located nearby. The first interview will begin at nine o'clock. Each of your interviews will be forty minutes or less in duration. This allows you at least twenty minutes to reach your next interview location. We will meet for lunch in the hospital cafeteria at twelve noon. Please don't be late. After lunch, we will take you by van to your afternoon interviews on the Homewood and East Baltimore Campuses. The afternoon interviews will begin at one o'clock. You should complete your final interviews by half past two and return to this office by three o'clock. I will begin reviewing the results as I receive them throughout the day. When you return, Robbie will schedule an appointment to meet personally with me on Wednesday morning and we will review your results. Do you have any questions?"

No one had any questions. Dr. Hobbs rose from his chair and again shook our hands, wishing us well. I took a look at my schedule. My first interview is with a Dr. Swanson of the Institute for Basic Biomedical Sciences located here in the BRB. My second interview is with the Institute of Medical History located in the hospital adjacent to the Center. I will be meeting with a Professor Robert James. Someone has done me a big favor. The original interviewer shown on the schedule was Dr. Altschul. His name has been replaced by Dr. James. I handed the list to Ramona and we headed for the Biomed Sciences Office on the second floor of the building.

Ramona looked me over in the elevator and straightened my tie.

"Do you think you are ready for Dr. Swanson?"

"Yes, I think I am. What's he like?"

"Dr. Hobbs does not want to introduce any third person bias into the equation. So, I'm not allowed to prep you on your interview professors. You will be meeting them one-on-one. I won't be in the room with you."

"The policy seems reasonable. I just hope I don't freeze up."

"I'm sure you will take it in stride. Just relax and remember they want you as a part of their team just as much you want them as well."

The elevator door opens and we walk down the hall to Swanson's office. I glance at my watch. Five minutes before nine. I made a mental note of the time. The receptionist asks us to be seated while she announces our arrival.

She returns shortly and says, "Dr. Swanson is ready to begin the interview. He is waiting for you in his office."

Ramona gives me a "thumbs up." I cross my fingers, say a quick prayer, and enter his office. The Doctor stands up from his desk and shakes my hand.

"Good morning, Jacob."

"Good morning, Dr. Swanson."

Dr. Swanson pulls up a chair next to his desk and says, "Have a seat Jacob and we'll get started. Do you object to me recording our talk this morning? It will help me complete my interview report."

"No Sir, I don't have a problem with you recording our conversation."

"Let me begin by saying I have reviewed your file and I must say I am impressed. I understand you are seeking a research position with JH. Is this correct?"

"Yes Sir, I want to pursue research in human genetics once I complete my doctorate."

"Actually, you will begin basic research as part of your training in your second year. What area of human genetics are you most interested in, Jacob?"

"My foremost interest lies in the area of human fetal development. I also have an interest in human memory. I'm also open for other areas of research."

"I see. At what level are you most interested in, macroscopic, microscopic, or molecular?"

"I believe my special skills lie in the ability to synthesize fields of knowledge, to see things as a whole. I believe most promise lies in a concentration at the molecular level, but I also have an interest in the historical record of mankind, the cultural big bang."

"What most attracts you to the molecular level, Jacob?"

"The dramatic difference between human gene expression from other primates is certainly interesting to me. The complexity of the gene types is truly amazing."

"Do you have in mind the complexity of the FoxP2 gene perhaps?"

"Yes, the RNA genetic process is also quite interesting."

"What are some of the researchers you have read and admire?"

"I have found papers by Nelle Lambert at the University Libre de Bruxelles in Belgium, and Courtney Babbitt from the University of Massachusetts to be quite interesting. Of course, I must add the fine contributions of those here at the Institute, Dr. Hobbs being one."

"Well, you have picked some giants in the field, Jacob. How do you think we can account for the rapid rate of evolution these genes demonstrate?"

"I really don't know. I try to keep an open mind on such mysteries. I don't think one should adhere too closely to a particular paradigm."

"I see. So, you might even be willing to question this proposed human accelerated rate (HAR) of evolution?"

"Yes, I can't accept any hypothesis based solely on political correctness. On the other hand, the great degree of uniqueness of the human cortex from all other animal species, including chimps, seems to argue against any such accelerated rate of evolution."

"Would you be comfortable working along the side of researchers who do embrace the HAR view of evolution?"

"I can do so, provided the person's point of view did not limit my freedom to pursue a path I felt has greater scientific merit."

"So, you come to us with a philosophy of the 'freedom to explore and publish' where the research leads you?"

"Yes Sir, I do."

"What do you see as difficulties with our present approach to this field, and what do you envision as possible new methods of research?"

"I believe the field is faced with some limitations on available analytical tools. Perhaps new methods of investigating molecular activities within the cell. We need the tools to delve more deeply into the cell, to better discriminate between cell types and genetic expression. Another concern I see for the particular field of inter-utero development is the lack of sufficient tissue to support the research. A new way of examining such tissue without harm to the embryo would be a significant advance."

"You are correct. The inter-utero tissue scarcity is a real impediment to research in this area at present. So, your position is we must not cross the ethical line of protection for the unborn?"

"Absolutely! Without naming names, I am concerned with the direction some research labs are taking."

"Yes, developmental genetics is facing a real moral dilemma. Jacob, you have answered my questions well. I'm going to conclude our interview at this point. I will review the tape of our conversation and make my recommendation in a timely manner to Dr. Hobbs. It has been a pleasure meeting with you and I wish you every success."

"Thank you, Sir. I hope to have a chance to visit with you again soon."

"Yes, I would like to sit down with you again. Goodbye Jacob."

We shook hands and I rejoined Ramona in the anteroom. I am shaking like a leaf, feeling very tense as we leave the Office.

Ramona asks, "How do you think it went, Jake?"

"I think it went well, but we won't know until the results get posted, will we? I believe the Institute of Medical History is next. Where is it located?"

"The Institute is located in the Bloomberg School of Public Health at the corner of East Monument and Wolfe Streets. It's just a short walk from here."

We arrived at the office of the Medical History Institute ten minutes before ten o'clock. The receptionist asked us to wait until Dr. James returned from an appointment. He arrived a few minutes later and ushers me into his office. Dr. James is quite cordial, but rather brief. Thankfully the Doctor avoided any discussion about the theory of evolution. Instead, he did most of the talking. He mentioned the notable contributions made by various other schools at JH. It is clear the doctor is very much centered on his field. I said very little. He described in some detail the recent advances made in the treatment of certain unusual cancers, such as Her2Neu. This cancer is an aggressive cancer of the breast. They are now treating it with a new drug, Herceptin. My concern is whether I succeeded in communicating my interest area complete enough to gain his approval. He no doubt will treat the other interviewers the same way. It should help balance out the scores, not that I am in competition with Joe and Craig. I am hoping all three of us succeed in being approved for admittance to the Institute.

My third appointment is with the Autoimmune Disease Research Center (ADRC) located two blocks to the east. We arrived ahead of time. I am scheduled to meet with a Dr. John McKay The Center is much smaller than the Bloomberg Building. As we enter the first-floor lab area, we see a large number of researchers using a variety of highly technical equipment. Ramona points Dr. McKay out to me. He is dressed in a white lab gown and working at the ultracentrifuge. The machine is emitting a high-pitched whirling sound. He motions for us to stand back behind the protective barrier located several feet from the centrifuge. I took Ramona's hand and led her back behind the safety barrier. She looks up at me with a mischievous smile on her face. She is squeezing my hand tightly and won't let go. Dr. McKay turns off the machine. It takes about five minutes for the high-pitched whirling sound of the centrifuge to degrade into a low whine. Finally, the machine comes to a complete stop. Only then does Ramona release my hand. Dr. McKay motions for us to join him in his office. I successfully avoid Ramona's attempt to grab my hand as we follow the Doctor into his office.

"Well I recognize Miss Bergman. How are you Ramona?"

"I'm fine, Dr. McKay. I volunteered as student guide for this good-looking man standing next to me, Jacob Cahn."

"Hello, Jacob. Can you give me a few minutes? I need to finish recording the spin."

"Sure, take your time. We are running ahead of schedule. I used a much smaller centrifuge in the immune research lab at Emory. I noticed the centrifuge still has a shipping label on it. Is it brand new?"

"Yes, we purchased this one just last month. It has the capability to accelerate to two million G's. I am running a preparatory study of viruses within cellular organelles at a very high rotation rate. You probably read the warning signs. Even though we frequently inspect these machines, they are dangerous. We are careful to protect our people against any mishap. You can't be sure these rapid spinners are going to stay in one piece. Also, it emits some rather nasty radiations. Did you note the radiation warning on the sign? The back side of the sign also is coated beneath with a lead shielding."

"I saw the radiation warning, but I didn't notice the shielding."

Just listening to the sound of the rotor, I fully understood his concerns. Dr. McKay finished recording his data and asked Ramona to wait in the lab while he interviewed me in his office. As we leave for his office Ramona is still smiling at me with that same mischievous look on her face.

Dr. McKay said, "I think you have made a good decision in choosing to come to JH Jacob. I assume they have informed you of the opportunity to participate in research activities during your second year."

"Yes, the opportunity to begin research early on is a major factor in my decision to attend JH."

"Jacob, I have made a point to review your application and your references. I must say I am impressed. I hope we can enlist your help here in the Autoimmune Disease Research Center. You will have the opportunity to get to know us this coming year. You might be interested in knowing we are on the verge of going to human trials with three new Autoimmune Disease (AD) drugs. One drug in final trials will better alleviate tissue rejection. Another one we are just now testing we hope will offer a cure for Crohn's disease and lupus. The third drug we are working with offers a cure for psoriasis. Given the cold hard facts, we know 50 million Americans are afflicted with more than one hundred different types of AD. AD is one of the top ten leading causes of death for women under sixty-five and these diseases cost the health system over one hundred billion dollars every year. The diseases are not well known. Less than one in five Americans can name even one AD disease."

"Those figures are just incredible, Dr. McKay."

"Jacob, how do you think we should proceed in the future, given these facts?"

"Well, it seems to me our Nation would benefit from a genetic screening and data bank to make available the health histories of family members."

"You have mentioned a worthy goal, Jacob. There is as yet no national data collection program to provide medical histories of AD cases, let alone any DNA database. We currently are attempting to reach the public with thirty-second radio and television spots, plus a website providing the latest information on these disorders. We are supporting the free distribution of the magazine Genome published quarterly by Big Science Media and distributed free of charge to major hospitals and medical centers nationwide."

"Dr. McKay, I remember seeing an ad a few months back while watching the Masters golf tournament."

"We are given some time by the networks for those ads. I think our discussion along with my review of your file has given me enough information to write up my report for Dr. Hobbs. By the way, your student guide is a very sharp young lady. She aced my immunology class last semester. She's not bad on the eyes either."

"She has been very helpful in getting me settled in. As you say, she's a beautiful woman. Thanks for spending some of your valuable time with me. I look forward to talking with you again soon."

"Yes, I'm sure we will have opportunities to talk again. You two have a good day."

I left the lab with a very good feeling. I felt confident Dr. McKay was favorably impressed.

It is ten minutes before twelve noon when we arrive back to the BRB cafeteria for lunch. Robbie is there already. I am hungry. I ordered a hamburger with fries, plus my usual coffee. Ramona's lunch consists of rabbit food. I paid for her lunch as well as my own. The other two candidates and their guides arrive a few minutes later. No one shared their thoughts about their interviews, but everyone seemed to be in a good mood.

We left for our afternoon appointments at a quarter to one. The van took us first to the Homewood Campus. My fourth interview is in Hackerman Hall with the Robotics Engineering School. Terry and I had visited it during our earlier visit to Homewood. I am to meet with the head of the Division of Biotechnology, Dr. Francis Glick. The van dropped off the two of us at the east gate on North Charles Street and we walked to Hackerman Hall. Dr. Glick's office is located in his worksite on the second floor, right next to the Divinci Machine where Terry and I had visited the other day. The reception room is on the first floor. We are greeted there by a Ms. Johnson.

"Hello. You must be Mr. Cahn."

"Yes, I am. This is my student guide, Ramona Bergman."

Ramona quickly leans over and whispers in my ear, "Jake, I'm Miss Ramona Bergman."

"Please have a seat both of you. I will see if Dr. Glick is ready to meet with you Jacob."

Ms. Johnson picked up the intercom and calls the Doctor. She tells me the Doctor is ready for my interview. I am to meet him upstairs in his office. Ms. Johnson offers to show me to his office, but I tell her I know where it is located. I took the stairs to the second floor and entered his work area. His office is located at one end of the room. I walked to his office and entered. The walls of Dr. Glick's office are covered with pictures and diagrams of various medical devices ranging in size from minute nano devices to large robotic machines. Interestingly, there is a nice photo of Jerusalem on the wall behind his desk.

"Hello, Jacob. I am Dr. Glick. You have picked a nice day to visit our facility. We are just now testing a new device. If you have time after the interview, I will give you a quick tour of our facilities here."

"I'd like very much like to see the facilities and the device you are working on Doctor."

"So, you come to us from Emory in Atlanta. Some of our graduates are now working at the Yerkes National Primate Research Center there. Have you visited their lab?"

"No Sir, I have not. I know they are performing some interesting research. Perhaps you are not aware undergraduates are not allowed in the lab."

"I didn't know this."

"Yes, the research lab is a high security facility."

"Do you have any experience along the lines of robotics, Jacob?"

"No Sir, I really do not. My major was pre-med at Emory and I only took a few courses outside of the major. I did take the usual required physics and chemistry courses, but I have not had any direct experience in robotics. I am very interested in the development of new research tools. I feel the pursuit of human genetics now is bumping up against some technological barriers. Our ability to study the embryonic development in the womb is greatly hampered by the moral constraint to protect the fetus. We could certainly use some new method of tracing fetal development with protection for the unborn."

"You do agree with the moral imperative to preserve the health of the fetus then?"

"Absolutely, Sir."

"I'm most concerned about this problem. If you were to suggest some means of accomplishing this goal, how then would you proceed?"

"My ideas may appear too absurd to be considered."

"No, don't let your lack of expertise stop you. I want you to share your thoughts. We often find some outlandish idea turns out to be the key to unlocking a new technology or a new application of an existing technology."

"Well, for safety to the unborn, any device or technique developed would have to remain physically outside of the womb."

"Yes, of course."

"Secondly, I believe such a device would have to make use of recent advances in physics, probably operating at the quantum level."

"Why do you believe quantum physics is necessary rather than utilizing the standard physics model, Jacob?"

"In order to determine changes in cell types based on protein fragments the device would have to be very robust, capable of tracing literally billions of protein fragments. If tools can be developed to manipulate matter and energy at the quantum level, this would allow researchers to access very discrete operations within the cell, such as alternate RNA splicing. Such a tool possibly could be used, not only to trace splicing, but also permit manipulation of fragments.

This would allow us to repair damaged cells. I have heard of some devices claiming to produce healing of cancers by exciting genes within the cell at specific frequencies."

"Yes, I have heard a few such reports, but I believe those high frequency machines are mostly ineffective inventions of rip off artists, Jacob."

"That very well may be, Dr. Glick."

"Your idea is to develop a device, not only of observing the cells, but also capable of manipulating RNA fragments by determining their response or resonance to specific frequencies, using a quantum computer capable of handling the immense number of neuron cells. Have I summarized your conceptual vision, Jacob?"

"Yes, you have described it exactly in one very long sentence. I know it is a shot in the dark, but I believe research toward developing such a tool might be possible."

"Let me ask, have you read or seen any research papers along these lines?"

"No Sir. As I have said, this is just some ideas I have been thinking and pondering over. I would appreciate hearing your opinion. Do you think this might be a goal worth pursuing?"

"Jacob, not everything we know is published in the journals. Let me just say I have seen some interesting attempts paralleling what you have described. Sorry, but I am not free to discuss them with you. Certain nameless governmental entities are hard at work developing the capabilities you have mentioned. At some point in the near future I may be permitted to discuss this topic with you further. Let me ask you a philosophical question. If you do not want to answer it, I will understand."

"I'm willing to answer any of your questions if I can."

"Fair enough. Jacob, would you be strongly opposed to altering human DNA at the germ level, possibly incorporating DNA from other species, effectively altering the DNA of future generations?"

"Sir, I would not be in favor of even testing the alteration of the Creator's handiwork."

"I see. Perhaps we will have an opportunity to discuss these issues with one another from time to time. I am disappointed you do not have the specific skill set to participate directly in our engineering activities. Our teams do require support from other fields, however. Do you have any skills with computer languages?"

"Yes Sir, I can write code in several user level languages. I have used the MUMPS and Oracle database systems. I can write in Fortran, C, and some machine code depending upon the type of machine."

"Do I assume you might be willing to help program some routines for our projects?"

"Yes Sir. Down the road, I can help with your programming needs, but I don't wish to make computer programming my career."

"I understand what you are saying Jacob, and I like your attitude. You have some original ideas. We need more of it around here. Let's take a walk downstairs to the Learning Center. I want to show you our newly developed ion chromatograph machine."

Professor Glick led the way to the first-floor Learning Center. He pointed to a newly developed ion chromatograph machine. "Jacob, this new ion chromatograph model separates charged compounds including anions, cations, amino acids, peptides, proteins, and some very unusual ions, such as hydrogen, helium 3, and lithium. It is used to purify proteins, but it can also reach below the protein level to analyze individual protein segments. I think this new machine could prove useful if you decide to pursue the field of fetal development."

"I'm a long way from starting such a project, Dr. Glick, but I will certainly keep this machine in mind."

"Do not put things on the back burner, Jacob. You will find time passes quickly around here."

The Professor continued to show me various student projects stored in workspaces in the Learning Center. All too soon, I had to leave for my final interview.

Ramona and I walked back to the east gate and boarded the van. My final meeting is with Dr. Steven Showers. He is in charge of the Emerging Technology Center, located in the East Baltimore Complex, one mile east from Homewood on 33rd Street. I am the only one scheduled to meet with Dr. Showers. The driver already had taken Joe Harmon and Craig Simpson back to the BRB. It took us just a few minutes to arrive in front of the newly renovated building next to the School of Nursing. We took the elevator to the third floor. The Center actually consists of a large number of offices, class rooms, and conference rooms. We located Dr. Shower's office and walked in.

The receptionist welcomed us, "Good afternoon. I assume you

are Mr. Jacob Cahn?"

"Yes, and this is Ramona Bergman, my student guide. Is Dr. Showers available?"

"Yes, he's waiting to see you. Go right on in Jacob."

I opened the door and entered his office. Dr. Showers is a short thin man with very intense, sharp piercing eyes. He wasted little time with informalities.

"Good afternoon Mr. Cahn. Welcome to the Emerging Technology Center. Our Center is a jointly run enterprise between Johns Hopkins and many public companies. Also, we are in collaboration with the University Space Research Association, or USRA. The USRA brings a needed unity to the whole enterprise of medical health in our Nation. The McKusick-Nathans Institute of Genetic Medicine and those other supporting research facilities have been very instrumental in helping us here carry out our function in this intricate system we call the new National Health Program. Our focus is on commercializing innovative technology from Hopkins as well as from other research institutions. We operate as a private company and are involved in placing the latest new technologies into the hands of medical practitioners. We do this by working with existing health providers, and also by launching new startup companies who then bring these new products to market. We feel it is important to make all incoming graduate students fully aware of our little group and what we do here on the East Campus. Our company is dedicated to providing a critical link between research and the application of new technologies and pharmaceuticals to the needs of our national health system. Advances in knowledge are of little value if they do not meet the needs of people. Need I say more?"

"Sir, I fully understand the importance of effectively applying new technological developments to our national health system. The free enterprise system must have access to the scientific sector of the economy. It is clear we have gone pretty much beyond the era when new technology emerges solely from individual entrepreneurs. My father is a general practitioner and instructor at Creighton University in Omaha. I have seen both ends of the spectrum, the developmental phase as well as the marketing of new products and technologies."

"It is big business Jacob. It sounds like your father's experience has given you a good understanding of this free enterprise system.

You are more aware of the importance of new medical technologies than many other medical researchers who believe their goals are simply to uncover new knowledge. The health business accounts for six percent of the Nation's total economy, and it is growing rapidly. Do you think you will be pursuing basic research, and if so, what area are you intending to concentrate on?"

"I intend to pursue research into embryonic development of the fetus at several levels, microscopic, molecular, and genomic. I also have an interest in exploring human memory and the many other aspects associated with it, such as inspiration, reasoning, and spiritual aspects."

"Very interesting. Speaking of spiritual matters, I heard through the grape vine you had a rather heated discussion with Dr. Altschul. Would you classify your interpretation of the human animal as the result of a Darwinian macro-evolutionary paradigm?"

"Apparently my discussion with Dr. Altschul has been widely banded about. No Doctor, I do not ascribe to the naturalistic view. I prefer to believe there is a higher, more plausible explanation."

"You may find some resistance to your world view, Jacob."

"Yes, I have quickly found this to be the case here at JH."

"Yes, even here."

The Doctor has a knowing smile on his face. He is non-committal, but I can tell he is on my side of the issue.

"Do you have any concerns about where this human genetic trail is leading to, Jacob?"

"Yes, of course. I'm not privy at this point to what is going on within the dark corners of genetic labs. The recent failure of prominent geneticists to restrict cross-species research at the germ level was very disturbing to me. I have no proof, but I believe there are powerful forces, hidden from the public view, which are promoting this unethical research."

"You are very perceptive, Jacob. I carry a very high security clearance permitting me to enter some of the labs you are referring to. There are some, shall we say, unusual agendas at work. As a result, I share your concerns. After I visit some of these labs, I don't sleep well at night. I'm sure you understand I cannot say more. As you continue in this field, you may find the need to discuss this problem further with me. Keep your eyes open and be careful, Jacob. The allurements of money, fame, and pride have corrupted many a bright, young scientist."

"I will, Doctor. I appreciate your candor. I will keep your warning in mind."

"I really don't have any further questions to ask you. I think I have covered the ground on how this system operates. If we had more time, you might want to discuss the effect of this new Affordable Care Act and its impact upon our fragile economy. It has been a pleasure meeting with you Jacob, and I wish you every success here at JH."

"I also have enjoyed our meeting today Dr. Showers. I may well have occasion to use your operation here as I make headway toward my research goals."

"Shalom Jacob."

"Shalom Doctor."

I left the final interview in a happy mood. Here is a man I can confide in without fear of betrayal. As we emerged from the building, the van was waiting for us in front. We returned to Homewood to pick up the others on North Charles Street. Joe Harmon and Craig Simpkin with their student guides Virginia Sinclair, and Joan Rodriguez are all waiting for us when we arrive. The van delivered us back to the BRB on Broadway.

On the way back, I asked, "Joe, how do you think you did?"

"It went well. It was kind of hard to break the ice, but once I started talking I finally relaxed. How did it go for you, Jacob?"

Ramona interrupted us. "Gentlemen, I need to ask you not to share specifics about your interviews. Let's just assume everyone has aced their visits today, okay?"

Craig added, "I think we need to follow Ramona's advice guys."

I wanted to object. "Ramona, are you placing a gag order on us?"

"Yes, I am, Jake. If you can't keep your big mouth shut, why don't you discuss the weather, or the last lousy movie you watched?"

I rejected her choices and switched the topic of conversation to sports. "So, do either of you play tennis?"

Joe says, "I don't play tennis; I play lacrosse, Maryland's favorite game."

"I'm wondering, Joe. Will my tennis game help me play this weird game of lacrosse?"

Joe laughs before answering. "No way are your tennis skills

going to help you in this sport, Jake. Lacrosse is a contact sport. I ended up with a broken arm last year. This kid from the University of Maryland slashed my forearm on purpose with his wicket in the last game of the season. Better stick to your tennis game."

We arrive at the BRB at 3:15. Robbie is waiting for us in the reception room.

"Welcome back everyone! I see you all survived the ordeal. No broken bones."

Joe answers, "It feels like my head may be broken. Just kidding."

"I know all of you are curious about the results of your interviews. We have asked the professors to email their reports to us promptly. Dr. Hobbs will require some time to complete his review before we can share the results with you. Also, you need to realize we still must assess quite a few additional applicants. You are three among one hundred twenty applicants. More than likely we will not be able to announce our final decision on your admittance until a week or so before the Fall semester begins. I will check with Dr. Hobbs how his review is going."

Robbie left the room. She soon reappeared with Dr. Hobbs in tow.

Dr. Hobbs is smiling. "Gentlemen, I am quite pleased with the results of your interviews. I realize you are very anxious to know how you stand with us. No doubt some of you have applied to other institutions. We don't want to endanger your chances to gain admittance to another school this Fall. I will be reviewing your interviews once again over the next few days. I should be able to share the results of your interviews by the weekend and give you at least some idea where you stand. I assume you understand our final admittance decisions cannot be announced until several weeks from now. We need to process additional applications before our final decisions can be announced. I have asked Robbie to schedule individual appointments with each of you. Again, I want to emphasize I am very pleased with your interviews. I regret having to keep you on pins and needles. I hope you understand. Thank you for your efforts today. We will be in touch. You are free to leave."

Having been left in never-never land, all three of us were pretty disappointed. I was just about to leave when Dr. Hobbs asked me to remain behind for a few minutes. I said goodbye to everyone, including Ramona. She looked disappointed, She, no doubt, was looking forward to another opportunity to get a ride back to her apartment.

I followed Dr. Hobbs back to his office and sat down opposite his desk. "Jacob, I want to say you did very well, exceptionally well in fact. I have little doubt you will be accepted to the Institute. In the meantime, I want to offer you a summer job. If you are willing. We have an opening for a lab assistant to work with the undergraduates in the Biology Department on the Homewood Campus. It pays a rather meagre salary of seven hundred and twenty dollars for each of the two consecutive summer classes. The classes are back to back, six weeks in length. Would you be willing to work this summer?"

"Yes Sir, I certainly would. Do you know who will be the instructor?"

"You will be working with Dr. Jerome Adelman. The text for the course is Introduction to Virology. I believe you took the same course from Dr. Miller at Emory?"

"Yes, I did. He was a great instructor. In the book store the other day I noticed he uses the same text."

"Did you also earn a top grade from Dr. Miller?"

"Yes, I did well in the course. It was a special blessing. The course clinched my appointment to the Dean's list for my last semester."

"The first session begins next week. Would you be able to meet with Dr. Adelman on Monday and begin work on Tuesday?"

"Yes, I don't have anything planned for next week."

"I do have one request. I'd prefer you not discuss this with the other two applicants. Of course, if they hear of it in some other way, it is not a problem."

"I understand. I may have to let my friend Terry know about it. I stay at his home from time to time."

"That's fine, Jacob. Just ask your friend not to talk about it on campus."

"Thank you so much, Dr. Hobbs."

"I appreciate your willingness to pitch in and help us, Jacob. Robbie will give you the schedule and how to contact Professor Adelman. I'm not certain who else may participate in the class. It may be Dr. Altschul."

"Dr. Altschul?"

"No, I was just teasing you. Do you have any questions?"

"No, Sir. I'm sure Ms. Capoletti will fill me in on everything.

Thanks again."

"You are entirely welcome. I will talk to you soon."

With those encouraging words, I left his office. Robbie gave me the information packet, and the phone number of the Biology Department.

Robbie asked me to complete and sign a W-4 form and a couple of other administrative forms. I asked her for a copy of the W-4 to frame on my wall. Then I left the office and boarded the shuttle for the Homewood Campus and my car parked in the visitor's lot. I arrived back at Timonium at 4:30 in the afternoon. Terry was just walking back to the house when I pulled up.

"Hey Jake! How did everything go?"

"Things went very well. I don't have the final okay yet from the Institute, but Dr. Hobbs said I scored high on the interviews. And I'm now employed."

"You have a job? Doing what?"

"Dr. Hobbs offered me a lab assistant job with the Biology Department this summer. They are offering two courses, one after the other. The first course begins a week from today."

Terry is impressed. "Wow! They don't waste any time, do they? What does it pay?"

"Seven hundred fifty bucks per course."

"Right on! So, I guess you will be moving into town soon."

"Yeah, I need to rent an apartment near Homewood. The classes will be held in the Undergraduate Teaching Lab building near the Rec Center."

"Ramona might let you move in with her?"

"I'm sure she would. You should have been with me today. When we were at the ADRC, she latched onto my hand and wouldn't let go."

"I told you she has her eye on you."

"Are you jealous?"

"Not really. By the way, Allen Schaeffer called looking for you. He tried several times to call you, but your phone must not be working. I told him you are staying here with us for a few days. He wants you to call him. He said you have his number."

"My battery died on my phone, Terry. This is kind of strange. I have been thinking about him the last few days. I wonder what he and Alice are doing this summer?"

"Use my phone. He has some exciting news to share with you."

We went into the game room. I plugged my cell phone into Terry's charger and called Allen. He answered right away.

"This is Jake, Allen."

"Hey Jake, how is JH treating you?"

"Everything is great."

"I've been trying to call you. Has your application been accepted yet?"

"It's not final yet, but the Institute Director, Dr. Hobbs, just told me today I did well on my interviews. He said it is likely my application will be accepted. Also, Dr. Hobbs offered me a summer job as a lab assistant in the Biology Department. So, things are cool. What are you and Alice up to this summer?"

"We have been studying at Youth with a Mission (YWAM) located in Lakeside, Montana. Alice is boning up on African languages. She will be teaching when we go."

"What language is she learning?"

"She is studying two, Ibo and Swahili. Swahili is mainly in Eastern Africa. There are over four hundred languages in West Africa alone. Right now, the Nigerian area is very chaotic."

"Yes, I have heard the terrorists are burning churches there. You need to be prayed up."

"Jacob, Alice and I recently have decided to be married before our trip to Nigeria with YWAM. I have invited Terry to be in the wedding party and I want you to be our best man. I realize it is short notice. Do you think you have time to come down here for the wedding?"

"Do you have a date picked out?"

"Yes. The wedding is scheduled for this coming Saturday. Because of the tight schedule, we are holding the wedding reception dinner on Friday evening. We want to be married before we get to Lagos. I realize it is very short notice for you and I will understand if you say you can't make it. Is this too soon for you?"

"No, I'll get there. I'm honored you have chosen me to be your best man. The lab assistant job starts next Tuesday and I have to meet with Dr. Adelman on Monday, so it will have to be a quick trip. I can fly down there on Friday and return on Sunday."

"Okay, Bro. Someone will pick both of you up at the airport. I have reserved two rooms at the Omni. The guys in the wedding party are wearing black tuxes, white shirts, crimson cummerbunds and bow ties. You and Terry will have to get measured and rent them there. We are going to hold the formal wedding dinner before the wedding on Friday night because Alice and will be leaving on a short honeymoon right after the wedding and a brief reception."

"No problem. I'm sure Terry and I will be able to rent them here. Do we need to rent patent leather shoes?"

"No, just black dress shoes will work. On another note, I assume Rachel has contacted you by now. There was a big splash in the Atlanta newspapers here. I'm not sure it was in your papers also."

"I don't know what you are talking about Al. I haven't been reading the papers here. I haven't spoken with her since I left Atlanta. Her father insisted we not contact one another for six months. Has she been in an accident or something?"

"No, she is fine physically. I don't want to say anything more. You need to call her. Perhaps the two of us can discuss it when you come for the wedding. We invited her to the wedding, but she hasn't gotten back to us."

"This doesn't sound good. I'll call her right after we hang up."

"Okay. Call me back when you two know your flight number. I'll have someone there to pick you up."

"Yes, I will. Take care Bro, and give my best to Alice."

"I will. Goodbye."

"Goodbye Al."

I am shaking like a leaf as I dial the Kaufman residence. Hannah Kaufman answers the telephone.

"Good afternoon. This is the Kaufman residence."

"Hello, Mrs. Kaufman. This is Jacob. Is Rachel able to come to the phone? I would very much like to talk to her."

"Hello Jacob. I have been expecting your call. Yes, she is here. Let me tell her you are on the phone."

There is a long pause. I hear some muffled talking. Mrs. Kaufman must have her hand over the phone.

"I'm so sorry Jacob. Rachel cannot speak with you right now."

"I realize I am calling before our time out period is over, but I just finished speaking with my friend Allen Schaeffer. He said I needed to call her. Are you sure she isn't able to speak with me?"

"Let me ask her again, Jacob."

I waited. I can hear Rachel in the background, crying her heart out. Hanna comes back on the line.

"Jacob, I'm very sorry. Rachel is not able to come to the phone right now. She asked me to relay a message to you. Rachel is now engaged to be married. My husband has approved her engagement to Mr. Jeremiah Levitt from New York. I believe you may have met him once."

"No, I have never met the man in person. The last time we spoke with one another Rachel told me she had no intention of marrying Mr. Levitt."

"Rachel is very upset by my husband's decision. I need to explain something to you Jacob. This is an arranged marriage. According to our Orthodox traditions, a father has the right to select a husband for his daughter, especially if she is the oldest sibling. In such cases, the daughter is obligated to acquiesce to her father's wishes. Do you understand what I am telling you, Jacob?"

I'm feeling totally betrayed and too shocked to answer her. How can this man, an Orthodox Jew, do this to me after our previous understanding?

"Jacob, are you there?"

I managed to utter a brief response somehow. "Yes, I am here. Thank you, Hanna. Please wish Rachel my best."

I hung up the phone.

Terry is staring down at the floor, clearly sensing my pain. "Jake, what's going on with you and Rachel?"

"Terry, my relationship with Rachel is over."

"I suspected as much. How can I help, Jake?"

"Thanks for your offer, but no. I will have to work through this on my own."

"I assume Allen told you he and Alice are going to be married next Saturday?"

"Yes. I'm planning to go. Allen asked me to be his best man. We will have to work fast to get our tuxedos and plane tickets."

"I already assumed you are going, so I booked our reservations on Delta flight 231 leaving from BWI Friday morning at 11:10. I also booked our return trip on Sunday morning aboard Delta flight 242 departing at 8:20."

"I really appreciate this, Terry. How much do I owe you?"

"We can settle up later. We both need to get our tuxedos ordered right away."

"I don't feel like going into town tonight. I'm sure we can rent them tomorrow and still make the flight on Friday?"

"I hope you are right. I will call Al back and give him the flight info."

"Okay. I'm going up to my room. Don't wait dinner on me."

"Sure, Jake. I understand. Tomorrow we can talk."

I slowly climb the stairs to the bedroom. My legs feel heavy, like I am wearing lead shoes. I shut the door behind me. This planned marriage of Rachel's will take me a long time to get over, perhaps a lifetime. I stretched out on the bed and went to sleep.

11

The Divine Calling

"These are they which are not defiled with women, for they are virgins. These are they which follow the Lamb whithersoever he goeth. These are redeemed from among men, being the firstfruits unto God and to the Lamb. And in their mouth, was found no guile, for they are without fault before the throne of God."
(Revelation 14:4–5)

My Destiny Revealed

I could not get to sleep Tuesday night. I tossed and turned. The night passes slowly. I keep staring over at the clock on the night stand. I finally drifted off about a quarter past two, only to be awakened again. The clock on my night stand is staring back at me. It is 3:30. I rolled over onto my back and stared up at the ceiling. About a minute later I realize there are words again floating through the air over my head, like I had experienced once before in Atlanta. The sounds of the words are echoing in my brain as their shapes pass overhead. It is Revelation 3:7–8:

And to the angel of the church in Philadelphia write; These things saith he that is holy, he that is true, he that hath the key of David, he that openeth; and no man shutteth; and no

*man openeth. I know thy works: behold, I have set before
thee an open door, and no man can shut it: for thou hast a
little strength, and hast kept my word, and hast not denied
my name.*

It is an amazing vision. I had read this passage addressed to the
Philadelphian Church just yesterday, describing the powerful keys
of the Lord. I'm not from Philadelphia, but I know these words from
Ha Shem are for me. I am beginning to understand just how different
a path my life is taking. Yes, Ha Shem has certainly opened a door
to me. After the vision is over, I just laid there, staring at the ceiling.
It is now crystal clear. God is directing my future. He has a special
purpose for my life. He has given me a preview of my destiny. So
why am I feeling this sad? It is because I now know I will never
marry Rachel, nor anyone else. In some miraculous way, the pain is
now being replaced by a deep supernatural love of my maker. His
peace is now filling my soul.

I woke on Wednesday to a beautiful morning again. It is exciting
to realize I am gainfully employed. I joined the Johnson family
downstairs for breakfast. Staring at the plates of scrambled eggs,
toast, and ham on the table, I faced an immediate dilemma. Yes, it
is ham, several slices of it, artfully displayed on the plate in front
of me. Mrs. Johnsons is watching me closely. No doubt my face is
giving me away.

"Jacob, is there something wrong with the food on your plate?"

"No, the eggs are just the way I like them. I'm sorry, but I can't
eat the ham."

Eldora looks puzzled. "My word, are you allergic to ham,
Jacob?"

Terry solved the mystery for her. "Mom, Jacob is Jewish and
Jews just don't eat ham."

"Oh, I didn't know. I'm so sorry Jacob. Let me fix you something
else. How do you like your bacon?"

Terry and Howie are laughing hysterically. Then Terry asks,
"Mom, where does bacon come from?"

"I buy our bacon from — Oh! It is from a hog. Jacob, I have
some left-over chicken?"

"Chicken is good, Mrs. Johnson."

Mr. Johnson said a quick prayer and we began to eat. Terry

usually has a lot to say over breakfast, but this morning he is not saying much else beyond asking for me to pass the hot sauce. I could see something is bothering him.

"Terry, you seem pretty quiet today. Is there something bothering you?"

"Well, I was just thinking you must be very good at hiding your feelings, considering what you learned yesterday. You seem to be handling it well?"

"I had help, Terry."

"Help? What kind of help?"

"I had a vision last night. I know the Lord has my life in His hands."

All three insisted I tell them more about my vision. When I finished recounting the vision of the floating words, Howie Johnson is bubbling over. He couldn't contain himself.

"Well praise the Lord and hallelujah. You saw all those verses float right over your head, Jacob?"

"Yes Howie, I did."

"And you had just read the same verses the day before?"

"Yes, right from the Book of Revelation. In the vision, the Lord showed me the very same words in the King James version no less."

"This is truly amazing. I truly wonder what God's plan is for your life Jacob."

"I do not know exactly what the game plan is for me, Howie. I will have to take this one day at a time."

After breakfast, I offered to help Terry with the horses.

"Terry, if you are going to muck out the stalls today I can help you."

"Are you sure, Jake? I need to put new bedding down. We won't be finished until ten or so. Do you have time?"

"I have enough time. I can contact Dr. Adelman after we finish the chores."

Terry and I excused ourselves from the table and went right to work. We didn't finish the stalls until a quarter past ten. Back at the house I called Dr. Adelman. He agreed to meet me on Monday at eight o'clock at the Undergrad Teaching Lab, just north of the Freshman Quad.

I asked Terry, "Say Bro, do you have time to go to town with me? We need to rent our tuxedos for the wedding. Also, I need to return a box of books I borrowed from the Emory library. Is there a UPS office in Timonium?"

"Yes, I have time. We need to rent the tuxes. There is a UPS office in Timonium also."

"If we have time after those stops, I want to look for an apartment located near the Homewood Campus."

"Sure, Jake. If you want to, we can order a sandwich for lunch at the Inner Harbor."

"Just as long as they don't wrap mine in sliced ham or fry it in bacon grease."

"You know the Orioles are playing a double-header today. I think it begins at three this afternoon."

"A little baseball right now is a great idea. I haven't been to the stadium to see a game yet."

"Okay then, it's a go."

A Yiddish Blessing

We both changed clothes and I checked on my bank balance while Terry looked up the address of the tuxedo shop. Then Terry went on his computer and bought two tickets for the double-header. We took Terry's truck. On the way through Timonium I dropped off my box of books. Terry pointed out the location of their church. We arrived at the tuxedo rental shop at 11:15. A salesman wearing a kippah met us as we entered the shop.

"Shalom, my name is Jacob Cahn and my friend here is Terry Johnson."

"Shalom, Jakub. Welcome to my shop. My name is Yehuda Sorenson, but my *khaver* call me Yudi. Are you here to rent some tuxedos?"

"Yes, Yudi we are. We have a bit of a problem. The wedding we are attending is this coming weekend. Is it possible for you to get our tuxedos ready by tomorrow morning?"

"Not a problem. What time can you pick them up?"

"Your sign says you open at 9 o'clock?"

"I'll be here. Is this wedding on Sunday?"

"No. The wedding is on Saturday in Atlanta."

"Saturday is Shabbat until sundown, Jakub."

"I am the best man, Yudi. My good friend in Atlanta is the *khosn*.

He's a goy Yudi."

"Aha! so your friend is a goy, is he?"

"Yes, he is a Christian goy and he is marrying a Jewish girl."

"Goy, smoy, what's the difference. I had a Christian friend once, but he gave me the brush when I told him he should get a bris."

We both laughed. Terry looked confused. I needed to explain it to him.

"It's a joke, Terry. Yudi said his friend had to get clipped."

"Clipped? What's so funny about getting a haircut?"

"Never mind. I'll buy you a Yiddish dictionary for Christmas and you can look it up for yourself, Terry."

Yudi laughs and says, "I tell you what Jakub, I'm feeling generous today. Since you are a son of Abraham, I'm giving you ten percent off, right off the top."

"Oy, mazel tov. *Vielen dank.*"

"Mekheiyeh."

Terry again looked bewildered. "Jacob, is there perhaps a Yiddish dictionary on my iphone? I feel like I am in a foreign land."

"Terry. Yudi is giving us a very generous discount because you are a goy, and a Christian goy at that!"

"Hum, I don't think so, Jake."

After we both stop laughing, Yudi asks, "Do you want black, white, purple, green, blue, or transparent tuxedos boys? I have all kinds."

Terry tells him, "They asked us to wear the following: black tuxedos, white shirts, crimson cummerbunds and bow ties to match. At least we don't have to wear those weird patent leather shoes."

"*Zer gut.* I will have everything ready for you the first thing tomorrow morning. Just for you two I will open an hour early at eight."

I patted this little Jew on the back and said, "You are a true friend, a real hamisch, Yudi. Since I'm getting such a deal, I'll foot the bill for both tuxedos."

Terry objected. "Jake, you don't have to pay for mine."

"After all you have done for me, Terry? I insist. Anyway, I'm now a working man. At least I will be very soon."

I gave Yudi the cash and we quickly finished the transaction.

"Well, boys, your single days are numbered. Enjoy yourself while you can. Just remember where you received the discount. I wish you well. Lekhaim."

We left the shop and made our way to the Inner Harbor for lunch.

"Jake, I really was surprised when Yudi took ten percent off the bill. It amazed me. I didn't expect a Jewish store owner to be so kind to a Gentile guy like me."

"Oh yes, Jews always do nice things for goy Christians like you. That's because Christians have been so kind to Jews through the centuries."

"That is not the way my history books read, Jake."

We found a parking place on Light Street and walked to the Harbor Place.

"Jake, Harbor Place is a seafood lover's heaven. What we'll do here is to just walk through the open market and buy whatever looks good to us."

We walked past booths filled with steaming pots of crab, oysters on the half-shell, fried fish, and more exotic offerings like fried eel and Alaskan prawn.

"Jake, have you ever eaten Maryland crab cakes?"

"No, I haven't, Terry."

"Well, I suggest you give them a try. Blue crabs are smaller, but much better than the Dungeness crabs you get in the South."

"Terry, I want to follow your suggestion, but we Jews are not supposed to eat anything from the sea which does not have scales. I admit I have slipped a few times since arriving in Baltimore, but I need to be more faithful on what I'm eating."

Terry bought two crab cakes and I settled on a bluefish sandwich. We ordered soft drinks, found a table, and sat down.

"Terry, the bluefish was good, but it is little on the oily side."

"You had your choice. Jake."

"Do you think we have time to look for an apartment before the game?"

"I think so. We should be able to get back to the ballpark by game time. Anyway, it is a double-header."

"That's true. Okay, let's give it a go."

Terry drove back to Homewood area and then cruised north on St. Charles Street. He turned into the West campus gate. He found a parking place in the visitor's lot next to the tennis courts. Terry suggested I take a look at the Bradford Apartments located one block north on St. Paul Street.

"Jake, the entire area along Charles and St. Paul Streets is known as Charles Village. The shuttle stops just a block southeast of the Bradford. The Bradford Apartments were completely renovated just last year. There's a shopping mall adjacent to the building with a food court as well as several other places to eat close by."

We walked over to St. Paul Street. I liked the look of the Bradford. It is separated from the other apartment buildings, yet still within walking distance to the shuttle and the campus. The big plus was the parking structure nearby providing free parking for Bradford tenants. Terry and I entered the Bradford office. The manager showed us around the facility, and we took our time to look at several vacant apartments. We followed the manager up to the fourth floor to look at apartment number 415. This apartment included a bedroom, private bath, living room and/or study area, and a small kitchen. It is all I really needed. What's more, it is fully furnished, including a TV twice the size of my old one. Wanting to get these details of my life in order as quickly as possible, I rented it. We returned to the office. The manager explained the house rules to me.

"Mr. Cahn, we do not allow the following: pets, loud parties, loud music, and cohabitation without permission."

I signed the lease agreement and wrote out a check for the deposit, plus the first and last month's rent. He gave me the key and a parking permit for my vehicle. I told the manager I wasn't sure exactly when I would be moving in. He said the apartment was ready for occupancy any time. Then Terry and I headed for the stadium.

The double header was with the Red Sox. Both games were great, although the Orioles blew their lead in the second game. We drank a few too many beers and ate way too many hot dogs. No parve on the wrappers, either. What do they put in hot dogs, anyway? It's too late to ask now. We arrived home very late. Eldora had been busy in the kitchen in our absence. Terry and I conducted a secret raid. There is a home-made apple pie sitting on the window sill. We each cut a large slice of pie, and topped it with vanilla ice cream from the freezer. We made short work of the pie al a mode. Afterward, we played a couple games of pool before turning in. This long but great day in Baltimore came to a close.

I awoke on Thursday with a headache. Apparently, my body

did not handle the beer from the night before well at all. I joined the family for breakfast. Howie is looking at the two of us with suspicion in his eyes. The black coffee helps to clear my head. Howie asks if we had forgotten to feed the animals yesterday. Terry confesses to the crime and explains we got back late. I said we will put off moving my stuff into the apartment until after the morning chores. This seems agreeable to Howie. After finishing our plates of hot cakes and fruit salad, we escaped to the barns. We finished the chores by half past ten.

"Terry, I guess I might as well move into the apartment today. I'm going to straighten up the room and pack my things. Right after lunch I will leave for town. You don't have to come along. It won't take me any time to unload my gear."

"Are you sure you don't need help? I think the refrigerator is still in your front seat."

"No, I have lots of time to get my things moved in and organized."

It is time for me to say goodbye to Mr. and Mrs. Johnson. "Howie, thanks so much for your hospitality, and especially the apple pies, Eldora."

"I'm glad you came, Jacob. I packed the rest of the pie for you to take with you."

"Thank you. I really appreciate everything all of you have done for me."

Mr. Johnson shakes my hand and says "I also have enjoyed your visit Jacob. Consider our home as your own. Whenever you feel the need for some company, or just a time to rest up, don't hesitate to come visit. I do want to take you to visit our church some Sunday in the near future."

"Yes, I want to attend some weekend. Terry pointed out your church to me just yesterday. I promise to take you up on the offer soon Howie, and I'll keep in touch."

"Godspeed, Jacob."

The three of them are waving goodbye as I leave. I got back to my new apartment at four o'clock. The manager furnished a cart for me to use. In no time, I succeeded in moving my worldly possessions from the car and up the elevator to my room, including my small refrigerator. I parked the car in the lot and walked back to the apartment. I gave the phone company a call and ordered a phone. They said I would have to wait a couple of days before the installation and gave me the new number. Once I finished moving in I felt rather lonely. I decide to call Dad on my cell and bring him up to date on my life in Baltimore. It rang only once before he answered.

"Hello Son. Why have you waited for so long to call? Is everything alright?"

"I'm sorry Dad. My phone battery ran down and I couldn't find the charger in my stuff. I didn't want to make a long-distance call from the Johnson's home. I found my charger just today. Do you remember meeting the Johnsons?"

"I remember meeting Terry, but not his parents. So, they put you up for a few days?"

"Yes, they did. They treated me royally."

"How are you and Rachel getting along?"

"Not good. She is history, Dad. Her father dumped me. He signed an engagement agreement with the artist from New York. It is one of those planned marriages."

"Oh, I'm so sorry Son. She was such a nice girl. You must be devastated."

"It's okay Dad. I'm getting beyond it. I am now free to pursue my career without having to fill the task of a new husband at the same time."

"Well, there are other fish in the pond, Son."

"Yes, that's true. I'm calling you from my new home away from home. I rented a fourth-floor apartment in the Bradford Apartment Complex. It is fully furnished and the rent is manageable. If you have a pencil and paper, I can give you my new address."

"Yes, go ahead."

"The address is 3301 North Saint Paul Street, Apartment 415. Baltimore, Md. 21218-3223. Also, I ordered a land line phone here. The number is 419-287-4341, but it isn't hooked up yet."

"Is the apartment located on the Campus?"

"No, but very close by. The Bradford is located just one block west of the Homewood Campus in an area they call Charles Village. Also, I now have a job. I've been asked to work this summer as a lab assistant in the Biology Department. It will pay my rent, plus little extra."

"How is your application proceeding?"

"My interviews went well. Dr. Hobbs, the Director of the

Institute, told me I have an excellent chance of being selected. They need to complete the other interviews before the final decisions will be made. The competition is stiff. They received a hundred twenty applications for the ten to twelve slots."

"I'm pleased things are going well with your schooling Son, but it sounds like you won't be coming home to visit us this summer?"

"No, I won't be. I won't have enough time."

"I want you to keep us better informed in the future, Jacob. Your mother has been kvetching for days on end."

"Is Mom there?"

"No, I'm at the hospital right now."

"I'm sorry about not getting back to you. I promise to keep you better informed in the future. By the way, I will be flying down to Atlanta next week. My friend Allen has asked me to be his best man. He is getting married to Alice Winters."

"Perhaps you will be able to talk to Rachel's father when you are in Atlanta?"

"No Dad. Mr. Kaufman is not one to change his mind easily. I will let you go now. You have my number in case anything comes up."

"Yes, thanks for calling. Goodbye Son."

"Give Mom a hug for me. Love ya."

I hung up the phone and took the elevator down to the street level. I wanted to check out the restaurant in the mall next to the apartment complex. The dinner I would grade as a 5 on a scale of 10. After dinner, I bought a newspaper before returning to my room. I read the paper until about nine o'clock and then went to bed.

I was up and going by six on Thursday morning. I need to call Terry early before he starts his morning chores. This day has a demanding schedule. Terry answers right away.

"Good morning, Jake."

"Terry, I want to check out our schedule for the day. I assume you need to finish your morning chores before leaving the farm. The tuxedo rental shop opens at eight o'clock. I can pick up our tuxedos and bring them back here to my apartment. What do you think?"

"Yeah, it would save us some time. Our flight leaves at 11:10 in the morning. We have to check in an hour beforehand. I probably should pick you up by nine o'clock. Can you be ready by then?"

"Yes, I'll be ready. No need for you to park. Just pull up in front of the apartments. I'll watch for you from the lobby."

"Okay. See you at nine o'clock."

"See you then Terry."

To save time, I opted to eat breakfast at the small café in the apartment building. I am hungry. I order pancakes, fried eggs, hash browns, and coffee. I finish eating by a quarter to seven and returned to my room. I planned to check my suitcase and carry the tuxedos in my suiter on the plane. I partially packed my suitcase. Then I left for Yudi's place. True to his word, my Jewish hamisch opened the shop an hour early for us.

"Good morning Jakub. I see you are anxious to pick up your tuxedos. I'm so sorry, but I can't give them to you yet."

I'm panicking. "Yudi, you said there would not be a problem having them ready? Terry and I leave for Atlanta tomorrow morning at ten minutes past eleven. Are you serious?"

"Calm down, Jakub … have a cup of coffee … everything is here. I just can't give them to you yet."

"Why not for God's sake? Oh, forget I said that."

"You are wanting to have a safe trip, right?"

"Yes, of course."

"Well … I haven't prayed for you and your friend yet."

I let out a sigh of relief. "Whew! You had me going there for a minute Yudi."

Yudi is laughing loudly as he reaches back and opens the cabinet behind the counter. Our tuxedos are hanging there. Then this spry old Orthodox Jew utters a prayer over Terry and myself in Yiddish. I could not understand that many Yiddish words in succession. I knew it wasn't the standard Priestly Blessing. I did recognize three words, Jakub, Terry, and Shalom. Seeing the questioning look on my face, he then repeated the prayer in somewhat better English.

"Hear my prayer, Oh Marvelous One for these two children of yours, Jakub and his friend Terry. Keep them both in the way you have destined for them before the world began. Give them bread to eat and raiment to wear, and bring them again to this house in peace. Amen."

"I am blessed," I told him. "Yudi, you are a treasure. You also are a bit of a wild and crazy guy, a true Meshugeneh. I know your prayers will follow us on our trip. Thank you."

"It's a pleasure doing business with you, Jakub. *Lekhaim*!"

　　　　　　　　　　　　　　　　　　THE DREAM GENE

"Thanks again, Yudi. May Adonoi also grant you a long life."

"Come back sometime and see me. I have two beautiful daughters still at home."

I couldn't resist giving him a big hug before leaving the shop, tuxedos in hand. When I arrived back at my apartment I called Terry to tell him we are on schedule and ready for departure. I put the tuxedos and my good suit in my suiter. I'm pumped up and at a loss what to do with the rest of the day. I decide to work off some excess energy by taking a quick run. I changed into my running duds and headed out for the Homewood campus.

The Wave

The run took me to an area on the far side of the campus I had not visited yet. The astronomy and physics labs are located there. I was hungry. I spotted a small café located in the main building. When I enter, there are just a few students in the room. I bought a hamburger, some chips, and a coke. I spotted two guys sitting at a table near the windows, leisurely eating their lunches. One very tall student is waving his arms back and forth in the air, apparently explaining some point of physics or perhaps astronomy. I decide to invite myself to join them.

"Good morning, guys. Mind if I share your table?"

The tall one invites me to sit down. "Not at all, have a seat. I'm Calvin Jones and my friend here is Jeff MacDonald."

"Hi. My name is Jacob Cahn. Jeff, are you using your arms to explain a bit of astronomy to your friend Calvin?"

"Good guess, Jacob. Yes, I was explaining the rotation of binary stars and their magnetic attraction or repulsion to Jeff. I don't recall seeing you before. Are you physics or astronomy Jacob?"

"Neither one. I'm genetics. I'm hoping to be accepted by the McKusick-Nathans Institute. I do have an interest in astronomy. Also, the Virology Department has hired me as a lab assistant for the summer sessions."

Jeff reached over to shake my hand and said, "Well, welcome to our little group here. Can I call you Jake?"

"Certainly."

As Calvin reaches over to shake my hand, he is intently studying me. "Jake, from what you have said, it sounds like you have a better than average chance of being selected to the Institute."

"Yes, I believe you are right. Dr. Hobbs, the Director, seems to have taken a liking to me."

Jeff asks, "Do you know anything about dwarf stars Jake?"

"Very little, Jeff. In my undergraduate course in astronomy, I remember the professor mentioned the various types of stars, but did not go into any great detail on them. I remember he stated most astronomers believe our star is a bachelor star, but some suspect we may have a connection with a brown dwarf, a failed star."

"You are right Jacob. In fact, quite a few astronomers now believe our sun is connected gravitationally and magnetically to a dwarf star, a brown dwarf. It has many names, some old and some new. NASA refers to this star as Planet Nine. Supposedly this dark star does not emit radiation in our visual spectrum, explaining why no firm imagery exists of the star. This strange body is known from ancient times as Nibiru, the Destroyer. Others reject the existence of the dwarf and believe the eccentricities of past orbits of planet Jupiter explains these ancient records, and the creation of the asteroid belts."

"Do you believe that our sun is magnetically and or gravitationally bound to a brown dwarf?"

"Yes, I believe so. NASA and a few others such as the Vatican say they have been watching the approach of a burned-out star for decades now, as well as it's associated moons and debris field. I believe it is headed our way, Jake."

Calvin adds a second bit of information. "We now have waves of particles coming our way from outer space, from beyond our heliosphere, Jake. These waves of negatively charged helium and hydrogen ions are reacting with protons from our sun, producing harmful ultra-violet and gamma rays which explains the highest temperatures we have records of around the globe. This is heating up the crust and mantle of the earth, causing the seas to warm, the ice caps to thaw, earthquakes and volcanism to increase, the death of thousands of birds, sea life, and land creatures. At this point, we aren't sure what is responsible for these waves of energy from beyond our solar system. It could be caused by the explosion of a far-away magnetar star."

"Calvin, I do recall the professor's discussion of Magnetars. They are humongous in size and energy. It is hard to believe I haven't heard anything on the news about this. Are you guys serious?"

Calvin's voice is deadly serious. "Jacob, we are now facing a major change in life on this earth. The National Atmospheric and Space Agency, as well as other governmental agencies in our country and other countries have put the lid on this to avoid public panic."

"Just how serious is this increased radiation, Calvin?"

"The most serious particles are the X-rays and Gamma rays. The X-rays can cause cancer and Gamma rays can damage your DNA."

"Is there a shielding to protect us?"

"Yes, you can go underground. The Hall Current travelling on the Earth's surface may protect you if it remains strong."

Jeff went to the bottom line for me. "Jake, thousands of people will soon perish. Only those who are prepared will survive, essentially the elite leaders of the world and a few others who have prepared underground shelters. The elite never waste an opportunity to use these events for their own nefarious purposes. For a very long time, they have used such events to install a one world government and economic system. They intend to reduce the population of the world to an optimum level they feel the earth's biome can support. The remaining lower-class populations will be enslaved. If you are having trouble getting your mind around this, you might want to check out some of the alternative media postings."

I gathered up my trash from the table and prepared to leave. "You guys surely know how to make a person uncomfortable. My close friend, Allen Schaffer, is getting married Saturday in Atlanta. I will have to revisit this with you when I return."

Jeff takes a card out of his wallet and gives it to me. "Sure Jake, any time."

Calvin also writes his cell phone number on his napkin and gives it to me. We shake hands and I leave the café, a bit wiser about this time I am living in. I spent the rest of the afternoon in my new apartment reading the old version of the virology textbook. The text emphasized the field of auto-immune diseases resulting from virus infections. I lost track of time. It is nine o'clock before I realize there is a gnawing pain in my stomach. It's time to lay aside the mind food and feed the rest of the body. I popped a pizza into the microwave, retrieved a slice of Eldora's apple pie from the refrigerator, and brewed a fresh pot of coffee. After dinner, I set my alarm clock for six and turned in.

12

A Different Road Taken

"Who can find a virtuous woman, for her price is far above rubies."
(Proverbs 31:1)

The Wedding

The alarm woke me at six. I took a quick shower, shaved, and dressed. I quickly went downstairs for breakfast, returned to my room and finished packing. It is half past seven before I am ready to go. I took a few minutes to catch the morning news. What I heard was interesting. The moderator was interviewing a Catholic Jesuit priest about the Vatican's astronomy research activities on Mount Graham in Arizona. According to this Jesuit priest, NASA and the Vatican share the use of an unusual telescope with the University of Arizona. The priest stated they are studying the heavens using a binocular infra-red telescope on the mountain. The name of this instrument is the Large Binocular Telescope Near-infrared Utility with Camera and Integral Field Unit for Extragalactic Research (LUCIFER). It blew my mind. The Jesuits have named this state of the art telescope after the Devil.

The interviewer pressed the priest for further information. "Exactly what have your astronomers been viewing?"

The priest was very non-committal, saying only "We are watching something approaching the Earth. I cannot say anything further at this time. We will reveal more on this at a later time."

I wondered if they were observing the brown dwarf? My time was up. I turned off the TV, gathered up my things, locked the door, and went down to the lobby. Terry showed up a few minutes later and we left for the airport. We took St. Paul's Street into town, turned onto Fayette, and caught the Martin Luther King Highway south. As Terry turned onto Interstate 295, I could not get the morning news program out of my head. The story of the impending cosmic wave and the Jesuit telescope LUCIFER is just too weird to talk about with Terry.

Terry finally asked, "Jake, you are pretty quiet this morning. Is anything bothering you?"

"No, I'm just slowly waking up."

Our journey takes us south on I-95, and onto the airport access road. We have plenty of time to get checked in. We bought a couple of sandwiches and coffee at the Burger King in the terminal to take with us on the plane. Terry purchased a farm magazine. We picked up our boarding passes, checked our luggage, and went to the gate. We are still an hour early. Terry sat down and began reading the Farm Journal magazine. I made a quick call to Allen.

"Al, they say our flight will be leaving on time. The ETA is 1:12."

"Very good, Bro. Hank Bowman says he will pick you up. He'll meet you in the baggage claim area."

"Okay, we will look for him there. See you soon."

Terry and I boarded the plane and I opened my old virology textbook I brought along to review. I found it hard to concentrate on the textbook. My mind kept drifting back to the meeting yesterday with Calvin and Jeff, the story about the Jesuit telescope, NASA's Planet Nine, and the wave of particles from outer space. What is the Catholic Church looking for in the heavens? I did manage to finish my review of the textbook by the time we descended to land.

When we arrived in the baggage area Hank is there, anxiously waiting for us "Welcome back to Atlanta guys. It's good to see you two again. You both are looking well."

We shook hands and shared a few war stories while we waited for the baggage to show up. I knew it would not be long before Hank asked about my sister Emily. I decided to pull his chain a bit. I didn't have long to wait.

"Jake, how is your sister Emily these days?"

"Emily is just fine, Hank. She is dating some guy from Creighton. It looks pretty serious. They will probably be married before the year is out."

The look of pain on Hank's face was priceless.

"Really? I was looking forward to seeing her again. Is she still planning to come to Emory this Fall?"

I didn't have the heart to punish him further.

"Yes, she is coming to Emory, Hank. She also is still very much unattached."

In a split second, I watch his face change from utter pain to a wide smile.

The lights began to flash on the baggage conveyor. It wasn't long before our bags showed up. Terry and I grabbed them and we all headed for the parking lot. As promised, Allen had arranged for both of us to stay at the Omni tonight and Saturday night. The wedding rehearsal dinner is planned to be held at Rose's Centennial Steak House on Marietta Street, just a block from our hotel. Everyone in the bridal party plans to meet in Salon One at seven o'clock tonight. After Terry and I checked into the hotel, Hank offered to show us around the campus. Terry is anxious to visit it once again, but I really wasn't in the mood. My mind now is dwelling on Rachel, wondering if I will be able to see her this weekend.

"Guys, I think I'll pass. I'll just take a walk around the downtown area. I need to spend some time alone. Why don't you two meet me at the reception dinner."

Hank understood my problem. "Okay, Jake, we will catch up to you at the reception dinner. I assume you can just walk to the restaurant."

"Yes, I can. I'll see you there."

Hank and Terry left me there to ponder my fate with Rachel. There are many attractions in downtown Atlanta. I had never taken the time to visit most of them. The Georgia Aquarium is located just two blocks from the hotel. I spent an hour at the aquarium, then I just wandered through Centennial Olympic Park. The walk gave me some time to sort out my thoughts. I wondered if Rachel will be courageous enough to attend the reception and wedding? If she does show, what will I say to her? I'm sure she will not be able to handle the reception dinner. Perhaps she might slip into the wedding chapel and we can talk after everyone leaves? Although she is gifted with a lot of courage, I really doubt she will show, even at the wedding. It is better if she doesn't show up. If I shared my recent dream calling to live a celibate life with her, it might very well push her over the edge. Then again, what kind of a future are we all facing? Are we on the cusp of a world-wide disaster?

I arrived back to the hotel at four o'clock. I turned on the TV and the Braves were playing. I watched the rest of the game. Then I took a shower and changed into my suit. I left a few minutes before seven for the dinner.

I arrived at Roses' Centennial Steak House at exactly seven. As I entered, the wonderful aroma of charcoal grilled steak and roasted onions wafted through the room. The maître de at the door ushered me into Salon One. Allen and Alice stood at the door greeting everyone. Alice insisted on showing me to the head table. The room is rapidly filling up with very noisy and happy people. The two families have gone all out for this affair. There are streamers decorating the ceiling and four walls of the room. Each table is decorated with a flower centerpiece. A crystal bell, inscribed with the name Mr. and Mrs. Allen Schroeder, decorates each place setting. There are pitchers of water, bottles of red wine, carafes of coffee, and a bottle of Champaign cooling in buckets next to each table. I looked at the place cards at the empty seats each side of me. The two seats to my right are reserved for Pastor McGuire and his wife. The place card to my left reads "Miss Rachel Kaufman." The chair is empty. I'm sure now I will not see Rachel this evening. Looking up I see Pastor Bill and his wife Marjorie making their way to our table. I stood up to greet them.

"Good evening Pastor. I'm so glad to see you again."

"It is good to see you also, Jacob. Let me introduce my wife Marjorie."

"I'm so pleased to finally meet you, Jacob."

"Mrs. McGuire, your husband has been a very important person in my life. I've looked forward to meeting you as well. I feel I already know you. Pastor Bill mentions your name often."

"How do you like Baltimore, Jacob?"

"I really like the city. Of course, every city has its unique character. I've enjoyed Atlanta as well. Allen and Alice have picked a perfect time for their wedding. The magnolia trees are in full bloom and their aroma is competing with the smell of the honeysuckle. How are things with your ministry Pastor Bill?"

"We are very blessed, Jacob. I believe Marj won't object to my telling you she is now in a family way. It will be our first."

"Congratulations. I'll venture to say you both must be on cloud nine."

"Yes, we are Jake. Marj has the spare bedroom all prepared. So, what is the status of your application at Johns Hopkins?"

"I won't know for certain until they finish the interview process, but they tell me I am almost sure to be admitted to the program. In the meantime, they have hired me this summer as a lab assistant in the Biology Department. The first class begins next Tuesday."

"Hiring you so quickly must mean the Institute is impressed with you."

Just then the head waiter came to take our orders. I glanced at the menu in front of me and selected a New York cut with all of the trimmings.

After Pastor Bill and his wife placed their orders, Mrs. McGuire asked, "Jacob, is there anything in your personal life you would like to share with us?"

"Yes, there is. I suppose you are making reference to the empty place setting next to me. Rachel Kaufman is now engaged to an artist from New York. Her father exercised the ancient Orthodox custom, a la *Fiddler on the Roof*, of an arranged marriage."

Pastor Bill nodded his head. "I read the article in the paper. I had asked my wife not to bring up the subject tonight unless you did, Jake."

"Bill, I did asked Jacob if he wanted to share anything about his personal life."

"Yes, Dear. I'm sure you had no intention on bring up the subject of Jacob's girlfriend, did you?"

"Not to worry, you two. I don't mind talking about it."

"Well, that being the case, how are you handling this Jacob?"

"I'm okay with it. I have to admit when I first heard about the wedding, I felt betrayed. Then I had another unusual vision, followed by the awesome experience of sensing Ha Shem's presence. So, I am okay with Sam Kaufman's decision. I understand his concerns."

Mrs. McGuire was hanging on my every word. "You saw another

vision? Would you mind sharing your vision with us? It sounds very interesting."

"No, I don't mind. I believe it was the day after hearing of Rachel's engagement. I remember waking up and glancing at the clock on my bedside table. It was exactly three o'clock. I rolled over and was staring at the ceiling when these words began floating in the air over my head. At the same time, the sound of each word was bypassing my ears and echoing in my brain. The words were repeating the prophecy over the Church of Philadelphia. I recognized the words from the Book of Revelation, Chapter Three because I had just read this passage the day before. After the vision ended, I felt completely at peace. My room was filled with a sweet aroma, something like lilacs. The awesome presence of Ha Shem remained for several minutes. My pain of Rachel's planned marriage was completely gone, replaced by an indescribable sense of His presence. If you remember Pastor, I experienced this floating word thing once before. One thing I know for sure. Ha Shem has my life in His hands. I now know my future will not include marriage, marriage to Rachel, or to any other woman. I now am married to the creator's purpose for my life. I still do not understand it, but I'm content to just follow his lead."

Pastor Bill was astonished by the way Ha Shem has moved on me once again.

"What an incredible experience, Jacob. Have you talked with Rachel since you heard about her engagement?"

"No, I have not. I tried to call her, but she would not speak to me. I think she is really hurting. Actually, I'm glad she isn't here. I have no idea what I would say to her or how I could possibly explain this latest vision to her. I feel very bad for Rachel. I know she is still very much in love with me, as I am with her."

Neither of them knew quite what else to say. It was an awkward moment. Mrs. McGuire kept trying to stop the tears from flowing, drying her eyes with her handkerchief.

Pastor Bill finally broke the silence. "Alice and Allen tell us they are on their way to Lagos Nigeria soon."

"Yes, Al told me they plan to leave right after a short honeymoon in the Smokey Mountains."

Pastor's face is grim. "Do they fully realize the situation they will be facing there? Northern Nigeria is in a state of total chaos.

Terrorists have declared a holy jihad there. Christian churches are their primary target."

"Pastor, I also don't have a good feeling about this mission trip, but they are determined to go. Perhaps they will listen to you. Northern Nigeria is a nightmare situation, for sure."

"Jacob, I advised both of them to reconsider, to choose a safer mission field."

"What did they say?"

"They told me God has their back. They are confident God will protect them. They quoted me this scripture from Second Corinthians, Chapter Two and verse eight: 'We are confident, I say, and willing rather to be absent from the body, and to be present with the Lord, that whether present or absent, we may be accepted of him.' How can I argue against God, Jake?"

"I hear what you are saying, Pastor. I've discovered one thing from these strange events in my life. I am not in charge of my destiny. We may think we are, but when we look back at our past, we see the evidence of his invisible hand, guiding our paths."

"That is a profound comment, Jacob. Over the short time I have known you, I have been impressed with your spiritual growth. Have you finished the research project I gave you yet?"

"As a matter of fact, I have Pastor. I applied the tools of statistical analysis to the problem. I carefully rejected paired cases where a dependency might exist, where the actual event could have been manipulated after the fact. I succeeded in identifying a valid number of discrete events, adequate to subject the data to a valid test.

The result of the test was quite amazing. I obtained a result of 0.001 at a confidence level of 95 percent. In other words, the probability is very low of that the matched events mentioned in the data could have occurred by chance. Even with these results I have remained stubbornly resistant to the acceptance of Jesus as the Messiah. I cannot truthfully dismiss the position of the new covenant scriptures. This Jacob will be wrestling with the issue for many days in the future."

"So, you still are sitting on the fence. Our little research study has failed to bring you to the point of asking Yeshua to be your Lord and Savior?"

"Yes Pastor, I am still torn between traditional Judaism and Messianic Judaism."

"Can I ask, do you know who owns the fence you are sitting on?"

"What fence?"

"I just said you were still on the fence, Jacob. Let me give you some information, Jake. Satan owns the fence. You will be required to come down on one side or the other. Commit your life to Yeshua Jacob?"

"It's not so easy. Even though I have been impressed by those who have shared their faith with me, I am struggling to place my trust in a God of three persons."

"You are operating from the intellectual part of your being. Our world views often overrule the obvious. Will you promise to continue pondering these Scriptures, Jacob?"

"Yes, I will. Sometimes I wonder, why doesn't Adonoi just make me decide? He must know I am too weak of character to make such a bold decision."

"Do you remember the great line Donald Sutherland uses in the movie *Kelly's Heroes?* 'Quit with those negative vibes. You gotta have a little faith, baby. Have a little faith!' That's your answer, Jacob. God wants you to use your faith. God has given us free will. Even though he knows what is best for us, he does not force us to make the right decisions. God won't ever do it all for you."

"Yes, I do know what you are saying, Pastor Bill."

Our conversation was cut short when Allen tapped his glass at the head table and asked everyone to take their places. While this was going on, I motioned to our waiter.

"Is there something you need, Sir?"

"Yes, there is. Please remove this place setting next to me. Miss Rachel Kaufman will not be attending the dinner this evening."

"I see. Do you mind if I seat someone else here, Mr. Cahn?"

"Not at all. Feel free."

"Thank you. We have a young boy who somehow was left off of the invitation list. I realize it is the head table, but can I seat him here next to you?"

"Yes of course. I will enjoy some company."

The waiter seats the young boy next to me.

"Hello, son. My name is Jacob Cahn. What's your name?"

"Tom Boone. I attend the church Mr. Schroeder goes to. My parents weren't able to come and someone left my name off the invitation list. I'm supposed to carry the ring to the best man."

"Well, you will be giving the ring to me, Tom. My name is Jacob Cahn. You must know Pastor and Mrs. McGuire?"

"Oh yes, Mrs. McGuire is my Sunday school teacher."

"Tom, will your parents be able to bring you to the wedding?"

"Yes, they will be back from their trip tonight."

"Well Tom, I hope you like steaks, because they are going to give you a big one."

"I'm hungry enough to eat an elephant, Mr. Cahn."

"Now that we are buddies, you can call me Jake."

"I like being your buddy, Jake."

While the waiters are serving our food, Allen reminds everyone of tomorrow's schedule. He asked the wedding party to arrive at the church on Elm Street by eight thirty for a short rehearsal. The wedding ceremony is scheduled to begin at nine o'clock. Then Allen asked Pastor McGuire to give the blessing over the food.

After the blessing, Mr. John Schroeder, Allen's father offered the first toast. He mentioned how proud he was of the couple. Then it was the father of the bride, Mr. Emil Sunderfeld's turn to address the guests. As soon as I heard his heavy Yiddish accent, I knew we were in for an interesting toast. I just hoped he realized he was in a church and would keep the humor within bounds. Mr. Sunderfeld stood up, raised his glass, and took a sip of wine. He then began his standup routine.

"A wise rabbi once said, if you don't have a sense of humor, you probably don't have any sense at all. Allen, since you were not raised in a Jewish home, I feel I need to give you some useful pointers. You are going to have to make some major adjustments to be able to live with a Jewish wife. For instance, there is only one place a husband can keep a bit of cash out of the reach of his wife. I recommend you hide your spare gelt under the vacuum cleaner. (Another swallow of wine.) It's the most secure place I know of. You see, growing up in a Jewish household gave me the advantage of early training. When I was a boy of just sixteen I remember coming home from school one day and telling Mother my teacher had given me a part in the school play. She asked, 'What part in the play is yours?' I told her, *Mame* I'm playing the part of the husband no less. My mother scowled and said, 'go back and tell your teacher you want a speaking part.' (More wine.) Of course, this is why Jewish mothers make such great

   THE DREAM GENE

parole officers. They never let anyone finish a sentence. (He has the waiter full up his cup.) Of course, all future grandmothers hope you two will be having a family soon. The sooner the *kleyn kinder* come along, the better. It says in the Torah, you are to teach these young ones with all diligence, when you sit in your house, when you walk by the way, when you lie down, and when you rise up. I suggest you stock up on vitamin pills. (two swallows of wine.) To diligently raise kids and keep the grandmas happy, you will need them close by. You will have to sleep in your car. (the wine again.) Allen, since you haven't finished medical school yet, I need to warn you. There is a controversy in the Jewish community about when these children become self-supporting. Traditionally the *kinder* are not considered viable adults until they graduate from medical school. This means you will be providing food and shelter far beyond the age when your children get their driver's licenses. There's only so much I can help you with before this wine evaporates from my glass again. So, I will close with this comment from my heart of hearts. May Adonoi bless you both with peace, happiness and many children. *Lekhaim!*"

As Mr. Sunderfeld sits down, his wife pushes a spoonful of mashed potatoes in his face. It reminded me a lot of home. Many of the other guests presented toasts to the bride and groom. The live band began playing. The couples left the Salon and made their way to the dance floor. Everyone had a wonderful time. The affair did not end until half past ten. I enjoyed seeing so many of my Emory friends again and visiting with Pastor Bill and his wife Marjorie. Terry and I walked back to the hotel together.

I watched as a golden moon rises over the city. The air is filled with the smell of magnolia blossoms. "Terry, this is another one of those perfect Atlanta evenings."

"Yes, the end of a perfect day. Do you see the tree with the bright red blossoms, Jake? It is a coral tree."

"Yes, I see it. Do you have a tree like it in Maryland?"

"No, the winters are too cold to grow those trees in Maryland."

It was a perfect evening. Terry and I both left a request at the registration desk for a wake-up call at six. Then we took the elevator to our rooms.

I had no difficulty getting to sleep. I didn't remember anything until the call at six Saturday morning. I quickly dressed, and knocked on Terry's door. We went down to the complimentary breakfast on the first floor. After eating, we returned to our rooms. I gave Terry's tuxedo to him and we both dressed for the wedding. The bow ties Yudi gave us were clip-ons. It took only a few minutes for me to get dressed. I took the time to call Dr. Michael Stein at his home on the hotel phone. I wanted to let him know the books are on their way via UPS.

"Hello, Dr. Mike, this is Jacob, Jacob Cahn."

"Hello, Jacob, how nice to hear the sound of your voice. What's new with you?"

"Well, I'm in Atlanta to attend a wedding. I'm the best man. I just wanted to let you know I sent the books I borrowed back to the library via UPS. They should arrive sometime next week. If I had known about this wedding sooner, I would have delivered them personally."

"I'll look for them, Jacob. How is your application to Johns Hopkins proceeding?"

"Quite well, although I won't know for sure if I have been accepted until they finish reviewing the other applications."

"I'm so glad you called. Do you have a minute or two? I need to share something with you."

"I'll make time. What's on your mind?"

"I assume by now you know Sam Kaufman has promised his daughter Rachel to Jeremiah Levitt. He is an artist living in upstate New York."

"Yes, I heard about it."

"Jacob, I need to confess something to you. I may have had a part in Sam's decision to arrange Rachel's marriage. I had a conversation with him a week or so ago and I mentioned your name. I told him I considered you to be a fine upstanding mench, and I really thought you would be making a real mark in your life. I did not expect his violent reaction to my comment."

"I don't see why that conversation would be a problem for him."

"There's more. I need to share a bit of the family's history with you. The Kaufman family lived in Berlin before World War II. His father, Hans Kaufman, was a well-known columnist and wrote a regular byline in Der Spiegel. He violently opposed the Eugenics Program proposed by Ernst Haeckel way back in 1868. It was adopted by Adolf Hitler under Reichskommissar Rüdin. There were other signs in the wind in the early 1930's, but the family couldn't financially manage to leave Germany before Hitler closed the borders. When the Nazis took power, Hans succeeded in smuggling Sam into Switzerland. He was taken care of by a Jewish friend of his living there. After the war, the Swiss family migrated to the US and took Sam along with them. The rest of the Hoffman family all perished in the death camps. There is one more detail to the story you need to know, Jacob. Hanz Hoffman's friend was my grandfather."

"I don't understand how this relates to me Dr. Mike?"

"There's more to the story Jacob. Sam asked me to tell him more about your chosen field, what kind of research is done, and so forth. I told him the field is exploding. I mentioned researchers are working on the ability to change our DNA, possibly resulting in ridding the world of many diseases and correcting defects in the human genome. I should have known better. The idea of altering inherited genes of course was anathema to him. He asked if such research could result in selective breeding. I admitted it was possible. At this point, he began to lose it. Swear words came rolling out of his mouth. I couldn't believe what my ears were hearing from this Orthodox Jew. He sounded more like a sailor than a lover of Ha Shem. His face was as red as a hot poker."

"What was he saying besides cuss words?"

"He kept yelling 'This is meshugeneh, utter madness, foolishness, craziness. No daughter of mine is going to marry a geneticist.' Finally, he calmed down and I said, 'Samuel, the Nazis are still with us. Our country did not win the Second World War. The Nazis did. We have allowed their scientists free reign in our Nation. Our government has sold us out to these wicked scientific survivors of the War.' I'm so sorry, Jacob. I should have known he would react violently to your chosen profession."

"No need to feel bad, Dr. Mike. I'm sure Rachel and I would have come across the same thing at some point in the future. It is better we not marry. Thank you for sharing this with me. I'm sure it has brought up some old and very painful memories for you. It certainly helps me understand the decision he made. I realize Sam and I truly live in two different worlds."

"Yes, the two of you occupy mutually exclusive realities, and Rachel is caught in the middle. I'm truly sorry Jacob. Please forgive me."

"I forgive you completely. Don't let this trouble your pure soul."

"Thank you, Jacob. I appreciate how you are taking this. Let's keep in touch. Oh, by the way, have you been watching the news lately? The alternate news folks are saying we will soon experience the arrival of an electromagnetic wave, and possibly also a brown dwarf. They say we can expect a lot of damage. I have seen some very weird things going on with my telescope recently. The planets are now much brighter. I have to put on a polaroid filter to observe Venus."

"As a matter of fact, I just watched on the local news channel an interview with a Jesuit priest about the very same thing. It is amazing. If you come up with some information about this, please keep me informed. Thank you for letting me know about your encounter with Sam, Dr. Mike. I very much appreciate it."

Jacob's Temptation

I just finished talking with Dr. Mike on the hotel phone when my cell phone rang. There was a minute of silence. Then I heard her voice. "Good morning, Jacob. This is Rachel."

"Oh, I'm so glad you called, Rachel. I have been quite concerned about you."

"Jacob, I just could not talk to you on the phone the other day. I hope you understand."

"I understand. I have been so worried, not knowing how you are taking this situation. I wonder if we can get together while I am in Atlanta?"

"Yes, of course. We need to talk. I just couldn't attend the dinner, and I won't be at today's wedding either."

"I understand. I'm flying out Sunday Morning at eight. Perhaps we can meet somewhere this evening for dinner after the wedding?"

"I would love to have dinner with you tonight Jacob."

"Shall we go back to the French restaurant?"

"No, I will cook something special here at the house."

"Are you sure? Isn't your home forbidden territory for me?"

"It is not forbidden territory this evening. We will have a quiet dinner and have some time to sort things out."

"Are you sure?"

"Yes, I'm sure. Can you take a cab or something? I'll be cooking."

"I'll take a cab or borrow a car. What time do you want me

there?"

"How does eight o'clock sound to you?"

"Eight should be fine. Are you sure this is okay with Papa?"

"I wouldn't ask you to come if there was a problem, Jake."

"Can I bring something?"

"No, just bring yourself."

"Okay, Rachel. I'm looking forward to it. Expect me at eight."

A flood of emotion is welling up inside of me as I hang up the receiver. I realize I'm not over this girl yet. Why is our relationship ending this way? Isn't love like this supposed to have a happy ending? As I look back on our relationship, I realized we are constantly in the midst of some kind of struggle, trying to connect two different worlds into one. I should have known from the start this conflict with Papa Kaufman could never be resolved. Somehow tonight we both must find a way to end it well.

There was no time to dwell on this any longer now. It is time to leave for the wedding. I met Terry in the lobby at 8:40 and we hired a taxi to take us to the church.

Terry asked the cabbie, "Do you know where Willow Creek Christian Fellowship is located on Elm Street?"

"Yes of course. I go to church there on Sundays when I'm not driving cab."

I asked, "When does your shift end today? I need to go see a friend this evening about seven-thirty."

"I just came on board. I won't get off until eight or nine tonight."

"Could you pick me up at the hotel later tonight?"

"Sure. Just in case you want to check on me, take my card. Just call the dispatcher and ask where Jason is. That's me. If he tells you I am on my way to the Omni, you will know you are covered."

"Thanks, Jason. I appreciate your help."

The cabby is smiling. "Important date with a girl I take it?"

"Yes, it's a very important date. She is cooking a special dinner for me."

"Well, we won't keep the young lady waiting. I'll see you at 7:30."

Terry and I arrived at the church for the rehearsal at exactly half past eight. The organist did not arrive until ten minutes later. Alice's sister Maudy stood in for Alice during the brief rehearsal. Then Maudy efficiently helped the guys insert red buttoners into their tuxedos, straightened bow ties, and made sure everything else was in order. The church sanctuary was filled by the time Allen and I walk to the front of the church. Some are standing along the side isles and at the back of the church. At nine o'clock the organ begins playing the wedding march. Nancy, a beautiful little girl of five, leads the bridal precession down the aisle, spreading rose petals out of her basket as she goes. Tom Winters, the boy I sat next to at the rehearsal dinner, escorts a second young girl down the aisle. They both carry the wedding rings on fancy pillows. The bride's maids, escorted by the guys come next. The girls are dressed in tan colored silk dresses, each carrying a bouquet of flowers. Then the bride, escorted by her father, comes into view. All the while this is taking place, I am thinking about tonight's dinner with Rachel.

I whisper to Allen, "Guess what Bro? I have a dinner date with Rachel this evening."

"Great, Jake. I hope things work out for you two."

"This will really be an attempt to find closure, Allen. We have to finish our relationship on an up note."

"I see. I have an important meeting tonight as well, Jake."

"Yes, you surely do. May Ha Shem bless you both with a wonderful wedding night."

"Thanks, Jake."

Allen kept nervously clearing his throat as he watches Alice walk down the aisle. Unlike my nervous friend, the bride is beaming and completely at ease. Pastor Bill takes his time leading them through their vows. At the proper time, I take Alice's ring from Tom and hand it to Allen. Pastor Bill asks Alice if she will take Allen as her husband. She is smiling as she says yes. Allen struggles to place the ring on her finger. Alice's maid of honor takes Allen's ring from the young girl and hands to Alice after Al quickly agrees to be Alice's husband. Alice places it on his finger. After the I dos and the rings, Pastor Bill pronounces them man and wife. It is a done deal. Light bulbs flash. They embrace for several minutes and then quickly exit the church sanctuary to the applause of everyone. Pastor Bill asks the gathering to allow the wedding party a few minutes for photographs before entering the fellowship hall. Then we walk back down the aisle. I accompany Trudy Walker, the maid of honor, to the church hall for pictures. I remember sharing a class with Trudy. After the photographer finishes taking the photos, the ushers open the doors and Allen and Alice greet everyone entering the hall. Of course, the meal is catered by Bernie's Delicatessen and served buffet style.

Terry asks Hank if he minds taking us back to the hotel. At half-past two the three of us leave. When I arrive back at my hotel room, I quickly got out of the monkey suit. I am happy for the couple, but very glad to have the celebration over with. I took a quick shower and then dressed into more comfortable clothes. I am restlessly waiting for time to leave. It seems like an eternity. I try to watch TV, but there isn't anything good on. I bought a newspaper and a Scientific American magazine from the magazine stand in the lobby. There is an interesting article in the magazine about a new revolutionary computer they are calling a quantum computer, or Qbit for short. The article states the design of this computer was first demonstrated on February 13, 2007 at the Computer History Museum in Mountain View, California by New-Wave Systems, Inc., based in Burnaby, British Columbia. The first commercially available quantum computer was a Model One operating with a 128 Qbit chip with quantum annealing. What is quantum annealing? I have no idea. Their Model Two expanded to 512 Qubits. The article said a third model is under production which operates on a 1024 Qubit chip. The company plans to ultimately produce a computer with 4096 Qbits. Already existing computers are capable of parsing a dataset containing every prime number in existence and storing the DNA information for everyone in the world. These incredible quantum computers are being used to launch research into a new field called Artificial Intelligence (AI). The new research is a collaborative effort with New-Wave Systems, NASA, and The Space Research Association (USRA) used by major universities. This new technology is now shared openly with all interested parties. It appears research activities rapidly are being centrally organized. This centralization effort extends beyond our national borders. What is this telling us about our future world? Who will be able to afford using these machines? I resolved to investigate this new device in greater detail. This technology is not without risk. Will it in concert with artificial intelligence lead to an a totally controlled world system? I made my way down to the hotel lobby a few minutes before seven-thirty. Jason the cabbie is waiting for me. I opened the door of the cab and jump in.

"Hello, Jason. Thanks for being on time."

"No problem. Where're we going?"

"We are headed for 1510 North Peachtree Avenue."

Jason drops me at the Kaufman home five minutes before eight o'clock. I walked up to the door and rang the bell. As I wait for the door to open, I am thinking back to the first time I stood on this doorstep, blessing the household by touching the mezuzah. This first visit went better than I expected. Her father and I communicated well with one another. This evening we will meet under much different circumstances. I wonder if Rachel's dad will still be cordial toward this supposed eugenics researcher? I may find myself facing a very angry man.

A smiling Rachel opens the door. She is dressed in a formal evening gown. She looks so beautiful she takes my breath away.

"Good evening, Rachel."

"Hello, Jake. Don't just stand there. I'm not going to bite you. Come on in. Here, let me take your coat."

"Rachel, I distinctly remember you telling me I should dress casually this evening."

"You are dressed just fine, Jake. How do you like this new dress I'm wearing?"

"Yes, it is very nice, but definitely not what I would call casual."

"I changed my mind and decide to dress up a bit. Do you think it is overly sensual, Jake?"

"No, I won't say it is overly sensual. It is a beautiful dress, Rachel."

"You still have a worried look on your face. Are you embarrassed by your casual attire? I remember you once had the same problem on a tennis court."

"Yes, I remember. My casual dress is not as much a problem as my concern about what your father's reaction is going to be when he learns I have been in his home."

"You don't need to worry about Papa. He and Hanna are in New York visiting my intended husband."

"In New York? Do your parents know you have asked me to dinner here, alone with you this evening, Rachel?"

"Do you remember the question you asked me when we first met? You asked me how I could play tennis on Shabbat. At that time, I told you, my Father doesn't have to know everything I do."

"Of course, I remember. It was the most wonderful day of my

life, Rachel. But this meeting isn't on an outdoors tennis court. Am I right concluding your parents don't know we are having dinner together tonight?"

"What a marvelous deduction, Jake. Yes, the two of us are going to have dinner together, just like an old married couple. Is this too difficult for you to handle?"

"No, but I know your father will be furious if he ever finds out about this. Does he happen to know any mafia hit men?"

"Relax. My parents need not ever know Jake. This will be our little secret, just like our ketuba."

"I'm facing another unknown danger."

"What danger?"

"I have never sampled your cooking before."

"Not to worry, It's all kosher. If you approve of my main course, you may enjoy the special dessert I have prepared for you."

"Dessert? Can I ask what kind of dessert you have prepared for me, Rachel?"

"You are a very bright boy. I'm sure you must have some idea what I have come up with. Please escort me to the dining room. Dinner is served."

I dutifully took her to the dining room, feeling like a rabbit caught in a trap. Rachel indicates I am to sit at the head of the table. This chair no doubt is where her father sits. She stands by the chair next to mine. I pulled out the chair for her and she sits down. I sit in papa bears chair. There is a bottle of Zinfandel wine on the table.

"Jacob, will you please open the wine while I get our salads from the kitchen?"

"Yes, of course."

Rachel disappears into the kitchen and promptly returns with two Caesar salads. She places a bowl of hot yeast rolls on the table with a butter dish and a tray of condiments including some red horseradish. This is followed by the main course, a ribeye steak, baked potato with sour cream and chives, and mixed vegetables.

"Rachel, this is a wonderful dinner you have prepared. How did you know I like steaks?"

"What man doesn't?"

She kept my wine glass full as we silently ate our meal. Neither of us knew quite how to continue the conversation. Finally, I broke the ice.

"Rachel, I talked to Dr. Michael Stein yesterday and he explained to me why your father has arranged your marriage. Your dad does not approve of my vocation, human genetics. He likens it to what the Nazis did during the war. He believes working in human genetics is tantamount to promoting eugenics."

"Yes, he said as much to me a week or so ago. He refused to listen to my explanation about your field. Did Dr. Mike tell you about my father's family background?"

"Yes, he told me Samuel's family managed to send him to Switzerland with a neighbor before the Nazis closed the border. Your father's entire family perished in the camps. It is useless to believe your father will ever consent to our marriage. As we have discussed before, I cannot ask you to abandon your family. I love you, but your father loves you deeply. Then there's your mother to consider. It seems we are condemned to live in different worlds."

"Jacob, I do not believe my marriage with Jeremiah will ever be a loving union. He is too into his own thing, his art. He is garnering quite a following in the art world. You know Papa is pretty well fixed financially. It is possible he has agreed to marry me only to provide a path into the Jewish elites of the art world. I cannot live such a fruitless married life. I will find love elsewhere."

"You cannot even consider such an option, Rachel, you are talking now like a crazy person. In time, the two of you may find love. It often happens in planned marriages. You will no doubt have children to love and care for."

"Jacob, I need you to understand something. At this point I am a hopeless meshugeneh. I am crazy out of my mind with love for you and no one else. What do you suggest, Jake? Do you want me to just kill myself?"

"If you truly love me, Rachel, promise never to take your life. Promise me!"

"I will not promise you, unless you give me a promise as well."

"Exactly what do you want from me?"

She did not answer. She quietly left the table and walked over to the buffet. Opening the buffet door, she pulled out a box wrapped in blue paper with a ribbon bow on top. She hands the box to me.

"Go ahead and open it, Jake."

I open the box. Inside is a carefully rolled parchment, our ketuba.

"Thank you. Rachel, I will treasure this always."

Rachel then asked me a rather curious question.

"Jake, I don't think I have ever heard you listening to even one song. Don't you like music?"

"I like music. I just don't often take time to listen to it."

"I want to play a song for you. Do you mind?"

"No, I don't mind."

"There is one song in particular I want to play."

"Sure, go right ahead."

Rachel walks over to the entertainment center in the living room. She selects a track on the CD player and pushes the play button, adjusting the volume down low.

"Do you recognize this song, Jacob?"

"Yes, of course, it's "The Way We Were" by Barbara Streisand."

"This is our song, Jake. We shared so many wonderful times together."

"Yes, I agree with you. I can't think of a song more expressive of our relationship. If you like, we can claim it as our song."

Rachel returns to the table and takes my hand in hers.

"Jacob, I want you to dance with me, just for a while. I want you to hold me."

My knees are feeling very weak. "You want to dance now?"

"Yes, right now."

"I believe it is against the traditions of the rabbis?"

"It is. Will you dance with me or not?"

I didn't know what to do or say. This is going far beyond my intentions for the evening, but how can I refuse her? I also must confess I am longing to hold her in my arms. How often have I dreamt of it? I stood up, took her in my arms, and we begin to sway with the music. It is wonderful. It feels like we are being catapulted into heaven itself. We danced to all of the songs on the CD. We danced around the dining room table, into the living room, and back again. We must have danced for the better part of an hour. Then Rachel takes my hand and begins leading me toward the stairway upstairs. It takes every ounce of willpower within me to force her to let go of my hand. I firmly pushed her back away from me.

"Rachel, this cannot happen."

"Jake, don't lecture me. Not now. I love you and tonight I need you to love me."

I looked up, away from her pleading eyes and said a silent

prayer to Ha Shem for strength. The answer comes. "You must tell her about Philadelphia."

I noticed a wooden bench along the wall under a window. I walked over to the bench and sat down.

"Rachel, we need to talk. Please come sit down on this bench with me. I need to explain something to you. What I have to say will be as painful for me as it will be for you. Just now Ha Shem spoke to me. He reminded me of a vision I had a few days ago. Actually, he gave me a scripture from the Bible, from the New Covenant book of Revelation. Chapter Three, verses seven to thirteen."

"So, you are telling me I am going to Hell now because I want to make love with you?"

"Not so, Rachel. It is a message to the Church of Philadelphia and it is about our lives right now. In verse ten it says the following:

"Because thou hast kept the word of my patience, I also will keep thee from the hour of temptation which shall come upon all the world, to try them that dwell upon the earth."

Rachel, I am being called by Ha Shem to serve in these last days. I don't fully understand it yet, but I know what I know, what I know. I must follow my destiny, not my desires. I know this seems unreal to you, but I believe Ha Shem will pour out a blessing upon your life if you will trust me in this. I will always love you and will cherish every moment we have experienced together. Can you trust me on this?

Her eyes are still pleading. "Jacob, don't you understand? I can't live without you. Tomorrow I will die. Do you want me to die, Jacob?"

"Rachel, with the help of Adonoi you will get beyond this. We both have been given strong spirits. We see only the back of the divine tapestry in this world. Our agony, in some mysterious way, is connected with his eternal plan."

Great drops of tears are now flowing down her face. She buries her head on my shoulder. I begin to carefully describe the rest of my vision to her. I tell her of my calling to a life of celibacy. It is as if I am plunging a dagger into her heart.

"Rachel, try to understand. Our song, "The Way We Were," is truly our song. Listen to the words. 'The misty water-colored memories, the pictures of the smiles we shared with one another.' Can you remember the laughter we shared when I lost those tennis games? Our special memories will always be with us. The song asks whether we would do it all again if we knew this painful outcome. My answer is yes, I would do it all over again. Don't you see? We will treasure these memories the rest of our lives. We will never let them go. But now we must pursue our separate destinies. Later in heaven we will be together with one another in the presence of Ha Shem."

"You are stronger than I am, Jacob."

Just then the headlights of a car flashes through the window and I heard what sounded like a garage door opening. Rachel jumps up from the bench and runs to the door. Peering out the small peep hole she says, "Oh my goodness, they are back a day early."

"Surely not your parents?"

"Yes, they are back. The car is just now pulling into the garage."

"This is not good, not good at all."

"What shall we do, Jake?"

"There is nothing we can do but remain calm and pray."

Hanna is first through the door. She stops short and her mouth drops open. She is literally speechless.

Rachel tries to calm her. "Hi Mom. You and Papa are home a day early, aren't you?"

Hanna finally is able to find her voice. "Yes, Rachel, we came home early. Your Father was worried about you. Now I know why."

Mr. Kaufman came through the door carrying their luggage. As he caught a glimpse of the two of us, he dropped the baggage to the floor. The look on his face turns from total surprise to livid anger. "What's this? What are you doing in my house, Jacob Cahn?"

He then turns his anger toward Rachel before I have a chance to answer.

"Daughter, what is Jacob doing here? It's only too plain to me. Knowing we would be gone, you have invited him here."

Rachel tries to explain, "Papa, I thought it would be a good opportunity for Jacob and I to talk about our relationship."

Papa didn't skip a beat. "The dress you are wearing tells a lot about the intentions you have planned. Do you remember? I told both of you there would be no contact for at least six months? Here it's been just a few days and I find you together again, alone. Is that music I hear? What has been going on here this evening?"

I seized the moment to try to rescue the situation. "Sir, Rachel invited me to have dinner with her and we have done exactly that, nothing more. When you arrived, we were in the process of discussing our separate destinies. I just finished explaining to Rachel I cannot marry her. In fact, I will never marry. I know this sounds incredible, but this has been revealed to me in a vision. It is the absolute truth. I have been called to live a celibate life. There is another thing I need to tell you. I spoke to Dr. Mike recently and I now fully understand your strong feelings about my chosen vocation. Let me assure you. I am not, nor will I ever be, a proponent of reducing the population of this planet through eugenics."

"Mr. Cahn, it was very poor judgment on your part to come to our home when Rachel was alone. You should have left just as soon as you found out Hanna and I were gone, unless of course you already knew she was alone."

"Sir, I did not know she was alone in the house. I assumed I would find the two of you here also. It didn't enter my mind you were in New York."

"Is he telling the truth Rachel?"

"Yes, Papa, it was all my doing. Jake did not know you both were in New York."

My explanation and Rachel's confirmation did not change Papa's attitude a whole lot. At least he did not vent his anger against me as he had done with Dr. Mike. He began to cool down.

"Well, Jacob, since you both have come to a rational understanding about your separate destinies, I don't see the need for us to discuss this matter any further. I insist you leave our home immediately."

"Yes, of course. Let me just retrieve the gift Rachel has given me from the other room. I wish you and your family the very best."

I walked back to the dining room table and picked up the box containing our ketuba. Then I walked to the entryway closet and put on my coat. Rachel was crying on Hanna's shoulder as I fumbled with the zipper on my coat.

Mr. Kaufman angrily demanded, "What kind of gift did my daughter give to you Jacob?"

I refused to answer him. I simply said goodbye, opened the door, and walked out of the house on North Peachtree Battle Avenue for the last time.

It is late. There is little chance I can get a cab anytime soon and waiting here on the Kaufman driveway does not appeal to me. I will make my way back to the hotel on foot. It will take me a little over two hours to jog the fifteen or so miles back to the hotel. The jog gave me time to settle my nerves. Perhaps the run will help me to get some sleep. As I reviewed this last conflict with Mr. Kaufman, I actually felt pleased with my ability to weather the storm without striking back. My small rebellion about refusing to show him the ketuba was the only time I failed to control my emotions. I arrived back at the hotel at half-past ten, took the elevator to my room, and went straight to bed. The long run did not help me get to sleep. My mind resorted to reviewing what the coming day would bring. Delta flight 242 is scheduled to depart at 8:20 tomorrow morning. Terry and I will take the MARTA light rail system from the hotel directly to the airport, a twenty-minute ride. Just to be safe, I put in a wake-up call at the desk for five o'clock. I will call Terry in the morning. I finally drifted off into never-never land.

The hotel woke me at five on Sunday. I called Terry and arranged to meet for breakfast at six. Then I dressed and packed up my gear. We both checked out of the hotel after breakfast and walked to the MARTA Station. We arrived at the airport an hour before the flight was scheduled to leave. I put Terry's tux in my suiter and promised to return them to my Jewish friend Yudi the first thing Monday morning.

As we waited in the airport for the plane to load, Terry cautiously asked, "So how did the visit with Rachel go last night, Jake?"

"Pretty much as expected Terry. There were a few surprises."

"How so?"

"Well, Rachel's parents were not at home. They were in New York meeting with Jeremiah Levitt's family. Rachel was looking forward to a more intimate evening than I was expecting or prepared for."

"So, how did the evening go?"

"It began with a great dinner. She prepared steak, and all the trimmings. Then she unrolled part two of her evening's agenda. Although Orthodox girls are not permitted to dance with the opposite sex, she put on a CD and pleaded for me to dance with her."

"Oh boy, what came next?"

"Not what you are thinking, Terry. About half past nine her folks unexpectedly showed up on the scene."

"Ouch! It sounds like the formula for an explosive situation. What happened?"

"Rachel and I were completely surprised when they arrived home early. Her dad lost no time in putting me on the hot seat. I tried to explain the situation to him. I said I thought he and Hanna were going to be at home. I said I didn't know about any trip to New York."

"Did her dad believe you?"

"Yes, I think he did, but it was a disaster. He did keep his temper down to a slow boil and we didn't come to blows. He simply demanded I leave his home immediately. So, I grabbed my coat, Rachel's gift, and left. I didn't want to wait in front of their house for a cab so I jogged all the way back to the hotel. It took me a little over an hour to get back."

"So, what is your status with Rachel now?"

"It's over Terry. There is no chance for any resolution."

"I'm sorry, Jake."

"Thanks. I really love Rachel. Given time I believe my pain will pass. How Rachel will deal with it is an open question."

My story is cut short by the announcement the plane is now ready for boarding. Terry took the window seat and was soon sound asleep. We arrived in Baltimore at 9:50. We picked up our luggage and took the shuttle to Lot B where his truck was waiting for us. It was half-past ten when Terry dropped me off at the Bradford. I took the elevator to the fourth floor, fumbled for my apartment key, and finally got the door open. It was good to be back in Baltimore. I took the ketuba out of the box, unrolled it, and carefully hung it on the wall. Rachel poured her heart into it. It is a wonderful keepsake. I resolved to pray for Rachel every time my eyes lingered on our engagement ketuba. I plopped down into my easy chair. What a strange visit to Atlanta this has been. I assisted in joining two wonderful people in marriage and destroyed the dreams of two others. Although my separation from Rachel is painful, at least this conflict in my life has been resolved. It is behind me now. I am free now to pursue my destiny as a human geneticist and servant of the most-high God. It all starts tomorrow morning."

13

Working for a Living

"Commit thy works unto the Lord,
and thy thoughts shall be established.
(Psalm 16:3)

Sacrificial Servant

Today is Monday morning, the day I go to work. After a shower, I put on my sweats and took the brochure I received from Robbie along with me to the food court located in the mall. I read it over my breakfast of scrambled eggs, a sweet roll, and a large cup of black coffee. The brochure contains a thorough description of the field of human genetics. There is a short summary at the top of the brochure.

It reads, "Medical disorders caused wholly or in part by a fault or faults in the inherited genetic material within a person's cells, that is, in the genes formed from the substance deoxyribonucleic acid (DNA), which makes up the chromosomes in a person's cells is known as a genetic disorder. The reason abnormal genetic material can lead to disorders or disease is that genes control the manufacture in cells of enzymes and other proteins playing roles of varying importance in the functioning of cells, and in the body as a whole."

This brief introductory description gives one some understanding of the intensity of study required in this field by a student of genetics.

The Johns Hopkins School of Medicine is unique in the Nation, bringing together experts from biochemistry, genetics, medicine and biostatistics to increase our understanding of how cells establish and maintain control of genes, also revealing what happens when cells lose their control. Work done at the various branches of medicine has implicated procedures for treating many medical conditions, literally the A to Z of medicine.

These are complex issues. There is so much I don't know about this field. After breakfast, I took a walk along 33rd Street, to the campus book store. I wanted to take a look at the virology section. I found the textbook used by Dr. Adelman. "Fundamentals of Molecular Virology" authored by Nicholas Acheson. This is a new edition of the same text I used at Emory. The paperback price is $84.95. I paid ten dollars less for the first edition I used in Atlanta. The work book required for the lab experiments is another $35.95. I can't help feeling sorry for the poor undergrad students trying to finance their education. I took a few minutes to browse through the new text. Dr. Acheson has included a lot of new material. He has included a review of recent contributions by prominent virologists, as well as a thorough discussion of the latest concepts in virology. I realize my work as a lab technician and general proctor is going to be a demanding task. I went ahead and purchased both the new edition and the lab workbook. My curiosity satisfied, I took the book and lab manual back to my apartment. Then I took a run across the campus.

I arrived back in my room at eleven o'clock. After another shower, I took time to catch the news on Fox. Chaos in the Middle East is continuing to make the headlines. This time another Coptic church has been burned to the ground by the Muslim Brotherhood. The situation in Syria is even more dangerous. Russian jets are coming into close contact with our jets and Turkey has taken control of a portion of Syria to obstruct the invading Kurds. I believe Israel and the US may soon invade Syria. A news report recently claims we are massing all kinds of equipment and men in Syria's neighbor Lebanon.

I like to keep tabs on what is going on in the world. Right now, I need to turn off the television and crack the cover of the new virology text. I spent a couple of hours reading the text before time to meet with Dr. Adelman. As I considered the task before me I felt the need to tell Ha Shem about my situation. In the few minutes remaining, I said a quick prayer to the Lord in heaven.

At a quarter of three I left for the Undergrad Teaching Lab located just west of the Freshman Quad. I arrived a few minutes before three. I did not want to be interrupted by any phone calls, so I turned off my cell. There is a middle-aged man waiting at the door. I assume it is Dr. Adelman.

"Good afternoon. Are you Dr. Adelman?"

"Yes, and you must be Jacob Cahn. I appreciate you helping us out this summer, Jacob."

"I appreciate the opportunity. I am looking forward to it."

"Well, let's get started. First, I want to show you around the facilities. This building, like several others on the Homewood Campus, is designed as a multipurpose facility. There are traditional lecture halls in addition to what we call smart classrooms. All of the course lectures given in the lecture halls are routinely recorded and made available to the students in these smart classrooms. In this way students are free to schedule their own time to review the lectures. They can replay them as many times as they need to. There are advantages to the professors and lab assistants as well. Since they are recorded, the lectures and lab reviews need to be given only once. The lab exercises will be held in our lab facilities here on the fourth floor of this building. Your lab reviews will be recorded and added to the database. You will still be required to hold open door sessions where individual students can meet with you face to face. Your schedule is going to be a busy one. Do you have any questions for me?"

"Will my lecture review session follow directly after your lecture?"

"Yes. I will leave the lecture hall and you can carry on with your review immediately afterward. There will be a student from the Audio-visual Department (AVD) to handle the recording, get you miked up, and so forth. Also, he or she will be in contact with the AVD to make sure the system is operating correctly. You need to instruct the students not to ask their questions until after you complete your lecture review. Questions from the students will be recorded and edited by the AVD people for possible inclusion with your taped review. The students will approach the standing microphone at the front of the room to ask their questions. Now, as far as the lab assignments are concerned, you will take them through the workbook according to the lecture schedule. Expect to spend your entire day in the labs. The labs will meet on Wednesday, Friday, and Saturday mornings. During these abbreviated summer sessions, we need to open the labs on Saturdays. Miss Bergman has volunteered to walk you through at least your first week of classes. I think you should take her up on her offer. Do you have a way of contacting her?"

"Yes, she gave me her phone number. I believe she gave it to me twice."

"Do I detect a bit of concern there Jacob?"

"Yes, you do Doctor. I already have established some parameters in our relationship. I do appreciate her offer and I'm sure it will help me get my feet on the ground."

"These single ladies do have more than one agenda constantly in mind, Jacob. I will leave it to you to keep your friendship with Ramona centered on the business at hand. By the way, if you do find some free time, feel free to explore what the other summer classes are doing in our Biology Department or other departments on campus as well. If you need assistance, I'm certainly willing to spend some time with you. There will be three exams. They will be on the second, fourth, and sixth Fridays. The final is a comprehensive. I believe I indicated to you the second course will begin right after the first session ends. Do you have any other questions?"

"Nothing comes to mind right now, Dr. Adelman. I brought the text and workbook along with me today."

"Oh, I guess I failed to tell you. You didn't have to purchase the text and workbook. The Institute furnishes all necessary materials for you. Your copies are in your mailbox in the Biology Department. You can take these books back to the book store for a full refund. If you have any trouble, let me know."

"I really don't need to return them. As an incentive, I may just donate them to a struggling student."

"A little bit of *corban* in the right place is a good way to begin, Jacob. Thank you for helping me with these summer classes. I look forward to getting to know you better."

"I feel the same. Thanks for your help, Dr. Adelman."

"I'm going to have to leave you now to your own devices. I'll expect to see you again in Lecture Hall 120 tomorrow at a few minutes before 9 o'clock."

"Very good, Doctor. I'll be there."

I took time after the Doctor left to take a look at a smart classroom. When I entered one, there were three medical students listening to a lecture on bedside manner. It is a quarter to six before I arrive back in the Bradford. I turned my cell phone back on.

Ramona had left me a message. I called her back. "Hello, Ms. Bergman. This is Jacob."

"Hello Jacob. You finally turned your cell phone back on."

"Yes, it and I are both charged up and ready to go. I just returned from my meeting with Dr. Adelman."

"How did it go?"

"Well. He showed me around the facility and we had a good talk."

"Did he mention my conversation with him?"

"Yes, he did."

"Am I on board or not, Jacob?"

"He encouraged me to accept your offer, but do you really want to do this, Ramona? It is going to be a busy schedule."

"I wouldn't offer if I wasn't willing to spend the time. You need my help. Summer sessions are grueling. You are taking your time getting to the bottom line, Jacob."

"Yes, you are on board. I appreciate your offer, Ramona. I do need to remind you there will be certain limits on our personal relationship."

"Platonic limits no doubt. We can discuss these at a future time. I will see you tomorrow, Jacob."

"Okay, Ramona, I'll see you at the lecture hall at a quarter to nine tomorrow, Room 120."

The growling in my stomach reminds me I haven't eaten since breakfast. It is about time to supply my apartment with food. I can't continue to pay restaurant prices every time I get hungry. The hotel clerk suggested I walk south on St. Paul Street. He said I would find two markets, the University Store and Eddies Grocery. I found Eddie's Grocery at 31st and St. Paul Street. I broke my cardinal rule. Never go to the grocery store when you are hungry. I filled a shopping cart with groceries and other unnecessary things I felt were necessities. When I arrived at the checker I realized there is no way I am going to be able to carry everything back to my apartment.

The checker took a look at my cart and quipped, "You are new to the campus aren't you."

"Yes, I just arrived last week."

"No doubt you are wondering how you are going to carry all this food back to your apartment?"

"Right! Do you have a solution for me?"

"Aisle five, top shelf."

"Can I leave the groceries here for a couple of minutes?"

"Sure, I'll wait for you."

I hurried back to aisle five. On the top shelf, I found several wheeled shopping carts. I managed to pulled one down and returned to the checkout.

I asked the checker, "Say, how do I work the logistics on this? I mean, where do I park the cart when I'm shopping?"

"You don't park it. You wheel it around the store and put what you need in your cart."

"That makes a lot of sense. I don't know what I was thinking."

"Yeah. Well don't feel bad. I have physics professors ask me the same question."

"I'm not physics, I'm human genetics. You know, microbiology."

"No, I don't know."

"Are you Eddie?"

"No, Eddie the owner now enjoys life in the Bahamas. I'm Charlie. What do they call you?"

"Lots of things, but my name is Jacob Cahn."

"Nice to meet you Jake. Our Jewish items are on aisle seven. We have some parve pizza on sale this week."

"Thanks, I'll pick one up."

I picked up a couple of pizzas and added them to the cart. Pizza will be my first home-cooked meal, or should I say home microwaved meal. I like this guy Charlie. He helped me transfer my groceries into the new cart and I wheeled it back to my apartment. I wheeled my new cart into the elevator and pressed the button for the fourth floor. Once in my apartment, I unload the groceries, stuck the pizza in the microwave, set the clock for six minutes, sat down at the table, and presto, dinner is served.

The seven o'clock news is just coming on. The lead story is about a large earthquake on the ring of fire. This one is located right on a major fault in the middle of Japan's southern Island of Kyushu.

This island is sitting next to one of the deepest subduction zones in the world, the Japan Trench. The quake was a monster, measuring 8.2 on seismometers in Japan, Europe, and the US. According to first reports, this long fault moved several feet apart. Suddenly, the alarm on my microwave sounds off. I almost jumped out of my chair. "No Jake, this is not an earthquake. It is your Hawaiian pizza ready to come out of the microwave."

I sat down at the table and ate pizza while listening to the news from the quake scene. The island of Kyushu has many industrial and commercial buildings. Most of the plant facilities were closed at the time of the quake, meaning the loss of life is not expected to be great. The report said serious aftershocks are continuing, hampering rescue operations. I wondered if these frequent earthquakes on the ring of fire are trying to tell us something? Are these earthquakes, tsunamis, volcanic eruptions, and monster hurricanes a wake-up call?

Scientists are now beginning to learn how to predict the occurrence of these catastrophes. Their relationships to our Sun's coronal holes, sun spots, filaments, and electric currents are enabling scientists to predict the occurrence, location and approximate strength of quakes with an accuracy now approaching eighty-five percent. The alignment of the planets to one another and to the sun also is an important factor. Scientists are now discovering the electromagnetic connections between the sun and the planets are even more critical than the effect of gravity. I wondered, are these geologic events harbingers of even greater disasters to come?

I needed to get some sleep before tomorrow morning's class. I finished the pizza by nine and turned off the TV.

As I was emptying my pockets into the nightstand drawer next to my bed, my eyes fell on my new Gideon Bible. I picked it up and began flipping through the pages. I turned to the back of the Bible where there is a section entitled, "How to Become a Christian." I remember Dr. Mike suggesting I read the entire Bible to gain direction from the Lord. He said Yeshua will confirm the truth to me. I have only a limited time to spend studying these Bibles I purchased, but it seems like Ha Shem is urging me to set aside some of my valuable time this summer to study this particular Bible. I'm wondering how different these Christian Bible translations are from the Jewish Tenach? They very likely are biased. Yes, I have many questions needing answers. I need someone to help me understand all of this. Perhaps Ha Shem will provide someone to help me.

I did not sleep well at all. I woke up at five on Tuesday morning. I turned on the TV, hoping to learn more about the quake in Japan. It was too early for the news, but there was a banner streaming at the bottom of the screen on Fox. It said over one thousand people died on the Island. I turned off the TV and quickly dressed in my sweats. I will work off some of this anxiety by taking a morning run. I jogged for over an hour. I got back to my cave a little before seven. After showering and getting changed, I sat down to a breakfast of corn flakes, a banana, and a cup of strong coffee. Then I pulled out the syllabus and the class schedule again.

Dr. Adelman was right when he said I would be busy. No time to study the Bible this morning. Each of the summer sessions is six weeks in length. The second session begins right after the first one is completed. The schedule for the classes is intense. Dr. Adelman's lectures are scheduled for Tuesday and Thursday mornings from 9:00 to 10:00. Afterward I am scheduled to review each lecture with the students from 10:00 until 11:30. After lunch, I meet from 1:00 to 2:00 with them in the tutoring room next to the smart classroom. After the tutoring class, individual students can schedule appointments with me for later in the afternoon. The labs meet on Wednesday, Friday, and Saturday mornings from 8:00 until noon. Each lab session meets for two hours. The labs are held in the Hands-on Learning Center on the third floor. My weekend begins at noon on Saturday and is over Sunday evening. I'll do this summer thing because I really need the gelt.

I arrived at the lecture hall on Tuesday at exactly 8:45. Ramona arrived soon after I did. The student manning the audio-visual equipment arrived minutes later. He turned the computer on the lectern on. The wide monitors each side of the room are displaying a series of screens showing the course title, instructor Dr. Adelman's name, the class schedules, and the synopsis for the course. Ramona and I sat in the back of the room, watching the students troop in. I estimate the room holds upwards of sixty students. Dr. Adelman arrives exactly at nine o'clock. He places a disc into the computer, and then motions for Ramona and I to come forward.

"Good morning, Jacob. Can I call you Jake?"

"Yes, of course."

     THE DREAM GENE

"I see your helper has arrived. Good morning, Miss Bergman."

"Good morning, Dr. Adelman."

"Jake, I am going to use the Microsoft Office Powerpoint program when lecturing. I will give you a copy of my presentation before each lecture session. Here is a copy of today's lecture. You can use the powerpoint copies to help you during your review. You don't have to use them. It is up to you."

"Sounds good to me. I know I will find them useful. Thank you, Doctor."

"The syllabus shows where we are going in the text. We will be moving quickly because it is summer session. Do you have any questions?"

"No, Sir. I have read the syllabus and the class schedule. I believe I'm ready for action."

"Okay. Let's begin."

Ramona and I returned to our seats in the back of the room and I took out my laptop and inserted the disk he gave me. His lecture followed the powerpoint presentation exactly, The Professor finished speaking fifteen minutes early. It gave Ramona and me enough time to take a break before beginning the review with the students. We waited in the room for the students to return. About half of them apparently preferred to skip my review, choosing to view the tape instead. I began my review exactly at ten o'clock.

"Well, it seems many of your classmates have chosen to skip my review this morning. So be it. Each of Doctor Adelman's lectures will be available to view on the AV system. My name is Jacob Cahn. I am your lab instructor and right-hand man. I hail originally from Omaha, Nebraska. I'm new to JH, having taken this course last year at Emory University in Atlanta. I must give you a warning. I often may unknowingly inject some Yiddish humor when speaking. I hope you will not be offended by my mixture of Yiddish and southern accent. Do you want to know why? Oui vey, I believe Jewish humor is the best humor in the world. If this presents a problem for you, I suggest you download a Yiddish dictionary from the internet. Yes, I'm Jewish. Are you aware Jews account for only one percent of the US population, but well over fifty percent of medical doctors are Jewish? Enough said. Let's get down to the business of learning about these little critters called viruses."

I ran the disc of Professor Adelman's morning's lecture, pausing it from time to time to emphasize a particular piece of information. The introductory disc was a nice summary of the first chapter in the textbook. After my review of the lecture, I opened it up for questions.

"I believe we have covered this morning's lecture by Dr. Adelman rather thoroughly, as well as the content of Chapter One in the textbook. I am now going to open the mike for Q & A. Raise your hand if you have a question. I will then ask you to come forward to the microphone. Please state your name and your question as succinctly as possible. The questions and my response will be recorded and will be available from AV. If I do not know the answer off the top of my kop, I will include it during our next review meeting. Please keep your questions related to this virology course. I do not give advice on auto repair, computer malfunctions, Pimlico race horses, or successful dating strategies. I do not edit these recordings, so if your question doesn't make the tape, you will have to take your problem up with the AV Department. Okay, who has the first question?"

I recognized a male student in the second row. He bounded up to the microphone. "Mr. Cahn, my name is Jonathan Logan. I have a philosophical question about this course, if you don't mind."

"No, I don't mind. Go ahead and ask your philosophical question Mr. Logan."

"Sir, do you believe human viruses are the end product of the evolutionary development of elementary life forms, or are they a uniquely created species of life?"

"Well, you have asked an interesting question, Mr. Logan. I will answer your question if you will give me an answer to my question first. Is this acceptable with you?"

"Yes, I agree."

"Very well. You are obviously quite a well-educated person, I assume you have some understanding of logic. Let's establish some rules first. Do you agree with the statement truth cannot be falsified?"

"Yes, truth is an absolute."

"Very good. My question to you then is this; Do you hold Neo-Darwinian Macroevolution to be a theory or a proven fact, in essence a truth?"

"Well, I will need time to consider your question."

"Please, take all the time you need, Mr. Logan. I believe you are considering three possible positions. Either you do not know how to answer my question, you would rather not respond because you are not fully informed about the topic, or you know there is insufficient evidence to prove beyond doubt the theory is completely truthful. In the latter case, you may not want to admit the theory is unproven, and just a theory. Since you are very intelligent, I am guessing the third choice is probably your position, and you do not want to admit this as it flies in the face of political correctness within the current biological science community. Have I correctly summarized the situation you are facing, Mr. Logan?"

"Yes, Sir. You have summarized the question quite well."

"And am I right in assuming you are in the camp of the third of these three?"

"Sir. I must refuse to state which of the three I am camped in at this time. I withdraw my question."

"Fine, I respect your reticence to reveal the location of your camp site. Very courageous Mr. Logan. You get a gold star today. You may sit down. Are there any comments concerning Mr. Logan's response? No? Therefore, I will not have to state my position concerning the theory of Macro-Evolution. I will say this. Atheism is a non-prophet, I say a non-prophet enterprise. Let us move on."

Only a few of the students seemed to catch my humor.

"Apparently, I have no other young lions in attendance today. Who has the next question?"

I can see from the student's faces I have instilled a sense of fear in them. The other questions presented by the students were much more mundane. They asked clarification of some of the terminology used by Professor Adelman and content from the first chapter of the textbook. I finished my first review session five minutes ahead of schedule. I reminded the students of the scheduled individual tutoring opportunity. I asked those interested in scheduling an appointment to meet me at one o'clock in the smart classroom down the hall. Then I excused the class.

Ramona walked up to the podium and whispers she has some comments for me. I suggested we meet in private in the smart classroom. When we arrived in the smart classroom, Ramona didn't waste time getting to the point.

"Jacob, I can't believe you started the review the way you did."

"Why, what was wrong with it? Are you referring to my southern Yiddish gag, or my description of atheism?"

"Neither one. I'm referring to your pro-Jewish comments. You don't know anything about who is sitting in the room, let alone those who will be viewing it on the tape. For your sake, I just hope no students decide to post a complaint on Twitter or some other place about your obvious Jewish bias."

"You are quite right, Ramona. No one appreciates humor these days, and not many Gentiles know anything about Jewish culture. Most think we Jews killed Jesus and neglect to review the Roman part of the story, but I will apologize to the class on Thursday.

"Our Jewish rulers did instigate the death of Jesus, Jake. I think you should just keep quiet and hope nothing comes of it."

"Where do you think the question from Mr. Logan came from?"

"I would guess from Dr. Altschul. I think he was trying to set you up."

"That's my conclusion as well. I will be more careful what I say in the future. It's nice to have you along to keep me out of trouble Ramona."

"Thank you. So, where are you taking me to lunch?"

"Who eats lunch? Yes, I'll buy lunch, but I'm not sure where to go. Do you want to walk across campus to Charles Street?"

"I don't want to walk across the campus."

"Why not?"

"Haven't you noticed? I have a bad leg."

"No, I have not. What is wrong with your leg?"

"My left leg is an inch shorter than my right one. I was in an auto accident when in high school. My whole body was a mess. My leg was broken in five places, my pelvis was fractured, and they had to fuse two of my vertebrae."

"Gee, I'm sorry Ramona."

"No need to feel sorry for me. The result of the accident put my mind on the right track. That's when I became serious about my education."

"Is there someplace close by we can have lunch?"

"Yes, there's a buffet in Building A's food court. It's right next door."

"Sounds good to me."

Ramona led the way to the food court. I paid for lunch and we sat down at one of the tables. She ordered Hawaiian pizza. I had a hamburger. I wanted more feedback.

"Other than my misplaced humor, and my brash delivery, how do you think I did, Ramona?"

"Just terrible! … No, I'm kidding. You did great, Jake. The lecture review went well and I am very impressed with your answers to their questions, especially the question from Mr. Logan. I just hope you will think again before you get off another one of those Jewish jokes of yours."

"Ramona, you made my day."

"I won't be able to attend the smart session this afternoon. I have a dental appointment downtown."

"No problem. I don't think I will have many students show up since this was the first lecture."

"You are probably right. What are you doing this weekend Jacob?"

"I may visit my friend Terry. Did I tell you his father owns a thoroughbred farm in Timonium?"

"I seem to remember hearing something like that when I called you on the phone. Do they have riding horses as well as thoroughbred race horses on the farm?"

"Yes of course, but I don't ride horses. They are flight animals you know."

"I just love horses. Perhaps you might ask if I can visit the farm some time?"

"I'm sure the Johnsons would love to have you visit their farm. It's really a beautiful spread."

"Please, Jake. Don't forget to ask them for me. I would love to go."

"I will ask Terry this very evening. Do you have anything planned for Sunday?"

"Nothing as of now. I never know just how many calls I'm going to get from male grad students wanting to go out with me."

"I will give you a call after I check with him. Perhaps you can squeeze a visit into your busy schedule."

"Great! I have to leave. You can finish my pizza if you want to."

I watched her leave and wondered how she would get along with Terry. Just perhaps I have solved my Ramona problem. I finished her pizza and then took a walk to the Rec Center.

The Rec Center is very well equipped with exercise equipment and has a nice eighth of a mile indoor track. There are showers and lockers as well. I just may start my days here every morning, if possible. I returned to the smart classroom a little before one o'clock. My recent antagonist, Jonathan Logan, is there waiting for me.

"Hello again, Jonathan."

"Hello, Mr. Cahn."

"Are we going to pursue your question about the theory again this afternoon?"

"No Sir. I do not intend to bring the subject up again today. Perhaps another time."

"Yes of course. We can have an honest conversation with one another about Darwin's theory as well as the positions of some of his critics. I have a small task for you, Jonathan. As the students enter, I want you to give each of them one of these handouts and have them sign in on this clip board. Can you help me?"

"Sure."

"Good."

It is five minutes past one. Only fifteen students have shown up. I closed and locked the door. The students will soon learn to show up on time for my tutoring sessions.

"Ladies and gentlemen, thank you for showing up on time. I appreciate your timeliness. The early birds do get the worms. We have an hour from one to two on Tuesday and Thursday for Q & A. I am willing to spend additional time as required, but I am not tolerant of people who make a habit of being late. Let us begin. These meetings are completely optional. You can decide whether to attend or not to attend. In the future, the clipboard will be available after my review of the morning's lecture. I want these afternoon sessions to be very informal. It is my wish to help you succeed in this course. No, more than succeed. I want you to excel here. Unlike the morning review, we will not be recording what goes on here. Ladies, you can relax. This should relieve some pressure for those of you who are uncomfortable with seeing their face on the AV system in the front of the whole campus. You should have received a schedule from Jonathan when you came in. Let's first review it. On the sign-up sheet is a box next to your name. Check it if you want to meet with me, one-on-one. These twelve fifteen-minute slots are available on Tuesdays and Thursdays beginning at 2:30 pm. If we need more time, I will schedule a second meeting with you. I will follow the order on the sign-in sheet, so if you are planning to take advantage of the individual tutoring, I recommend you get your name on the list as soon as possible. Please enter the time you want to meet from the remaining open slots. You are free to trade times with other students if necessary. Does everyone understand the plan? … Good. Now who has a question?"

The students seemed to like my informal approach to the "Q & A" meetings. They were curious about how the labs would be handled. I told them we would follow the lab workbook to the letter. Microscopes are provided, but I reminded them they also will need to purchase the lab activities package containing the prepared slides. No one signed up for a personal meeting with me, so the rest of the afternoon was mine. I got back to my apartment at half past two and sat down to study the lab workbook.

My life now is taking on a regularity I have not experienced since I was in pre-med at Emory. The labs on Wednesday are going well, and the undergrads are serious about their work. Professor Adelman gave them another thorough lecture the next Thursday. Before I realize it, my first week is over. I walked out of the lab at 1:30 Saturday afternoon.

One thing I do miss is my Saturday tennis games. I walked over to the Rec Center to see if I could pick up a game with someone. When I got to the tennis courts I saw there was a game being played on the lacrosse field. Today the JH team is playing against the University of Maryland. I quickly understood why Joe Harmon advises tennis players not go out for the sport. The players are hacking away on one another with those sticks. It didn't matter if a body part is between the ball and their sticks.

I recognized Joe Harmon sitting on the bench, so I sat down on the bleachers to watch the game. During halftime, I walked down to say hello.

"Joe, I didn't know you are on the school team. Didn't you warn me about the hazards of playing lacrosse?"

"Yes, true, but I actually love the game. I still have one more year of eligibility. I warm the bench more than I play. The coach hasn't seen fit to put me in the game yet today. How is your summer tutoring going?"

"Great. I am enjoying it. "I wanted to check out the tennis courts at the Rec Center. Perhaps I can pick up a tennis game with someone. I need to get back into shape. Is this a regular season game? I don't see anyone in the stands."

"No, this is just a practice game with our arch rival. Despite what I told you about the mayhem, I actually love this game, at least when I get to play."

"What position do you play?"

"Usually left wing."

"I don't see anyone working the scoreboard. What's the score?"

"So far, it's zip. But I think we are wearing them down. We should score in the second half."

"So, have you heard anything about your application?"

"Nothing yet. Have you?"

"Not a word."

Just then Joe's number came up. The coach stood up and pointed at Joe.

"Hey Harmon, you're in on left wing."

I watched the balance of regulation time for the game. Neither team scored. I didn't stay to watch the overtime. It is five o'clock when I arrived back at the apartment. I heated up another pizza and then sat down to read the commentary by Pastor J. Vernon McGee.

14

The Healing

"Jesus saith unto him, "Rise, take up thy bed and walk. Behold, thou art made whole: sin no more, lest a worse thing come unto thee."
(John 5:8–14)

A Surprising Miracle

In the middle of reading the McGee commentaries, I suddenly remembered Ramona's request to visit the Johnson farm. I gave a quick call to Terry to see if we might be able to visit the farm on Sunday. I left a message for Terry to call me back. My phone rang an hour or so later. It was Terry.

"Hello, Terry."

"Hi, Jake. I got your call. What's up?"

"Terry, do you remember me telling you about my student guide Ramona?"

"Yeah, I think I warned you about her."

"Well, it seems she likes horses. I told her about your farm and she asked me if we can pay you a visit?"

"I don't see why not."

"I will be the cab service. I wonder, can we do it tomorrow? Does this work for you?"

"You do remember we go to church on Sundays."

"Yes, I know. Ramona said she doesn't have a problem with your Sunday schedule."

"We don't get back until around one-thirty. You could decide to come early and have breakfast with us before church. We leave at 9:15."

"I will ask her. I seem to remember you sit down for breakfast about eight o'clock?"

"Yes, but if you are coming, you will have to let Mom know ahead of time."

"Okay. I will get back to you on this."

"How did your first week in the trenches go?"

"It went well. I enjoyed dishing out the work instead of being on the receiving end."

"Yes, putting them through the ringer like they did to us. Jake, I'll wait to hear from you about Sunday."

"I will let you know right back, Terry."

I hung up the phone and returned to the commentaries. Then I decided to take a final look at the lab session materials. I finished preparing for the first lab session by a quarter to five. Ramona no doubt is back at home from the dentist by now. I gave her a ring.

"Hello, Ramona. This is Jacob. I thought I would just give you a ring."

"How wonderful. So, you are finally proposing?"

"No, this is not the kind of ring you are looking for."

"How did your open session go?"

"It went well. Who do you think arrived in the room ahead of me?"

"I would guess it was Mr. Jonathan Logan."

"Bingo! How did your visit to the dentist go?"

"It was a white-knuckle time for me, but I survived."

"The other day you mentioned you would like to visit the Johnson farm. I talked to Terry and he said you are more than welcome to visit their farm. I can take you tomorrow if you are free?"

"Are we talking tomorrow, Sunday?"

"Yes. Here's the deal. The family goes to church and gets back about half-past one. Here are your options. We can arrive about one-thirty and have lunch with them, or we could arrive a bit later after lunch. The third option is the one I favor. Terry said we are welcome

 THE DREAM GENE

to come early and have breakfast with them."

"Would we have to go to church?"

"No. They won't pressure us to go to church. You can go if you want to or you can stay with me on the farm."

"I've never gone to a church before."

"While they are gone, I can take you on a tour of the farm. You can size up the horses in the pastures and walk through the barns. We might even give them a hand with the morning chores."

"H'm. So, Terry is inviting us to have breakfast with his family?"

"Yes, he is. It will be kosher. Mrs. Johnson knows not to feed you ham or bacon."

"The breakfast is tempting. When would we have to leave to arrive for breakfast?"

"The family religiously sits down for breakfast at eight o'clock on Sundays."

"Religiously? I take it that's a pun."

"If you decide to do the breakfast thing, we need to arrive a few minutes before eight. I can pick you up by seven-fifteen or so. Do you want to think about it and get back to me?"

"No, I don't need to think about it. I'll take Terry up on his offer of a country breakfast."

"A country breakfast sounds good to me as well."

"What should I wear?"

"Do you know how to ride?"

"Yes, of course."

"Will your leg give you any problem?"

"I don't really know. I haven't ridden a horse since my accident. I guess we will just have to see how well I do."

"Wear some jeans. To fit the occasion, I will be wearing jeans and my blue plaid shirt with the snaps instead of buttons."

"Are you going to ride?"

"No way José. Not if I can help it."

"Will Terry's girlfriend also be there?"

"I've never heard Terry speak about any girlfriend."

"Well, this is absolutely cool, Jake. I am looking forward to it. You've got class. You're a real hamisch."

"Yeah, I know, I'm Mr. Wonderful. I'll be at your place bright and early tomorrow morning."

"Okay Jake, I'll look for you about seven."

I hung up the phone. So, Ramona is asking me about Terry's girlfriend. That's interesting. Terry actually may solve my Ramona problem. This trip will be interesting.

I'm now in survival mode. I took an inventory of my cupboard and came up with the makings for a chili size. I put together a hamburger patty from the fridge, a can of chili, and an onion. I started my coffee brewing, diced up the onion, heated up the can of chili, and topped the whole thing off with hot sauce. Dinner ended with a piece of the pie Eldora sent home with me. Yes, I can cook a decent meal. I called Terry back and told him to expect to see us at eight in the morning. Well fed, I settled back in my easy chair and watched an old movie on the Turner channel. I fell asleep in my chair for an hour or so. Then I crawled into bed. I would have to leave early in the morning to pick up Ramona.

I woke up Sunday morning to the sound of thunder. Will Ramona still want to visit the farm? It is too early to call her. I am hungry. I need a little something to tie me over before the breakfast at the farm. I brewed a cup of coffee and scrambled a couple of eggs. At six I made the call.

"Good morning Ramona. This is Jacob."

"Good morning, Jake."

"Ramona, have you looked outside yet?"

"No, why?"

"It's raining. I'm wondering, do you still want to visit the Johnsons this morning?"

"I don't know. What is the weather man saying?"

"I haven't a clue."

"Just a minute. Let me check on my cell phone."

I waited while she checked on the weather.

"Jake, they say this squall line will pass through quickly. It's going to be sunny for most of the day. I'm willing to go if you are."

"Okay, Ramona. I'll see you in a half hour or so."

I hung up the phone and took a quick shower. I put the plaid shirt on, a pair of decent jeans, and my tennis shoes. Then I put on my raincoat and headed for the parking lot.

Ramona is outside waiting when I arrived at her apartment. She is really decked out for the occasion. She is wearing a fringed leather jacket, fancy jeans, cowboy boots, and a western hat.

"So, you are a country girl. I never would have guessed it."

"Just because I don't have a southern drawl, I suppose you thought I don't know the front from the back end of a horse. I'll have you know I used to ride horses every day when I visited my uncle's farm in Virginia. He raised tobacco and had a few horses. They weren't thoroughbreds. They were plain ordinary cow ponies."

"Cow ponies? What's a cow pony, a cross between a cow and a small horse?"

"C'mon Jake, do I need to explain what a cow pony is to you?"

"No, you don't. I'm not strictly a city kid either. We had horses in Omaha, even thoroughbreds. I used to frequent the race track in Omaha occasionally."

"How often?"

"Oh, maybe once every couple of weeks in the summer."

"Okay. So, let's hit the trail, pardner."

The weather report was accurate. When we arrived in the fair city of Timonium, the sun is shining. We pulled up to the house a few minutes before eight o'clock.

Terry was waiting for us. "Good morning Jake."

"Good morning, Terry. It looks like we picked a good day to visit the farm."

"Yes, the storm is over. This must be Ramona. Welcome to our ranch, Ramona."

"Thanks for inviting me Terry. I am looking forward to today."

"You two go right on in. Mom has breakfast ready."

"Sounds great, Terry. Follow me Ramona."

I led Ramona into the house and to the kitchen. Howie stood up from the table to welcome us and I did the introductions.

"Ramona, this is Terry's dad Howard Johnson, and his mom Eldora."

Eldora walked over to Ramona and gave her a hug. "Welcome, Ramona. It's nice to have some feminine company around here today. Jacob didn't mention my other boy, Ted. He is on the road most of the time. Are you a student at JH also?"

"Yes, I am Mrs. Johnson. I am a student enrolled at the McKusick-Nathans Institute of Genetics, the same program Jake is applying entry to."

"I think it's wonderful to see young women aspire to seek higher education. It's been a long time coming."

"Yes, it has. I'm fortunate to be living in this day and age."

"Yes, I can remember when girls were not even allowed to enroll in universities. Some would just sit out in the hallways and listen to the lectures of the professors."

Terry quipped, "Just how old are you Mom?"

Eldora frowned and ignored his question.

"Well, sit yourself down. We are having French toast, fried apples, and fresh beets. I pulled the beets just this morning from the garden."

Howie is watching Ramona with a smile on his face. "Everyone calls me Howie, Ramona. I'd wager from your outfit you have ridden a horse or two before."

"Yes, I have, Mr. Johnson, err Howie. My uncle has a tobacco farm down in Virginia, along the Rappahannock River. I used to enjoy visiting his farm and riding the horses. They weren't as classy as the horses I see in your pastures."

"Well, we still have a few cow ponies here on the ranch. When was the last time you rode?"

I can tell from Ramona's hesitation she is having a tough time answering him.

"Howie, Ramona was injured in an auto accident while she was still in high school. She hasn't ridden since then, but I know she is really looking forward to giving it a try today."

"Later today I'm gonna get you back in the saddle again, Ramona."

Terry starts singing, "I'm back in the saddle again. Out where a friend is a friend—"

Eldora had heard enough of his antics. "Terry, you are hurting my ears. If you want a piece of my pie you better stop right now."

"Ma, you know I like to sing."

"Well, why don't you just go sing 'Far, Far Away' if you need to."

Howie turns my way and asks, "So, Jacob, do you ride?"

"Only automobiles, Howie. One thing I do know about horses, they are unpredictable."

Terry is shaking his head. "You won't get Jake on a horse, Dad. He knows horses are flight animals and he has no interest in giving them a chance to demonstrate it."

"We may have to stick him on a buckboard, or the mule. After breakfast, we are going to go to church Ramona. Would you and

Jake like to come along with us?"

"I'd like to go, but I don't think this outfit I have on would be fitting."

"No need to worry about your jeans. People here are pretty laid back. A few of our good old boys come every week in bib overalls. How about you, Jake? You haven't attended church with us yet."

"Well, I didn't bring my *tallit* or my kippah."

"Same thing applies to you. Nobody is going to look twice at your get-up."

"Okay. Since Ramona wants to go, you can count me in."

Howie looks at his watch. "We will leave here by 9:30."

I wondered what I am getting myself in for as we all cram into Terry's truck. Terry is careful to seat Ramona in front. I took the window seat and Howie and Eldora climbed into the back. We headed out.

A very plain sign in front of the small country church reads "Timonium First Methodist Church, Pastor John Grafton." As we pull up in front of the church, the pastor is standing at the entrance, welcoming the faithful. Terry lets us out in front of the church and then parks the truck in the parking lot. As we enter, the pastor greets us with a "Good Morning" and shakes our hands. The usher hands each of us a church bulletin. Howie and Eldora are busily greeting their friends as we walk to the front of the church and file into the third row. Ramona saves a seat for Terry next to her. Looking around I estimate the church seats about a hundred people. By ten o'clock almost every seat is taken. Pastor Grafton and Terry are walking down the aisle together. Terry takes a seat next to Ramona. Two minutes later I notice them holding hands. The pastor takes a seat in the front row next to his wife. The worship team takes their places on the platform. A young girl with a guitar walks up to the microphone. Terry leans over and tells me she is their worship leader and her name is Gretchen Hoffsteader.

This girl Gretchen begins the service. "Good morning, church. Let's all stand as we enter into God's presence this morning."

Her greeting takes me back a bit. These people evidently are expecting God to show up. Everyone stands up, including myself. The words for each song are projected onto the screen above the music team. The drummer sets the tempo for the first song, and the band plays a very lively country gospel song by Neal Haggard.

Everyone in the church is singing their hearts out as the band plays the set of five country gospel songs. I am surprised. The group is no professional band, but they are very good. I watch the young people in the first two rows excitedly raising their hands in the air and swaying to the rhythms of the music. Following the worship service, Pastor Grafton asks the ushers to come forward to take the offering. He also makes it clear those attending for the first time are not expected to give an offering. Then these so-called ushers pass plates down the rows. I reached into my wallet and put a twenty in the plate. Next, one usher comes forward and prays over the offering. After the offering, Pastor Grafton's wife comes forward and makes a few announcements. She says they will hold a baptism service the following Sunday. Pastor Grafton comes to the pulpit and welcomes the visitors, namely Ramona and myself. He invites the congregation to greet one another and to take a few minutes to get a cup of coffee and a cookie or two. The snacks and drinks are spread out on a table at the rear of the sanctuary. I poured myself a cup of coffee and took a couple of homemade cookies. Howie takes the opportunity to introduce Ramona and I again to Pastor Grafton and his wife. The snack break lasted about fifteen minutes. Then the band plays another song. The pastor begins his sermon at eleven o'clock. The title of his sermon, "The Great I Am," appears on the screen. He asks the congregation to turn to the Gospel of John. The pastor presents his message by reading several verses from the Book of John where Jesus repeats the words "I Am" numerous times. The pastor's comments center on the claims Jesus made for his divinity. As the pastor continues his message, he becomes more and more animated. Near the conclusion of his sermon, he is walking up and down the aisle of the church, repeating the claims Jesus made of his fulfillment of the prophecies of the coming Messiah.

"Brothers and sisters, Jesus our Lord and Savior proclaimed he is the 'Great I Am.' He is the bread of life, the light of the world, the door of the sheep fold, the good shepherd, the resurrection and the life, the way, the truth, and the Son of God Almighty, the Alpha and the Omega, the all-powerful eternal Son of God, and our soon coming King. Can somebody say Hallelujah?"

The Pastor strides back to the pulpit and raises his right arm, saying, "My beloved, Jesus summarized it all when he declared to those Jews present, 'Before Abraham was, I am.' No Jewish son of

Abraham could miss what Jesus was declaring. It was crystal clear to them. Jesus was claiming to be Yahweh, the Messiah. And the religious leaders killed him for it. This holy name of God has four consonant Hebrew letters, JHVH but no vowels. We today don't even know how the word was pronounced. These four letters have a technical name. The word is called the Tetragrammaton. It is a compound word, 'tetra,' meaning four, and 'grammaton,' meaning letters. But what is the meaning of this word? Brothers and sisters, it means the 'Always Existent One,' The 'Great I Am!' Can we say 'Glory to the Great I Am' this morning church? Yes, glory to you Lord Jesus: our Savior, our Lord, our Healer, and our soon coming King."

The congregation breaks out in applause and praise. It is as if the roof has blown off this little church in Timonium. The meaning of the sermon was not lost on Ramona. When the pastor asked if anyone wanted to accept Jesus as their Savior and Lord, Ramona calmly rose out of her seat and walked to the front of the church. I couldn't believe my eyes. Ramona is giving her life to Yeshua. She is accepting Jesus. She has become what Alice described so clearly to me when we were at Bernie's several months back. Whether she realizes it or not, Ramona is now a completed Jew.

Why I resisted this pastor's powerful invitation I'm not quite sure. I really wasn't feeling any sense of rebellion. I just could not follow Ramona to the front of the church. Questions still keep running through my mind. What would my family think of me if I were to accept Yeshua? Is Ramona's decision the real thing or just an emotional reaction to the pastor's emotional message? No, even though my head is telling me this Jesus is the real thing, I'm not ready to accept him as my Messiah. He might end up sending me to India, China, or some other god-forsaken country. Perhaps there is an enemy speaking words of doubt into my soul.

We arrived back at the farm at 1:00. I watched Terry take Ramona's hand as we walked toward the house. Then I saw it.

I was the only one noticing how Rachel was walking, walking without her limp. Everyone gathered in the family room. The Johnson's are absolutely excited about Ramona's conversion to Christianity. While Eldora goes to the kitchen to prepare lunch, they are sharing their thoughts about the morning service.

I got up from my chair and quietly walked over to Ramona.

Leaning down I asked, "Ramona, can I have a private word with you?"

"Sure, Jake, what is it?"

"I happened to notice something as you walked into the house."

"What something?"

"Ramona, you were not limping. What's going on with your leg?"

Ramona jumped up to her feet. "Oh, my goodness, Jake! You are right. My limp is gone. Look everyone, my left leg is the same length as my right leg. It's grown! Hey everybody, look at my leg!"

Terry is looking befuddled. He asks, "What are you talking about, Ramona? Your two legs look the same to me?"

"Yes, you are right, Terry. They do look the same, don't they?"

Ramona is so excited, she can't say anything more. She now is crying crocodile tears. I had to clue Terry in.

"Terry, don't you remember me telling you Ramona was in an accident years ago?"

"No, I don't remember you saying anything about an accident. What accident?"

"Terry, Ramona has suffered with a bad leg for years. She was involved in an auto accident when she was still in high school. She broke several bones, including her pelvis, two vertebrae, and her left leg. Her leg was broken in several places. The doctors put her back together, but the left leg ended up one inch shorter than her right one. Before today I have watched her every day walking with a limp, especially when she is tired. A few minutes ago, she walked into the house without the slightest limp. The only conclusion I can make is Ramona has been healed, more than likely at church this morning."

The Johnsons and everyone else were absolutely amazed and overjoyed. Howie decides everyone should celebrate by taking a ride after lunch together along the nearby equestrian trail.

I naturally am curious about Ramona's healing. "Howie, does this kind of thing happen often in your church?"

"No, Jake, it doesn't happen often. I sure wish it did. No, this is a very special miracle Sunday."

Given everyone's excitement, I couldn't refuse to go along with them on the ride. They put me on a twenty-five-year old horse who is feeling every bit of those years. On the way home, my horse decided he was a young race horse again. He wanted to be first back to the hay bin. I did manage to stay on him. Even so, we didn't get back to the farm until almost five. By the time the horses were washed down and curried, we all were ready for dinner. Eldora just assumed Ramona and I were staying for dinner. She had a meatloaf in her oven. The smell of the meatloaf is permeating the entire house. I originally intended to leave right after the ride, but the smell of Eldora's meatloaf changed my mind. I noticed Ramona and Terry are both "leaning" the whole time they are together. I can see Terry likes this city girl turned cowgirl. It is half-past eight before the meatloaf and Eldora's apple pies are consumed. I excused myself saying I need to get ready for next week. Terry invites Ramona to stay the night, but she explains she needs to help me in the morning. After taking Ramona to her apartment, it is nine-thirty before I get back to my cave.

Just as Ramona previously warned me, this summer session is passing quickly. On Friday of the second week, I gave the students their first exam. As expected, Jonathan scored the highest grade, an A-plus. I am pleased with their results. The median grade is a B plus. There were two scores below the C. I called both of those students in for tutoring. I put the exams in a sealed envelope and carried them to the Biology Office. Professor Adelman wasn't in the office, so I left the package in his mailbox and sent him a text message telling him I had completed the first exam grades. He answered an hour later. He had picked up the grades. On Saturday, I let the students go a bit early. I felt they deserved a little time off. Once back to my apartment, I called Ramona.

"Hello, Ramona. How's things?"

"Hi Jake. Things are simply wonderful."

"I was just wondering if you would like to take in a ball game with me this evening? The Orioles are playing Boston."

"No can do, Jake. Terry is picking me up in an hour or so and I am spending the weekend at the farm. I am going to be baptized on Sunday."

"Oh. I see. I must now be further down on your list now."

"Do you want to go also?"

"No, I don't think so. I will just check in with you later."

"Okay, Jake. Oh, Terry is just now pulling up in front of my apartment. Sorry, I have to go."

"Okay. I'll talk to you soon."

I felt abandoned. I tried to invite Joe to the Orioles game, but he said he was visiting his brother in New Jersey. I went ahead and went to the game on my own. The following weeks I heard very little from Ramona or Terry. Sure, she was there at class, but her mind was elsewhere. I felt truly abandoned. I began to realize how important friendships are.

15

Enemies in the Camp

"Thou art my King, O God: command deliverances for Jacob,
Through thee will we push down our enemies:
Through thy name will we tread them under
That rise up against us."
(Psalm 44: 4–5)

The Spy

The first Summer Session went quickly ahead with no problems. Before I realized, it was over. There was no need for a lab on the final Saturday. Professor Adelman called and said he wanted to meet with me at the Learning Center at eight in the morning. When I arrived, he was already there.

"Jacob, I have a check for you. I'm sure you can put it to good use."

"Yes, I can certainly use it. I want to thank you for this opportunity, Professor."

"You did well, Jacob. I know these abbreviated sessions can be demanding, but you have done a first-rate job."

"Thank you, Sir. I did try to help the students. I remember how difficult this course was when I was in their shoes."

"Jacob, I'm looking forward to your services during the next summer session.”

“Is there anything you want me to improve on for the next session?”

“No, there’s nothing I want you to change. The second session schedule is not much different from the first one. It begins on Monday. Do you want to meet before we begin, or do you feel you are ready to go?”

“I’m ready. I will see you on Monday morning.”

“Very good, Jacob. Will Ramona be helping?”

“No, she has been seeing my friend Terry and seems to have lost interest in my company.”

“I see. This makes your task a bit less complicated.”

“Yes, it does.”

“Jacob, I have one concern I want to talk to you about. It has nothing to do with your performance in the classroom. It is just a bit of friendly advice. I am hearing on the grape vine you had a bit of a confrontation with Dr. Altschul at the applicant’s dinner. Did I hear correctly?”

“Yes, you heard right. Dr. Altschul took issue with my position on the theory of Neo-Darwinism. I simply said it was an unproven theory. He seemed to be very upset by what I said and actually attacked me. I believe he harbors a bad opinion of Jews.”

“You mean he made some antisemitic remarks?”

“Yes, his remarks were blatantly antisemitic.”

“Jacob, I want you to be very careful about what you say about the Theory. You touched a sensitive nerve in Professor Altschul, and he wields a lot of influence and weight around here. Do you understand what I’m saying?”

“Yes, I understand. I will be more careful in the future.”

“Good. Bear in mind, Professor Altschul has ways at his disposal to discredit a young applicant or student.”

“Like what kind of ways?”

“Perhaps he might use others to set a snare for you, such as one of his protégés.”

“I will be on the lookout for spies in my midst. Since you mention it, there is one student who seemed bent on baiting me along those very lines.”

“Oh? What is his name?”

“His name is Jonathan Logan.”

"Oh yes, Jonathan Logan is a favorite of his. Welcome to the realm of professional intrigue, Jacob. By the way, you might want to take a look at what they have scheduled at Shriver Hall. It is a debate and Altschul is one of the participants."

"Yes, Joe Harmon told me about the debate."

"Okay. I will see you on Monday morning, Jacob."

After our conversation, I walked down past Wyman Quad to Shriver Hall to check out this debate. Joe said it he thought it would be a sell-out. I stopped to read the billboard at the Shriver. The debate is being sponsored by the Department of Space Technology and Science. The listed debaters include two Christian apologists, an agnostic and an atheist, the atheist being Dr. Altschul. The date of the debate is scheduled for Friday, the week before the Fall semester begins. The ticket office was open, so I purchased two reserve tickets. The gal at the window also gave me two brochures about the debate. When I arrived back in my apartment, I called Joe and told him I planned to attend the debate and had purchased a ticket and one for him as well. I said I had a brochure to give him also. Then I took time to read the debate brochure.

They have scheduled two Christian apologists to participate. The first is from a ministry called "Faith and Reason (FAR)." The brochure lists their website, so I pulled it up and took a quick look at the scholars on their staff. Their backgrounds are quite impressive. All the scholars have published articles in leading journals. They all appear committed to good science. This ministry conducts a heavy schedule of forums on university campuses and churches, primarily in the US and Canada. FAR also supports a network of local chapters located in the US, Canada, and a few other countries.

The second Christian apologist ministry to participate is Punjad International Ministries (PIM), It is named after the founder, Dr. Sangra Punjad. I pulled up their website as well. This ministry is truly global. There are offices in: The United States, Canada, United Kingdom, Romania, Turkey, the Middle East, India, Singapore, and Hong Kong. Purespring, the humanitarian arm of the ministry, assists other internationally positioned charitable organizations. The website states these partnerships help rescue, rehabilitate, restore, and train disadvantaged individuals for reentry into the local economies. They also work with women and children. The descriptions of both of these Christian ministries impressed me.

Next, I turned my attention to the opposing team. Dr. Carlos Gonzalez, from our very own Bloomberg Center for Physics and Astronomy, will be presenting the case for agnosticism. The scholar assigned to present the case for atheism is none other than Dr. Frederick Altschul. I'm looking forward especially to this one. I will be finished with my second summer session before the debate is held.

I served myself the left-over chili, with a lot of onions over the buns and hot dogs for dinner. I have an unproved theory explaining my love of onions. I may have inherited in my DNA the love of onions from one of those Hebrews who wanted to return to eat onions in Egypt. I spent the evening reading from the virology text, preparing for the second summer session before turning in.

The second session of the virology class went even more smoothly than the first one, even without the help from Ramona. I am impressed with the attitude of these students. They have their fun times, but most of them are hard workers. In my opinion, they are more committed to their careers than the undergrads in Atlanta. Perhaps the different cultures of North and East may explain it. I met with Dr. Adelman on Friday following the final exam of the second class to give him the grades. I thanked him again for the opportunity and for the second check. Now I am looking forward to the Fall semester. I still have not received confirmation of my acceptance to the Institute.

I decide to give Joe Harmon a ring to be sure he is on board for the debate.

"Hello, Joe. This is Jacob."

"Hi, Jake. What's up?"

"Are you ready to attend the debate at the Shriver this evening?"

"I'm looking forward to it. Do you have our tickets?"

"Yes, I have them. I think I will dress up a bit for the occasion. It might help me get a word in."

"Not really, Jake. What they usually do is take questions via cell phones. They won't see your ugly face."

"Thanks for telling me. I won't worry about combing my hair. I will check to be sure the battery on my phone is fully charged."

"I'm just looking at mine now. I plan to be there a little early."

"Okay, I'll meet you at the door at 6:30."

I hung up the phone. My cell phone needed a charge. As I am

plugging it into the wall, the screen is blinking, telling me there is an unread message. It is from Robbie. It is close to the beginning of the Fall semester. Perhaps they have made a decision on my enrollment? My hands are shaking as I call her back. "Hello Robbie. This is Jacob Cahn returning your call."

Hello, Jacob. I have a message from Dr. Hobbs. We are holding a press conference this coming week to announce our selection of applicants. We want all of our selected applicants to attend, if possible. I will have to get back to you on the exact day, but it probably will be near the end of the week."

"Does this mean my application has been accepted?"

"Yes, Jacob, your application has been accepted. Congratulations! The meeting will be held in the Turner Auditorium at the BRB. As I say, we are not sure what day next week when the announcement will be held."

"Robbie, this is a great day for me. I want to take this opportunity to thank you for all you have done for me."

"My pleasure, Jacob. I wish you the very best in your studies. I know you will do well. I'll see you at the press conference. Enjoy the day, Jacob."

After such good news, I had to take time to thank Ha Shem. I stood up, and raised my arms heavenward. I prayed the Lord's prayer, which was really the disciple's prayer, thanking him for this blessing in my life.

The Debate

It is almost time to leave for the debate. My cell phone is now fully charged. I left at 6:20 for Shriver Hall. Joe is waiting for me at the front door when I arrive. Our two reserved seats are in the third row from the front. The title of the event, "Scientific Evidences for and Against the God of the Bible," is drawing a big crowd. By the time the meeting begins, the auditorium is completely filled. Dr. Hobbs is the acting moderator. He introduces the speakers and gives a short summary of their backgrounds. Then he explains the format for the meeting. It is divided into two parts, arguments from physics and astronomy and arguments from the biologic sciences. Each presenter would be allowed twenty minutes to state their case. There will be a twenty-minute Q & A from the audience at the end of each topic. First up is Dr. Gonzalez, a well-known proponent of the quantum origin of the universe and a member of the JH faculty. Following his presentation, Dr. Robert Hughes from the FAR Ministry is scheduled to present the argument for a divine origin of the cosmos. The second part of the program begins with a presentation by Dr. Punjad, a well-known Christian apologist. My devious friend, Dr. George Altschul then will challenge Dr. Punjad's statements. I can't wait for the opportunity to contest Altschul's position. It is pay-back time. Dr. Gonzalez began with a description of the mysterious world of the quantum. His argument repeated the position of Stephen Hawking. He asserted a horizon formed by light rays can emerge from a bounded region, a singularity need not be assumed. He postulated these light rays emerge from a region containing spacetime loops. Hawking believes our civilization at some point will build a machine with the capability to create a spacetime loop. His theory, the warping of spacetime, could permit some entity to come back to the place and time they started from, or to some other spacetime position. Although fantastic to the laymen, Dr. Gonzalez stated the Hawking theory has convinced many well-known physicists of the possibility. I remembered reading a research report issued by the Large Hadron Collider Project at Cern Switzerland. Cern researchers claim to be on the verge of opening a stargate into another dimension. I find all of this highly disturbing. I consoled myself with the knowledge our world is still under the direction and control of its Creator. Dr. Hughes began his criticism of the presentation by Dr. Gonzalez by criticizing the quantum physics argument. He considered this view of physic, even at the quantum level as myopic. He referred to the extremely fine-tuned features of our universe and world. The chances of such a universe coming blindly into existence consisting of such an intricate sequence of favorable events for life points to the existence of a causal agent existing beyond even the abilities of the Quantum. He asked, "What could explain the birth of our innate sense of good and evil?"

I have to agree with Dr. Hughes. Why would survival of the fittest result in a beautiful butterfly, not to mention those beautiful feminine coeds on campus. If such a path to another dimension inhabited by alien creatures does exist, there must be one who has orchestrated it all. With my weak understanding of the topic I did not want to display my ignorance by texting any question to Dr.

Hobbs about part one of the debate. I left it open to questions by my true friends from the Departments of Astronomy and Physics.

The second discussion topic seemed almost an extension of Dr. Hughes' position. Dr. Punjad began by going right to the juggler, presenting hard evidence for the resurrection of Jesus Christ from the grave. He placed his argument firmly on factual information. He presented the hard evidence of the matching blood patterns from the Shroud of Turin and the Sudarium of Oviedo, Spain. The Shroud first shows up in history in the Fourteenth century. The Sudarium is traceable further back, to the sixth century. Next, Dr. Punjad pointed out the accounts of the resurrection cannot be laid to the doorstep of mythology since myths require a significant period of time to develop. The accounts of the resurrection are established very quickly and attested to by a sizeable number of witnesses, believers, unbelievers, and prominent men like the Jewish historian, Flavius Josephus. Then the Doctor went over ground I already have examined, namely the prophecies in the Torah and the Tenach. He rested his case against the theory of Macro-evolution on the evidence of man's uniqueness, his consciousness, his appreciation for beauty, his creative abilities, and, most importantly, man's innate sense of a supreme being.

It was Dr. Altschul's turn. He began by questioning the facts presented about the Shroud and the Sudarium, but did not present any real evidence falsifying the information presented by Dr. Hughes. Altschul then launched into a discourse on the tendency for Christians to explain any unknown phenomenon by attributing it to their God, the so-called "God of the Gaps" argument. His next argument was the problem of evil. How could an all-powerful and all holy God permit suffering in the world? Either he is not all powerful, or not all holy. I watched Dr. Punjab as he struggled to keep his composure. At the end of Dr. Altschul's presentation, Dr. Punjab asked for an opportunity to rebut Dr. Altschul. Dr. Hobbs reminded the speakers of the agreed-upon format disallowing rebuttal arguments. He explained there simply was not enough time for them. Rebuttal arguments could take us into the wee hours of the morning. Then the Director opened the floor for questions from the audience. This was my golden opportunity to confront Dr. Altschul. My strategy was to challenge Altschul's position indirectly by directing my question to Dr. Punjab. I quickly sent a text message to Dr. Hobbs.

For a few minutes, Dr. Hobbs reviewed the messages on his tablet. Then he said, "I have quite a few questions for you gentlemen. I have one here from a Mr. Jacob Cahn. Jacob is directing his question to Dr. Punjab. It reads as follows, "Dear Dr. Punjab, how do you resolve the problem of evil in our fallen world with a holy, all powerful God?"

Doctor Punjab has a smile on his face as he begins his response. "Thank you for your question, Mr. Cahn. The problem of evil is the most frequent battleground between people of faith such as myself, and atheists and agnostics. There is a basic underlying error made by the majority of agnostics and atheists who argue the existence of evil in this world means God is either unloving, or not all powerful. They fail to recognize one simple fact. People of faith are not imprisoned by the 'either/or' view of their protagonists. Dr. Altschul and many others of his persuasion erroneously make the assumption this universe is all there is. The question of the problem of evil is most eloquently answered by the Judeo-Christian world view. This view is not in conflict with scientific evidence. Quite the contrary. Earlier this evening we heard evidence from physics of possible other dimensions. Atheists and agnostics confine their argument to their naturalistic world view. They mistakenly believe in the finality of this existence. The Bible clearly teaches the Creator has a very high purpose for mankind. Only a loving God would take 13.7 billion years to prepare a home for humankind. So, what does this loving God attend to accomplish with this Cosmos? The first purpose is to call out a people to his name, to call them into a perfect relationship with himself. This plan of the Almighty is far superior to Doctors Hawking's loop of spacetime or Cern's piercing of the veil into other dimensions. Just what dimension is Cern attempting to open, the pit of Hell? They hold in high regard the goddess Shiva, the destroyer. The Judeo-Christian God declares he will create an entirely new creation, an eternal perfect kingdom with totally new physics for his children. Jesus said, 'I go to prepare a place for you. If it was not true, I would have told you.' God's second objective is to defeat Satan, the source of all evil in this present world. It is foolishness to accuse God as the source of evil. He has allowed evil to exist, but is not the source of evil. Here is a very important difference between angels and mankind. Both had been given the choice to honor God or to rebel against him. Both angels and the souls of men are eternal. Angels are already living in the eternal spiritual dimension and have made their decision, for or against God. There is no salvation available for them because they have made that decision living beyond this temporal existence. In the case of mankind, we have a different situation. Although the DNA of mankind carries the sins of Adam and Eve, their progeny had not yet had the opportunity to choose between right and wrong, good and evil, obedience and rebellion, for they exist in this finite realm of time. Their souls have not been given the opportunity to choose until the act of conception. The Scriptures teach Adam and Eve were tempted in order to test their obedience to their God by the Fallen One, Satan. Since their failure, mankind has been separated from his Maker. It was necessary for a holy God to send Jesus Christ to this earth to redeem fallen mankind and provide a way for man to choose God. Jesus is wholly God and wholly man, suffering on the cross for all the sins of mankind. Only a completely holy and loving God could offer fallen mankind the grace of forgiveness through the death of his son Jesus. He is the spotless sacrifice for the sins of this world."

The crowd gave Dr. Punjab a loud and favorable response to his answer to my question. During Dr. Punjab's response, I was observing Professor Altschul's reactions. He is helpless to respond. He became increasingly agitated, his face contorted in pain. Then I realized he is glaring directly at me with murder in his eyes. I am looking at the face of evil.

Dr. Hobbs kept to the schedule. The debate concluded at nine o'clock. People did not want to leave. They gathered in small groups, discussing their personal assessments of the debate. Dr. Adelman waved for me to come over to his group. I told Joe I would catch up with him outside. I join Adelman's group.

The doctor takes me aside. "Jacob, I would like to share a few words with you if you have the time."

"Certainly! What's on your mind?"

"Jacob, do you perhaps remember our previous conversation about Professor Altschul and the warning I gave to you?"

"Yes, of course I do."

"Jacob, I need to repeat my warning to you. Any more shenanigans like tonight will doom your chance of getting a degree here at JH. As I explained before, this person you are sparring with is a heavyweight. He wields a mighty punch here at JH. He is responsible for securing major funding for the University, especially to the soft sciences of biology and genetics. The money flows from his grant applications, not his research. The man has ingratiated himself by bringing in a lot of grant money to the school. The Administration looks very favorably on those like him who have contacts with major funding sources."

"I see. Does this man have strong relations with any particular sector or group in our economy?"

"Yes, he does, but I'm not free to divulge the specific entities involved. Let me just say, the well the water flows from is deep, very deep."

"My Uncle Sam has a deep well like that."

"You are a very bright boy, Jacob. Please be careful. I don't want to see you taken advantage of. It's best we put these confrontations to rest once and for all."

"I understand what you are saying. I appreciate your willingness to share this with me Dr. Adelman. I have to confess the motivation for my question was clearly wrong. I saw the opportunity to strike a blow and forgot the man, although fallen, is still a creation of the Almighty."

Dr. Adelman gives me a slap on my face and says, "This will help you to remember, Jacob."

The Doctor rejoined his group. I left the auditorium with my head down. Joe is waiting outside. As I joined Joe, I spotted Jonathan Logan waving at me. I waved back at him.

"Who's the guy waving at you, Jake?"

"He's Jonathan Logan. He took the virology course during the first summer session. I heard he is a protégé under Altschul's wings."

"You certainly find a way to keep things lively around here. Your question set Altschul up beautifully. You gave Dr. Punjab a great opportunity to present his rebuttal to Dr. Altschul."

"Yes, but I have set myself up for a real bruising sometime in the future."

"The truth did him in. The truth is always there to rescue us."

"Yes, you are right Joe."

"What's wrong with your cheek? Did you bump your head against something?"

"No. I just received a little correction from Dr. Adelman. I will live to regret what I did this evening, Joe. Did Robbie call you today? I received a call from her."

"Yes, she did. She told me they are planning a press conference next week some time. Do you think this means I made the cut?"

"Of course. She wouldn't tell us about the press conference unless we have been selected."

"I wonder about Craig Simpkin. Do you think he made the cut also?"

"We may not know until the press conference. I hope we see his face there. I gotta go. Take care."

"You also, Jake."

I arrived back in my apartment feeling pretty apprehensive about the Altschul thing. I needed to do something physical. It is late, but I changed into my sweats anyway. I left the Bradford for the Rec Center. The door is standing open. I went inside. The only person there is a security guard. The name on his jacket is John.

"Good evening, John. I assume the Center is still open. Can I take a few laps on the track?"

"Sure, be my guest. I will be here until midnight. Are you timing your laps?"

"No, I just need to get some exercise. I find running is the best way to unload my anxiety. I won't set any records tonight."

"I understand. Just don't overdo it. You look a bit tired already."

I have the track all to myself. After the fourth mile, I am feeling better. I said goodnight to John and made my way back to my apartment.

Reconciliation

I slept soundly. I didn't wake up until seven Saturday morning. It is amazing how a good night's sleep can rejuvenate you. After a shower, I dressed in my tennis gear. Then I fixed a great breakfast of scrambled eggs, leftover chili, and buttered toast with strawberry preserves. With my tennis racket and bag in hand, I left for the courts. I hope to find someone to pick up a game with. When I arrived, all of the courts are in use. I put my name on the chalk board, found a bench, and sat down. I'm just watching the games in front of me when I feel a tap on my shoulder. Turning around I see Jonathan Logan is standing behind me.

"Hello Mr. Cahn. I didn't expect to see you on the tennis courts."

"Hello Jonathan. Well, I'm waiting for a court to open. How are you?"

"I'm fine. It seems our paths continue to cross."

"Yes, it does. You have aced the virology course. Perhaps we should end the formality. You can call me Jake if you want to."

"Okay, Jake. My full name is Jonathan Curtis Logan, but people who know me usually call me by my nick-name, JC. I assume you enjoyed the debate?"

"Yes, I liked it a lot. How about you, JC?"

"I found it quite interesting. You do know I come down on the other side of the debate. I hold to a naturalistic explanation."

"Yes, I know. You are in Dr. Altschul's camp, even some say a favorite of his."

"Yes, well from your question to Dr. Punjab, no one is in doubt where you stand on this issue."

"I found Punjab's position very sound. Of course, you knew this before the debate."

"It's true."

"So, are you open to investigating this issue further, or is your position pretty much set in stone?"

"You may be surprised. I can't give you an easy answer. Before last night, I have been an inveterate believer in the naturalistic explanation for life on this planet. I must admit I am now wrestling with some of the things that were touched on in the debate."

"Good for you JC. I'm going to give you another gold star. Perhaps we can find an opportunity to have our own debate?"

"I am open to a further discussion, but don't get your hopes up too high Jake."

"No, not at all."

About this time one courts came open and my name was next on the board.

Looking at my tennis bag I asked JC, "Say, I have an extra racket. Do you play tennis JC?"

"Yes, I try to play a few games through the week, but the courts are always crowded.

"Do you want to play a singles match with me?"

"Yes, but I don't have tennis shoes on."

"It looks like your loafers will do the job."

"Okay, I'll play a couple of sets."

"Shall we say a prayer together first, JC?"

"No, I will let you do the praying."

I had to smile about his response. I easily won the first two sets. I then eased up on him in the last one. I felt I needed to encourage him after the way he challenged me during the virology course.

After the last set, JC says, "Jake, you didn't tell me you are an expert at this game."

"I'm no expert, but I did manage to make the tennis team at Emory."

I decide to invite him to lunch. Perhaps he might share some interesting information with me about Dr. Altschul.

"Are you hungry JC?"

"Matter of fact, I am Jake. Shall we get some lunch?"

"Yes, let's do. I'm buying."

"But you won more games than I did?"

"It's my treat. Where should we go?"

"Well, there's a food court close by in the Brody Learning Center (BLC), next to the Embassy Library."

"The BLC food court sounds good to me. I haven't visited the Embassy Library yet either."

I engaged him in conversation as we walked over to the BLC food court. "Where is home for you, JC?"

"I'm from Philly. My family traces their history all the way back to William Penn."

"Really! A true blue-blood. Tell me about your family."

"There is a park in Northwest Philadelphia named after my ancestor James Logan. He was the personal secretary for William Penn before Pennsylvania was a state. Originally called "Northwest Square," the park was used for public executions and burial plots until the early nineteenth century. In 1825, it was renamed Logan Square after my ancestor. Where do you hail from, Jake?"

"My family lives in Omaha, Nebraska. My father is a doctor there. My ancestors came to the US much later than yours, JC. Dad grew up in Valdosta, Georgia. He met my mom while an intern in Buffalo. Do you have any siblings?"

"Yes, I have two sisters. They are older than I am. They have families of their own. Do you have any brothers or sisters?"

"I have a younger sister Emily. She is just entering Emory this Fall. Say JC, I have a question for you. You don't have to answer it unless you want to."

"Okay, what is it?"

"Do you remember during the very first Q & A segment of the virology class you asked about my position on Neo-Darwinism?"

"Yes, how could I forget. You cleverly derailed my question."

"True. Do you know why I skirted around your question?"

"I'm not sure."

"Let me explain. Having a lot of chutzpah is part of who I am. I enjoy what I call intellectual wrestling matches. I believe I have a gift for it. I won't suggest to you where I think I got the gift. Anyway, I like to get things out in the open. So, I have a question for you. Did Dr. Altschul ever discuss our disagreement about Darwin's theory with you?"

"You mean did he ask me to try to trip you up in some way?"

"Yes, I would like to know if he asked you to get me to state publicly in front of the students, with the cameras going, my position on this controversial question of evolution."

"He didn't in so many words say he wanted me to pin you down on your position. You might say he encouraged me to pursue the topic in greater depth with you."

"I believe he wanted to get my position on the tapes. So, it sounds like we have a conspiracy of sorts underway here?"

"Jacob, all I can say is he is giving you more attention than he normally gives to new grad students."

"I have one more question JC. Do you intend to continue to follow Altschul's suggestions about me in the future?"

"No, I won't be doing any more of his dirty work for him. Then again, you need to realize I am not the only playing card in his deck, Jake."

"Yes, I'm sure you are right. Thanks for your honesty sharing this with me JC. I appreciate it."

I took a look at my watch. It was five minutes before two.

"I need to take off now, JC. I have some reading to catch up on this afternoon. Let's exchange phone numbers. Perhaps we can schedule some tennis time together."

"I would like to."

We exchanged cards and then went our separate ways.

The Letters

After lunch with JC, I returned to my apartment at the Bradford. I hadn't remembered to check my mail box for a couple of days. There were three letters in my box, one from the Institute, another from Rachel, and a third from Dr. Mike in Atlanta. The Institute envelope contained a brochure listing the class schedules for the coming year. The second letter was an invitation to Rachel's wedding with a rsvp at the bottom. The wedding is set for the first of next month to be held at a synagogue on Long Island. What with my classes starting, there is no way I can attend, nor would it be wise. I'm sure she knows as much and simply wanted to keep me informed. The envelope also contained a list of suggested gifts from Sax Fifth Avenue. I pulled up the website and selected a nice-looking table lamp from the list. I arranged to have it gift-wrapped and sent with a card to her home address in Atlanta. After taking care of the wedding gift, I opened the very large envelope from Dr. Mike. It contained a very interesting document. I decide to call Dr. Mike at the Library in Atlanta to ask him about it and to tell him the good news of my acceptance by the Institute.

"Hello, Dr. Mike. Do you know who this is?"

"Of course. How are you Jacob?"

"I'm just fine. I called to let you know they have accepted my application here. They are holding a press conference next week to announce the new grad students."

"Great. Have you heard from Rachel since our last conversation?"

"Yes, I talked to her on the phone a while back. I just received an invitation to her wedding today. She is getting married to an artist in New York the first week in September."

"How do you think this will work out for her?"

"I don't think it is going to turn out well, Dr. Mike. She thinks this artist, Sam Leavitt, is probably marrying her for financial reasons, for her father's money. I would really like to read their ketuba."

"That's dreadful news. Well, I have something to share with you. Do you have a few minutes?"

"Yes. I assume you are calling about this document I receive from you today in the mail."

"Oh, you already have received it. Have you read it yet?"

"No, I decided to call you and ask about it as well as my acceptance to the Institute."

"Good timing then. I have a story I believe you will find interesting. This story begins in Bagdad, Iraq. During the Jewish diaspora, Iraq was one of the largest and most thriving Jewish communities in the world. The community existed for more than twenty-five hundred years, since the Babylonian captivity. The area has been a treasure of ancient Jewish history gathered from generation to generation, including historical works, copies of the Torah, rabbinical commentaries, and artwork from homes and synagogues. Over a hundred thousand Jews lived in peace among their Syrian, Persian and, most recently, Islamic neighbors. Then in the early 20th century everything changed. The tide of antisemitism sweeping across Europe came to Iraq as well. Arab leaders from many nations, including Iraq, attracted to Hitler's open hatred of the Jews, aligned themselves with Germany. The Jews in Iraq bore the brunt of their hatred, especially during the Shavuot celebration of 1941. The pogrom in Iraq left hundreds of Jews dead and destroyed homes and synagogues. Their properties were dispossessed. Over one hundred thirty thousand Jews fled to Israel, leaving their treasured possessions behind. But, as always, there were a few Jews who stayed. They were called 'Shadow Jews.' Saddam Hussein, in an effort to raise his position within Islam as the reincarnation of Nimrod, sent his troops to loot these thousands of books, manuscripts, and other valuable Jewish artifacts. They remained hidden until 2003 when Saddam was overthrown. The US Military discovered the treasures in a flooded basement of the Secret Police. The Military brought the treasures to the US and asked the Smithsonian to restore them and place them in their archives. That's the good news. The bad news is our Government now has agreed to return them to the Islamic leaders in Iraq. I have been working feverishly with several Jewish historians and archaeologists to copy the most important Jewish materials before they are sent to almost certain destruction in Iraq."

"This is very interesting, but how does it relate to me, Dr. Mike?"

"I'm getting there. The Woodruff Library is now in possession of copies of these manuscripts. They are not written in English of course. Some are in Hebrew, Farsi, Arabic, even Acadian. I have been translating a few of the most interesting documents into English and modern Hebrew. I came across one very curious document. In this document, a Babylonian scribe presents an account of the 'Fallen Ones.' I think you will find it very interesting. The copy I sent to you is in English."

"Whoa, you are giving me a front and center seat on this."

"Yes I am. Let me know what you think about it."

"I certainly will. Well, I will let you get back to your research. We'll talk again soon."

"Very well. Goodbye Jacob."

Dr. Mike's story has me curious. I can't wait to read it, but I will set it aside for now. I cooked up a meal of spaghetti and then began to read the brochure from the Institute, but I can't seem to concentrate on it. I keep falling asleep, reading the same paragraph over and over again.

The Looming Apocalypse

After dinner, I decide to turn on the TV and see what was going on in the world. I first turned to the Fox Business News channel. The moderator is discussing the implementation of the REAL ID Act of 2005. It is scheduled to be introduced no later than 2017. The man interviewed is Mr. John Talbot, the CEO of "Identity Retection Group (IRG)." IRG is a start-up security firm. The US Government has given this company a multimillion-dollar contract to produce a new type of Radio Frequency Identification (RFID) chip. The CEO's description of the new chip is scary.

Mr. Talbot stated, "Our revolutionary innovative RFID chip implant will integrate several new technologies, permitting the Nation to advance to a new level of efficiency and integration. Biometric authentication of a person's identity using a harmless implant beneath the skin will allow us to eliminate ID theft and other crimes, crimes costing individuals and our financial system billions of dollars each year. Not only will this chip yield positive identification, it also will establish the geographic location of the individual. A recent agreement with the Military to utilize their newly enhanced version of the Geographic Positioning System (GPS) will provide us with much improved locational ability. I can't reveal the actual level of accuracy, but it will be significantly less than the previous forty feet. The advanced chip also will include a complete informational archive, including the following; date of birth, gender, residence, race, medical history, financial information, State driver's license number, any criminal records, and an in depth cultural profile. The advantages of this chip are endless. To cite one example, a person injured in an auto accident with an imbedded chip will furnish instantly to first responders the person's complete medical history. Our chip will eliminate the need to carry cash, further reducing the occurrence of theft. The point of sale device will communicate with the imbedded chip, automatically sensing product codes for the purchased items and immediately transferring payment from the individual's bank account to the seller. For the first time, this chip will permit conversion to a totally cashless economy."

The interviewer smilingly asked, "Sir, will this ID system be mandatory for everyone, or voluntary?"

"It will be completely voluntary. We believe any resistance to the implant will quickly be overcome once the population realizes the advantages it affords. Of course, complete conversion to a cashless economy ultimately will require mandatory compliance. We believe total compliance will be achieved by the Spring of 2020. Some segments of our society no doubt will view the system with suspicion. We believe the gains far outweigh the negligible loss of personal freedoms. Let's face the facts. Our Founding Fathers created a wonderful system of government, but those documents are no longer relevant to our modern economy and way of life. Our Government must become more efficient if we are going to be a contributing member of the emerging community of nations."

There it was. The real objective; the implementation of a "One World Government." This entrepreneur doesn't seem to share my concern about personal freedom.

While Fox broke for another commercial segment, the neurons in my memory bank are lighting up. My recollection of the back of our dollar bill and my course in Latin brought to mind the translation of the ideas this CEO is professing: *Annuit Coeptis Novus Ordo Seclorum*, the birth of the New World Order. Is this New World Order a completely new idea, or have we known about such an undertaking long ago? I remember just recently reading about just such a system in a much older source.

I picked up my Gideon Bible, and read from the Book of Revelation, Chapter Thirteen and verses sixteen through seventeen:

And he causeth all, both small and great, rich and poor, free and bond, to receive a mark in their right hand, or in their foreheads. No man might buy or sell, save he that has the mark, or the name of the beast, or the number of his name.

The Fox guy asked another great question for him, one which has been rattling in the back of my brain for some time.

"Sir, where do you believe this explosion of technology will end?"

"I believe we are on the cusp of a major change in human history. Some have referred to it as the New Age of Human Hybridization, literally the transformation of humankind. As we continue to develop biometric devices, we very soon will be able to change man's genetic makeup, replacing defective genes, and inserting entirely new genetic material. Many researchers believe we will soon be able to rid mankind of disease, and extend the human lifespan to perhaps five-hundred years. Transhumanism is our goal. We intend to create a completely new kind of human being. No longer will cyborg beings exist only on the screens of science fiction movies. We will create human hybrids capable of amazing talents and skills."

I heard enough. This fellow is hoping to become a god. I turned off the TV. These new technologies promise so many good things, but are they simply dangling a delicious carrot at the end of a very crooked stick? This conspiracy is far greater than the one I am experiencing in the small world of the Institute. This could be the ultimate conspiracy. Where will it all end? Only God knows. I picked up my Gideon Bible and Concordance. I began reading every passage I could find on the End Times. After midnight, I put the Bible down, set the alarm for six, and went to bed. As I lay there, I realized something quite scary and foreboding is taking shape in our Nation and in the world. These malevolent groups intend replacing our great Republic with a one world dictatorship run by a small group of powerful elites. These people are completely evil, involved in the grossest kind of sin. They are the enemies of everything good and decent. Good has become bad, and bad has become good in the warped minds of these elite monsters. In the near future, I expect to face wolves in sheep's clothing.

16

From Darkness into The Light

"Jesus cried and said, 'He that believeth on me, believeth not on me, but on him that sent me. And he that seeth me seeth him that sent me. I am come a light into the world, that whosoever believeth on me should not abide in darkness."
(John 12: 44–46)

The Scroll

I did not sleep well. Monday morning finally came. I now have to transition from master to slave. At least my course work will not begin until next week. After dressing in my running clothes and a light breakfast, I took a long run around the Homewood Campus. When I arrived back to the apartment, I took a shower. Then I took some time to read the material Dr. Mike had sent to me. The name of the document is *Sherebiah, The Servant of Ezra Whose Eyes Are Open.* As Dr. Mike earlier explained, the document found by our troops in Iraq was sent to the Smithsonian Archives for restoration and safe-keeping. Our government then made the crazy decision to honor the request of the war-torn Iraqi government to return these ancient documents to them. Dr. Mike has been working feverishly to translate them before they are returned to an unknown fate in Iraq. The English translation sent to me is in Dr. Mike's beautiful handwriting. He has added commentaries in the margins of the document. Few people write like Dr. Mike does anymore. Certainly, I don't even come close. His annotations identify many Hebrew and Babylonian names known from other sources. The first comment reads, "Possibly the Levite mentioned in the Book of Ezra, Chapter Eight and verse eighteen." He certainly is a man of *sekel*, a man of wisdom and understanding. Could this ancient document actually date back to the time of Ezra in Babylon, 445 BC? I continue to read from the translation, comparing the writings to the accounts in the Books of Ezra and Esther. The names coinciding with the Tenach give strong evidence for the authenticity of the scroll. The scroll translation goes on to describe how the remnant of the Jews had taken the daughters of the Canaanites and other people groups for their sons, thereby corrupting the seed of Israel. Then I came to a remarkable passage. Dr. Mike's translation reads, "For the Fallen Ones corrupted the seed of the people. The Rephaim, giants of great renown, the progeny of forbidden unions, are as beasts of the field."

It all begins to come together for me. The Fallen Ones are the rebellious angels who followed Lucifer. Dr. Mike's translation agrees perfectly with Chapter Six in Genesis, verse four of Chapter Two in Second Peter, and verse six, in the Book of Jude. The Genesis Chapter Six, verse 4 reads: "There were giants in the earth in those days, and also after that, when the son of God came unto the daughters of men, and they bare children to them, the same became mighty men, which were of old, men of renown."

These are the angels who left their first estate, responsible for the sin of procreation with humans, producing the Rephaim, Nephilim, and other hybrid beings. Satan succeeds in deceiving the people of the world as mankind worships these creatures as gods. People have a distorted view of history. They fail to see the unseen battle in progress between God and Satan. The majority of people base their lives on a purely secular view of history. Few see the chess game being played out between the Devil and Almighty God in the heavens.

Dr. Mike's comment in the margin states, "Note this period was a crucial battleground between the Lord and Satan's effort to corrupt the royal lineage of Yeshua through the sons of Shealtiel (Ref: Ezra 3:2-3, Neh. 12.8-9). It was necessary to remove this royal line from contact with the Babylonian culture."

Dr. Mike suggested I read *The Armageddon Strain* for more information on the Biblical Nimrod, believed to have lived in the present City of Mozul. The DNA of the progeny of the union between the Fallen Ones and human women was irreconcilably changed. Some human DNA was changed even after birth. Dr. Mike is suggesting Nimrod's DNA may have been genetically altered after he rose to leadership, becoming a mighty Nephilim.

Our DNA can be altered during an infection in the cell, or by bombardment of gamma rays and X rays, or mutations at the time of conception. Today the DNA within cells can be altered by microbiologists in several ways. A skilled geneticist can easily alter DNA by injection a single corrupted RNA retrovirus strand into the cell. This single-stranded RNA then can access nucleotide pools in the host cell and form a double-stranded corrupted DNA copy. This bad DNA can then be incorporated into the host chromosome using a viral enzyme called an "integrase." The new fake gene then can order the cell to make more RNA copies of the original corrupted RNA. These viral RNAs then infect additional cells in the body. The alteration of DNA/RNA strings now can also be easily done using the new CRISPR technology. This technology is capable of cutting the strands at specific points, allowing the geneticist to produce strands specific designed to achieve desired objectives.

Another annotation by Dr. Mike refers to Genesis, Chapter Eleven and verse six where it says the following: "Nothing will be restrained from them, which they have imagined to do."

Nimrod's tower very well may have been able to access the second heaven, Satan's realm.

Dr. Mike's translation of the ancient document contains a third interesting passage: "Every high place is given over to the images of Moloch and Chiun. Thou shalt put these people away, lest the day of the Lord be darkness and not light for you, even very great darkness, and no brightness in it. Separate yourselves from them, my children."

This passage is quite similar to the passage in Joel Chapter Five, verses twenty-five through twenty-seven. The truth now is dawning on me. I now understand why El Shaddai commanded Joshua to kill every Canaanite, men, women, children and even all of their animals. This command was not capricious or unjust. Instead it is the clear expression of Jehovah's great love for mankind and his desire to protect the human seed. Only in this way could he preserve a pure strain of human DNA to enable Yeshua, the God Man, to bring redemption to the human race.

After reading Dr. Mike's translation and viewing the program on TV last night, I'm not sure whether to turn on the TV or not. I finally did so. I searched the channels and came up with a History Channel program on UFOs. The information shown was obtained from the Mutual UFO Network (MUFON). A retired man from McDonald-Douglas was speaking at their annual convention. He began by sharing photos of McDonald's experimental activities in the Antelope Valley, California. The man talked about the existence of a secret base just west of Lancaster called "The Ant Hill." He said the underground facility carried out flight testing of UFOs developed at Groom Lake, Nevada. I know this base does exist. I happened to meet a man in Omaha who had lived in the Antelope Valley. He once told me he helped construct a secret base on a mountain west of Lancaster, on land owned by the Tejon Ranch. He never did tell me what the base was working on, or even the name of the base. Although the information from MUFON is amazing, I believe the History Channel's interpretation of the MUFON information is seriously incorrect. Thanks to the information sent by Dr. Mike, I know they purposely are omitting several important facts. They never mention the existence of sinister demonic activity related to these strange craft, nor the agreements our government and others have made to secure advanced technology from the Fallen Ones.

Casting the Dreidel

These revelations are having a very big influence on me. I now am aware of the ongoing conflict between Satan and Ha Shem. The reality of this loving God-Man Jesus. This is now passing from my brain to the depths of my soul. My heart is pounding as I open the drawer of my night stand and take out my Gideon Bible. I turned to the back pages where I previously had seen a page entitled How to Become a Christian. There are three passages of Scripture on the page. I slowly read each of them. Just three easy steps to be saved. I must confess my sins. I then must declare my commitment to repent of my sins. Then I am to ask Jesus to come into my heart, to save me and become my Lord. As I considered these three steps, my mind is replaying the conversations I had with the three guys in the pizza café going to summer camp, and the waitress Judy Morgan at the restaurant in the Inner Harbor. I am wondering if they are even now offering up prayers on my behalf. I know this issue in my life needs to be settled, once and for all. I must come to a decision, either accept Yeshua or reject him. I can no longer live with one foot in secular Judaism and the other as a potential but uncommitted Christian. My heart is telling me "This is your time Jacob Cahn. You must make a decision. You must cast the dreidel." I mentally threw it down. The decision has been made. I dropped to my knees and prayed aloud, "Yeshua, I believe you are the Son of the Creator God of this universe. Today I am placing my trust completely in You. I now confess my sin to you, each and every one of them. You know all of them, and I repent. I repent of my failure to open my heart to you over these many months. Please forgive me and become my Lord and Savior. I will seek to follow you for my remaining years and make you the Lord of my life. Amen."

As I spoke these words there comes a peace I have never felt before. It is welling up inside of me. I'm not able to contain it. I begin shouting out loud, over and over again: "Jesus, you are the Son of God Almighty, Jesus, you are the giver of new life in the Holy Spirit, Jesus, you are my healer and defender, Jesus, you are my soon coming King."

Time seems suspended. I remained on my knees for the better part of an hour, just praising the Lord. Now I know what I know from the inside out. The Lord and his Holy Spirit are real and living within me. My fear, uncertainty of the future, and the heavy load of anxiety he has replaced with a profound sense of inner peace and joy. The conflict I have struggled with for many years between my secular Jewish background and Yeshua's calling upon my life, is now totally resolved. Hallelujah! This son of Abraham is now a humble servant of Yeshua Ha Messiach.

I turned again to the instruction page in the Gideon Bible. The page urges new converts to tell their family and friends about their decision. This will be very difficult for me. To tell my father may possibly result in a permanent breach in our relationship. I realize Dad has to be the first person on my list.

I picked up the phone and dialed his cell. "Hello, Dad."

"Hello, Jacob. It's nice to hear your voice. Is everything going well for you?"

"Everything is great, Dad. I have been accepted by the Institute. They are holding a press conference tomorrow evening at six."

"Mom and I would be thrilled to death to see you, even for a quick trip. Do you have time to visit us?"

"No, I really don't have time Dad. I just called to bring you up to date on things here. I start my classes next week. They are putting my stipend directly into my checking account, and all of my tuition is fully paid for the year."

"That's great, Son."

"Yes, and I have made some great friends here. I really like Baltimore."

"I'm glad to hear you are making a place for yourself there. Are any of these new friends a serious prospect for marriage? Your mom keeps wondering how soon she is going to be able to brag about her grandchildren."

"No, Dad, there's nothing on my radar screen about marriage. I have another important announcement to tell you about."

"What kind of an announcement?"

"As you know I like to do a lot of reading. I have been reading from the Tanakh, and from the Christian New Covenant. Dad, I have come to the firm conclusion Yeshua is 'the real deal.' Just minutes ago, I cast the dreidel and accepted Yeshua as my Messiah."

"What did I hear you say? You are now a believer in 'that man' Jesus?"

"Yes Dad, I am now a completed Jew."

"There is no such thing as a completed Jew. You are simply a Jew, nothing more and nothing less. You are a son of Abraham and the inheritor of over three thousand years of Jewish history."

"You are correct, Dad. I am a son of Abraham, but also a child of Ha Shem."

"Oui vey! First you decide to become a geneticist, and now this. I don't want you to say anything more about this to Mom or Emily. Let me know when you come to your senses. Do you understand me?"

"I understand what you are saying Dad, but my decision has been cast. I will never change my mind."

"Goodbye, Jacob."

As expected, Dad hung up the phone. He refused to hear me out.

I know of no way to heal this breach in the relationship.

I begin to get the blues, but I don't want to go there. Then I remembered Pastor McGuire saying Allen and Alice were scheduled to arrive back in Atlanta today. Perhaps they will take my news better. I dialed Al's number. His familiar voice came on the line.

"Hello."

"Hi, Al. It's me."

"Hey, Jake, great to hear your voice. I'm so glad you called."

"I take it you both are safely back in Atlanta."

"Yes, we arrived just last night. We are happy to get back. Things are very chaotic in Northern Nigeria. The State Department advised us not to go back until the area quiets down. We witnessed unbelievable atrocities toward Christians. It is amazing how the Lord blessed us faithfully with his protection in the midst of total chaos. We had the experience of witnessing many salvations and miraculous healings. What's going on in Baltimore?"

"Well, I have been accepted to the human genetics program at the Institute. They are holding a press conference tomorrow evening at six."

"I knew you would be, Jake. I'm sure it won't take you much time to get your teeth into the program."

"No, it won't. I already have challenged the status quo here."

"How's that?"

"I locked horns with a professor over Macro-evolution. He wields a lot of power in the Institute. Even so, I went ahead and challenged his naturalistic view of the origin of life."

"You did?"

"Oh yeah, I certainly did. My zeal may have put my career in jeopardy, but I intend to stand up for the truth and hold my ground. Political correctness over the sacred ground of Darwinism is an affront to good science."

"I agree with you, but you need to remember you are only a lowly grad student. You may have jeopardized your tenure there. You need to be wise in how you communicate your positions."

"Yes, I know. Everyone is telling me the same thing, but I will not yield on this. Anyway, I have more news to tell you. I have accepted Yeshua as my Lord and Savior."

"Wow! Wait a moment while I put Alice on the phone."

I can hear Al telling Alice about my salvation as she comes to the phone.

"Hello, Jacob. What's this Al is telling me? You have accepted Yeshua?"

"Yes, it's true, Alice."

"That's so cool, Jake. I am just thrilled for you. Given time, I knew you would come to accept him. God told me you are his when Rachel and I were arguing at the restaurant."

"Really? Well your word from Yeshua has come to pass. I hoisted the white flag and surrendered to Yeshua."

"Al and I had the opportunity to talk with Pastor McGuire on our way home from the airport. He brought us up to date on Rachel. We are so sorry things didn't work out for you two. I feel a little responsible. After all, I am the one who brought you two together."

"Don't let it burden you. Rachel and I shared a wonderful time together, but marriage was just not in the Lord's plan. He is doing something new in my life, and he has called me to remain single."

"You will never be lonely now, Jake. You have Yeshua with you twenty-four-seven."

"Very true, Alice. I do worry about Rachel. I'm not sure what the future holds for her."

"We will continue to pray for both of you, Jake. Please keep us informed on your current situation at the Institute. I have to rescue our dinner on the stove, so I'm going to let Al finish our conversation. Take care."

"Goodbye, Alice. Are you there, Al?"

"Yes, I'm listening."

"I'm not going to keep you any longer. I just wanted to give you the news. I need to call Terry and Ramona with the news also."

"Ramona? Who's she Jake?"

"Ramona is a second-year grad student who has been helping me get oriented here. The Institute appointed her as my student guide. I introduced her to Terry and the rest is history. They are now engaged. Ramona accepted the Lord just last week. She even experienced a miraculous healing. Her left leg was an inch shorter than the right one. She injured it in an auto accident when she was in high school. The Lord miraculously healed her last Sunday. Ramona and I will probably get baptized together this Sunday at Terry's church in Timonium."

"Wonderful. It seems the Lord is working overtime in Baltimore.

Thank you for calling us, Jake. We love you and will be praying for you."

"I will be doing the same for you two, Al. God bless."

I hung up the phone and called Terry. He answered right away. "Terry, this is Jake."

"Hi, Jake. I was just now thinking about you. Is everything going well?"

"Yes, thank you. The press conference is tomorrow evening at six. o'clock. I will be studying these course materials all week, trying to get a running start on my classes. Are you saying Ramona is getting baptized this Sunday?"

"Yes, she is. Do you want to come along?"

"Actually, I am wondering if the pastor might be able to squeeze another person into the tub on Sunday?"

"Well I'm sure he can. Who is asking to get baptized?"

"I am! I finally surrendered to Yeshua."

"Wow, Jake, this is awesome! Everyone here will be thrilled when I tell them the news. I will call Pastor Grafton and arrange everything. The baptism will be during the service. I assume you remember where the church is. The service begins at ten o'clock. Pastor Grafton will want to have a brief talk with you before the service. Bring a spare set of clothes and a towel."

"Okay. This will give me a chance to see if my watch is really waterproof."

"Also, be sure to take your wallet out of your pocket before he dunks you."

"I'll will hand it to you. I guess I've covered everything. I'll meet you at the church tomorrow about half past nine."

"One more thing, Jake. We will be inviting our friends to the farm after the service to celebrate both of your baptisms. We're going to barbeque a pig. Not really. Mom is marinating a brisket. Also, Ramona has something to show you, so plan to spend the afternoon with us."

"I would guess Ramona will be showing off her diamond ring to me."

"Yes, but she has something else to show you as well which will totally amaze you."

"I can't imagine what she's come up with. Do I get a hint?"

"No. You are too good at hints. Wait until Sunday. Gotta go

Jake."

"See you on Sunday, Terry."

I am very keyed up. I need to take another run. As I made my way to the Rec Center, the sun is just setting. My past days without the Lord are now over. I am aware of my brand-new life in Yeshua. He has taken all of my pain, all of my shame, all of my uncertainty and buried them like this setting sun. My spirit now is at peace.

It is a quarter past six when I got back to my apartment. I needed to put something together for dinner. I did a quick inventory and came up with the makings for another size. I thawed out a hamburger patty in the microwave, opened another can of chili and beans, chopped up a Spanish onion, and spread out the roll, catsup and mustard bottles on my kitchen table. It took me longer to cook the size than to consume it. Then I sat down and enjoyed watching the Orioles game. The game wasn't over until eleven when I turned in.

New Brothers in the Lord

How do I know there is going to be something special happening tonight? Perhaps it is because I am a brand-new completed Jew. The vision began as soon as I closed my eyes. I am sitting beneath a tree in a beautiful meadow. The sun is just coming up over the hills on the horizon. Already everything is brightly illuminated: the clouds, the flowers in the meadow, the brook, lined with trees winding through the meadow. I don't see anyone, yet I know I am not alone. I hear myself saying, "Hello, is anyone here?"

I watch the vision slowly zoom in on the area of the small brook. There are two men wrestling with one another. Yes, it is the patriarch Jacob wrestling with the Angel of the Lord. I'm at the ford Jabbok. There is no sound. The vision then fades slowly away. I know the message. I have wrestled with this Angel of the Lord also. The ancient contest has been mine also. This vision has confirmed the climatic change which has taken place in my life as it was in the life of my ancient namesake. The Lord is with me, of whom shall I fear?

My thoughts returned to the time when Pastor Bill told me I was a seer, a prophet of the most-high God. I now face the dawn of a new chapter in my life. Until yesterday, I only knew about this man

Yeshua Now I know him intimately. I recalled one of the songs I heard in the little church in Timonium, *I Am a Friend of God*. Yes Jesus, I am your friend.

This new page in my life is very different. I now am able to hear the Lord's voice. He is spending time with me, telling me I am destined to know deep things. He also is continuing to give me stern warnings of a great deception coming upon the world. He tells me the Evil One will attempt to use my knowledge and abilities to aid him in this great deception. I must resist him. I must not yield to the temptations of fame, wealth, power, or authority. Not only must I pray, He tells me I must learn to spend more time with him to maintain this new intimacy. I cannot succeed without this close relationship with him.

My heart is pounding as the reality of this calling sinks into my consciousness. I am shaking with fear. I tell him, "Lord, I am not fit for this. I am an imperfect human being. Are you calling me to spend time conversing with you like Moses? Surely this must be above my pay grade."

The Lord answers me with an encouraging word and a strange prophecy: "Have you not read what my Apostle Paul wrote? My grace is sufficient for you? My power is perfected in weakness. Do not fear. Walk in unity with your brothers. You will meet three men. One I have sent will mentor you. He will open your eyes to understand what must be accomplished before the End. You will meet him in the Promised Land. His name is Zvi. You will meet other brothers there also. The end is coming soon. The great harvest is ripe. I have sealed you as my servant for this time. Be encouraged. You will prosper in the days ahead. My Spirit will pour out divine wisdom and understanding upon you. Do not fear. I will reveal all when it is time. Peace be upon you."

The incredible vision quickly melts away. The message is over. I wonder when and how these things will take place. Has the Lord also spoken to this man Zvi about me? Will he know he is my brother and mentor? I am shaking like a leaf. I remember reading in Chapters Seven and Eight in the book of Revelation about the 144,000 sealed for the time of the end. I now know I am one of them, destined to do mighty things, to remain single and celibate, humbly serving the King of Kings. The thought is overwhelming. I slept peacefully the rest of the night.

The Baptism

I awake, energized, and ready for the day of my baptism. I seem to have been given a special touch. The clock on my bed stand reads 5:10. After my shower, I put on jeans and my cowboy shirt. I cooked up some oatmeal and brewed a pot of coffee. It was still an hour before I needed to leave for the church. I turned on the morning news. There has been another earthquake, this time in New Zealand. Unlike the previous quakes where little damage had been done, this one was a real shaker. It measured 7.9 on the Richter scale. A tsunami is due to arrive in the Philippines in six hours, but it is not expected to cause extensive damage there.

At nine o'clock, I turned off the television, packed some extra clothes and a towel in my tennis bag, and left for Timonium. Thirty minutes later I arrived in front of the church. Pastor Grafton is there waiting for me. He shows me into his study. The room is quite roomy. There is a large oak desk, and a rawhide leather chair. There are four other chairs in the room. Tall bookshelves cover three of the four walls. This man is no itinerate country preacher. There are at least five-hundred books in the bookcases. Pastor Grafton invited me to take a seat.

"Jacob, I would like to hear how you came to the Lord, how you came to faith, and placing your trust in Jesus."

"Pastor, I have known the truth about the Messiah for some time, but I was not able to move this understanding from my head to my heart until recently. I did not know what it means to intimately know Yeshua until just the other day. A good friend of mine in Atlanta, Dr. Michael Stein, sent me a translation of a scroll discovered by our troops in Iraq. The scroll was sent to the Smithsonian for safe-keeping, but our government is going to be sent back to Iraq soon. Dr. Mike is trying to translate this and some other ancient documents before they are returned. The scroll is quite old and contains information linking it to passages in the Tanakh."

Pastor Grafton was very interested and asked me to share the scroll information with him at a later time. I promised to mail him a copy. Then I showed him the Gideon Bible I used as a guide in accepting the Lord. I told him I had followed the help section, humbling myself and praying a repentance prayer. I told him how I had promised to make Yeshua the Lord of my life and how I experienced his filling me with the Holy Spirit. This experience has giving me a joy I have never before experienced. I told him I feel like a brand-new person.

"Jacob, you are a new person. I am overjoyed with the story of your conversion. I will be proud to baptize you today. Let's go get you baptized."

Terry and Ramona drove up outside as we were leaving the pastor's office. I asked the pastor if he wanted to talk to Ramona, but he said he had already met with her. The pastor stood at the door welcoming the faithful as Terry, Ramona and I entered the sanctuary. Ramona and I sat down together in the front row next to the three others being baptized. Everyone in the church stopped by to congratulate Ramona on her conversion. The congregation is in a happy mood, looking forward to the baptisms. None of the congregation knew beforehand of my own conversion, so they were very surprised when Pastor Grafton announced my name along with Ramona's and the three others to be baptized.

The pastor brought a sermon from Chapter Seventeen in the Gospel of John. The Pastor pointed out the fact to the congregation the prayer in Matthew Chapter Six, widely known as The Lord's Prayer, should be called the disciple's prayer. Only then did I realize how wonderful this passage is. It truly is the disciple's prayer and a great model for my own requests of the Lord.

After the sermon, Pastor Grafton asked all those being baptized to come forward to receive communion. Then he released us to get ready for the baptism while the rest of the church received their communion. The band begins playing songs as the pastor walks into the baptismal pool. Ramona is the first to get baptized. She gives her wonderful testimony of how the Lord touched her. There isn't a dry eye in the place. I am the last one to be baptized. I repeated what I told the pastor earlier in his study. As I came up out of the water, the congregation is wildly clapping. Then the Pastor prays a blessing over the congregation and concludes the service.

When we arrived back at the farm, I was anxious to find out what Ramona's big surprise is. Terry tells me I will have to wait until after dinner. About forty or so people from the congregation have shown up for the celebration. I tried my hand at horseshoes while everyone waited for Eldora to get the dinner on the table. I learned the game of horseshoes is a major pursuit here in rural Maryland. Two of the men brought their own set of horseshoes. One of the men pulled out of his case a set of silver-plated horseshoes. I lost every game, but I didn't mind. When Eldora rang the triangle, everyone sat down to a dinner of brisket, potato salad, fresh peas, and choice of non-alcoholic beverages. I sampled something they called elderberry juice. I switched to coke.

As we are finishing our meal, I noticed Ramona was missing. "Terry, I don't see Ramona. Is she hiding someplace?"

"No, she is down at the training track getting ready to ride. Follow me."

Almost everyone walked with us down to the training track. Terry pointed to the starting gate located in the chute on the far side of the training track. "Jake, do you see who is up on the Black Gold over there?"

"Yes, I see Ramona. She is there behind the starting gate, perched on top of the horse like a jockey."

"Yep, she's fixing to give Black Gold a good workout. He is our best three-year old stallion, Jake."

"But she is riding on top. Isn't it too dangerous for her to ride like a jockey?"

"No, she is a natural. I've never seen anyone ride any better, not even Ron Turcotte. She knows how to get the best out of a horse, Jake."

I watch Ramona calmly walk the horse into the gate. Howie is standing at the side of the gate with his hand on the starter button. Suddenly the bell rings and the horse and rider come charging out of the gate. They fly around the first turn and down the home stretch in front of us. Ramona takes the horse around the second turn and down the backstretch. I watch as she rounds the far turn for the second time. She is leaning forward, perfectly balanced above the horses' withers. She really gets into the horse with her whip down the home stretch. They flash past us. I watch as she pulls the young stallion up smartly and walks him slowly back to where we are standing.

Terry is smiling. "She did seven furlongs in 1:27 flat, Jake. Not bad for a quarter mile track. Did you see how fast she broke out of the starting gate?"

"Terry, I don't believe this. Aren't you afraid she will take a spill?"

"Not this girl. She loves to ride on top. To ease your mind,

Dad and I taught her how to dismount safely in the case the horse stumbles or breaks down. We plan to give her a mount at Pimlico this year, maybe when Black Gold makes his first start as a maiden."

"Really?"

"Sure, why not? She can work out with some horses in company in the mornings right away. She should have it all down by the time the meet gets underway."

"You two are meshugenehs Terry, simply crazy."

"If you mean we're crazy about one another, you are right. Horses are a part of our lives, Jake. You will just have to trust us and the Good Lord to know what we are asking Ramona to do."

"I suppose so. Well, this is certainly some surprise. I would never have guessed it."

A few minutes later, Ramona has Black Gold on the walker. Her face is beaming as she joins us.

"Ramona, I have never seen you this happy. What happened to all those blues you used to carry around with you?"

"They are all gone, Jake. I am gloriously happy."

"I should say you are. I'm very impressed. And your hair matches the color of Black Gold's mane."

"Of course, Jake. When I ride him, we are one and the same, a powerful machine."

We all walked back to the house where Eldora and the other women had replaced our dirty plates and silverware with clean plates and new silver. An assortment of home baked pies is displayed on the dining room buffet. Ice cream is available in five flavors. I decide on a piece of luscious-looking lemon meringue pie. I followed this with a smaller piece of pecan pie, and topped it all off with a sliver of Eldora's apple pie a la mode. I am now too full to continue any further pie testing. The men are standing outside now, around the fire pit telling windies. The flood gates are open and I am being treated to a lot of country stories about dogs, pigs, chickens, and every other critter you can think of.

Howie proceeds to make his own contribution. "Guys, I was ridin' with Ed the other day over near Andover Creek and we saw this big hole in the ground. Thet hole was so deep we couldn't see the bottom of it. We were both curious how deep thet hole was. Ed found this old transmission sittin' nearby. It didn't look like it was any good, so we jist threw it down that thar hole and see how long it'd take to hit the bottom. Just after we threw it in, this dawg come flying by doing sixty miles an hour. Darn if he didn't just jump right down thet same hole. We didn't know what to do. A few minutes later this old guy comes by. He wants to know, 'Did y'all see my dawg?' I told him, 'Maybe so. A few minutes ago, we seen a dawg fly right by us'n. Darn if he didn't up and jump right down thet hole over thar.' This ole guy thinks a minute and then says, 'Well thet can't ha been my dawg, 'cause I dun left him tied to an old transmission I found around here someplace.' Thet's the truth, ain't it Ed?"

Remembering how quickly I learned the proper use of 'y'all' in Atlanta, I thought I had better leave before I picked up any of these Maryland country words.

It took a while, but I finally was able to get a word in edgewise. "Everyone, I want to thank you for helping me enjoy one of the best days of my life, especially you ladies who baked all those delicious pies. I only wish I had more room to try out a few more of them. But it's time for me to leave. Thank you and may the good Lord continue to bless you all like I've been blessed today."

Eldora handed me a whole cherry pie and insisted I take it back to the apartment. A young woman also insisted I take one of her pies along as well. I noticed she had written a nice greeting and signed her name and phone number on the paper plate covering her rhubarb pie. Rhonda Sutherland is her name. Another young and pretty woman with a name starting with R. Why do I seem to attract these R girls?

Terry and Ramona saw me off. I leisurely drove back to the Bradford. Tomorrow at eight o'clock I will begin the seminar class for new students. My first in-depth course is Molecular Biology and Genomics. I sampled a piece of Eldora's pie and then went to bed.

17

The Training of a Geneticist

*"If a man therefore purges himself from these, he shall be a
vessel unto honor, sanctified, and meet for the master's use, and
prepared unto every good work."*
(2 Timothy 2:21)

Out of the Starting Gate

It is Monday morning. My training as a geneticist is now
officially under way. I have worked hard to see this day. The new
student seminar is scheduled to start at eight o'clock in room 305
of the BRB, and will include students from several schools, not
just the Institute. The first regular class, "Molecular Biology and
Genomics," begins at ten. This course is given by the Medical
School. In the afternoon, the course, "Fundamentals of Genetics,"
will meet at 2:00. Then at six tonight is the press conference. I am
still thinking about yesterday's experience at the farm. I am totally
amazed by the skill Ramona displayed, especially how quickly she
came out of the starting gate. I want to do the same thing today
at the seminar. I want to start this year breaking in front of the
herd. I left the apartment twenty minutes after seven and arrived
at the BRB by a quarter to eight. Dr. Hobbs was again present to
welcome the incoming students. There were students from several

other departments besides the Institute. I counted 30 students in the room. Then Dr. Hobbs introduced Professor Arnold James. Dr. James began the Seminar by asking each individual student about their major, family, and outside interests. He is making notes about each student. Dr. James called Joe Harmon's name right off the bat. He didn't call my name until half-past eight.

"Mr. Jacob Cahn, where are you?"

"Right here Sir."

"Can you give me some idea why you are here in this seminar this morning, Jacob?"

"Yes. I am proud to say I have been accepted to the McKusick-Nathans Institute of Genetic Medicine. My BA was pre-med at Emory University. I succeeded in making the honors program there. It was there I first became interested in human genetics. I am looking forward to a future in basic research."

"Where do you hail from, Jacob?"

"My home is Omaha, Nebraska where my father is a general practitioner and teaches at Creighton University Medical School. My father graduated and interned at Emory. The family expected me to attend Emory also, but I decided to change my major."

"Have you a particular research area in mind, Jacob."

"Yes, I do. I hope to work on fetal development. It is my hope to develop a methodology for recording precursor neuron development from outside of the womb, avoiding any harm to the fetus. I also am considering research into human memory, its location in the brain and the relationship of memory to creativity, and other related attributes."

"Well, both of those are worthwhile objectives. Thank you, Jacob."

It took another hour before he reached the last name on the list. He then said we were free to leave. I had just a half hour before the class in Molecular Biology and Genomics begins. When Joe and I arrived, the room is almost full. We found seats near the back of the room. According to the course listing, Professor Dr. John Silvers is the course instructor. The title of his first lecture is "A Summary of the History of the Department of Medicine." Dr. Silvers began his lecture precisely at ten o'clock.

Ladies and gentlemen, this course, Molecular Biology and Genomics, introduces the student to the field of biotechnology. We will be concentrating on the techniques used in the study of genetics and molecular biology. There's a lab component in this course teaching the techniques used to map genes, sequence gene sets in the DNA, and map complete genomes of selected organisms. The course is one of eight basic science courses offered by Johns Hopkins University School of Medicine. Throughout the history of the Department, it has been at the forefront of basic biomedical research, contributing to the training of many leading molecular scientists. The Department of Microbiology was established in 1959, with Dr. W. Barry Wood as its founding Director. The early faculty of the Department of Microbiology included virologists, bacteriologists and immunologists. Daniel Nathans and his colleagues first used the restrictive enzyme discovered by Smith to divide the genome of the tumor virus SV40 into specific fragments whose functions then are individually studied. This new methodology greatly assisted the analysis of gene function and proved to be an important foundation for the subsequent recombinant DNA revolution. During the 1960s a number of the staff became Department Chairs at other universities. The Department continues to be an international leader in biomedical research and training. Our fourteen faculty members in the Department are versed in a broad range of research topics. You may be interested to know the son of Daniel Nathans is on our team. He has been awarded a Nobel prize for his research.

Our areas of principal emphasis includes the following: cell cycle control, gene expression, DNA identification and replication, signal transduction, and developmental biology. We also are leaders in the fields of microbial pathogenesis, immunology, DNA transposition, telomere structure and function, RNA catalysis and the molecular biology of vision and olfaction. Since I have now given you a bit of background of our department, is there a brave soul here who would like to give me a brief definition of this complex field?"

No hands are raised. I realized this is another golden opportunity to make a good impression. I racked my brain, trying to recall the definitions I read the previous week. It is a great opportunity, too good to pass up. I raised my hand, hoping the words will come back to me.

The Professor says, "I see the hand in the back row. Please come down to the microphone, state your name and your major."

I made my way to the front of the large room. "Good afternoon.

My name is Jacob Cahn, first-year graduate student at the McKusick-Nathans Institute."

"Good afternoon, Jacob. Please proceed with your definition of this field?"

"I want to define the term genomics first. The genome includes the complete DNA set within an organism. The field of genomics is a three-part discipline applying recombinant DNA, DNA sequencing methods, and bioinformatics to sequence, assemble, and analyze the function and structure of genomes. The field includes efforts to determine the entire DNA sequence of organisms and fine-scale genetic mapping. The field also includes studies of intragenomic phenomena such as heterosis, epistasis, pleiotropy and other interactions between loci and alleles within the genome. In contrast, the investigation of the roles and functions of single genes is the primary focus of molecular biology. A segment of DNA involved in producing a polypeptide chain can include regions preceding and following the coding DNA as well as introns between the exons. The chain is considered the unit of heredity. Formally genes were called factors. The study of single genes and gene expression makes use of very sophisticated high-tech tools of biotechnology. This branch of molecular biology also studies the use of micro-organisms to perform specific industrial or medical processes, using altered bacteria as surrogate hosts to create beneficial and useful new cell types."

"Thank you, Jacob. Very well done. I'm sure many of you are unfamiliar with the terms mentioned by Mr. Cahn. Let me assure you. By the end of these six weeks you will become very familiar with them. Judging by the reluctance to raise hands today, I'm going to let you go early so you can hit your textbooks. Tomorrow we will meet in the genetic sequencing lab to demonstrate some of these tools to you. The lab is located just down the block at 725 N. Wolfe Street, room 418. You are dismissed and I will see you tomorrow at nine o'clock at the lab. You are now excused."

I am elated to have been given a second opportunity to gain an early start on this course also. But I knew I would have to study even harder to stay ahead of the class. The syllabus for this class requires a couple of book reports, and one oral presentation. The lab for the class is scheduled for Friday afternoons from one to four-thirty.

After the Molecular Biology and Genomics class, Joe suggested we have lunch together. "Jacob, there's a small pizza parlor across Broadway and down a block. How does pizza sound to you?"

"Fine, Joe. Let's do it."

We walked to the pizzeria and ordered a large pizza. We found a vacant table and waited for the pizza to arrive. I am anxious to hear what Joe thinks will occur in the afternoon session.

"Joe, I have to admit I am a bit nervous about this afternoon's session. What direction do you think Professor James will take in his Fundamentals of Genetics course?"

"I'm not sure, Jake. I have the feeling he won't launch into teaching mode yet. I believe he will continue to size up his class."

"You are probably right. I think he will continue to ask questions. Perhaps his questions may center on the history of human genetics rather than family backgrounds. To use an analogy, this will provide a golden opportunity to get a quick beginning out of the starting gate, Joe."

"Yes, I think you are onto something there, Jake. Dr. James will want to get a handle on our depth of understanding and knowledge of genetics, human genetics in particular."

We continued to pass ideas back and forth over lunch. After the last piece of pizza, we walked back to the BRB. We arrived early, and it worked out well for us. Joe and I agreed to sit as close to the front of the room as possible. We found two seats in the middle of the second row. Once the last stragglers entered, Professor James began the afternoon class.

"Ladies and gentlemen, I am going to ask you a series of questions about the history of human genetics. If you wish to answer the question, please raise your hand. When I select a person, stand up, state your name, and answer the question as succinctly as possible. Here is my first question; Who would like to tell me about the namesakes of this Institute?"

I raised my hand along with perhaps seven others. Dr. James has a good memory. As he recognized a student, he called them by their first and last name. Several students are versed on who Dr. McKusick was, but none said much about Dr. Nathans. He finally called my name. "Mr. Jacob Cahn, do you have something to share with us?"

"Yes Sir, I do. I believe we have pretty much exhausted the information about Dr. Victor McKusick, so I want to say a few words about his co-researcher, Dr. David Nathans. Dr. Nathans came from very humble beginnings. His parents emigrated from Russia and settled in Wilmington, Delaware. His father lost his business during the depression and the family struggled to put food on the table. David was the youngest of eight children, and the only child to seek a higher education. He graduated from the University of Delaware in 1950, and earned his medical degree from Washington University in St. Louis in 1954. My own father graduated in the same year from the Emory University medical school in Atlanta. After a brief period working as a family physician, Dr. Nathans made the decision to pursue a career in basic research. He began working at the National Cancer Institute in Bethesda, Maryland as well as at Rockefeller University. In 1962, he was invited to come to Johns Hopkins. Dr. Nathans's early research focused first on viruses causing tumors in animals and then on cellular responses to growth factors, the mechanisms causing cells to grow and multiply. In 1969, Dr. Nathans spent six months at the Weizmann Institute in Israel working on animal cells and viruses. There, he received a letter from a Johns Hopkins colleague, Dr. Hamilton Smith. Dr. Smith had discovered a restrictive enzyme enabling the researcher to cut the DNA of a bacterium at specific sites along the double helix. At the time, Dr. Nathans was working with SV40, a small tumor virus causing cancer in primates, enabling cells in tissue culture to become cancerous. This letter from Dr. Smith stimulated him to consider using restrictive enzymes to analyze the genes of SV40. Dr. Nathans also used the enzymes to break up the DNA found in cancer cells in an effort to learn what made them different from normal cells. He found the restrictive enzyme cuts the simian virus in ten places, creating eleven well-defined fragments. Then he found new ways to use the cutting technique to help determine where genes begin and end in the DNA of the virus. This helped him find a gene in the virus giving the order for production of a tumor-making protein. Without his restrictive enzyme contribution, we might not yet be able to break the DNA sequence into subparts."

"Well said, Jacob. Can you add anything more?"

"Yes Sir. In 1978, Dr. Nathans shared a Nobel Prize in Medicine with Drs. Smith and Arber of Basel, Switzerland. Dr. Arber discovered the type of restrictive enzyme to dismantle genetic material at specific points. The Nobel selection committee at the

Karolinska Institute in Stockholm predicted their work would help in the prevention and treatment of malformations, hereditary disease and cancer. Restrictive enzymes led to the now-standard prenatal tests for cystic fibrosis, sickle cell anemia, muscular dystrophy, hemophilia and many other genetic diseases. In 1993, Dr. Nathans won our Nation's highest scientific award, the National Medal of Science."

"Jacob, are you reading this off of your tablet or do you have a photographic mind?"

"The latter Sir."

I am pleased. I accomplished my goal of getting out of the starting gate fast. Dr. James continued to assess the students in the class, asking a variety of questions about the history of human genetics, its heroes and its important contributions to the field of medicine. Joe came through with a thorough presentation on the status of research on the Ebola virus. He also clearly explained the two major processes used to sequence genes. At the end of our first meeting, Dr. James passed a paper around asking what topic we wished to present to the class. Not wanting to stir the pot any further than I had already, I titled my paper simply "The Evidence for Evolution." Dr. James reminded us of the traditional student/faculty retreat scheduled for the end of the first semester. We would be spending an entire week at Bar Harbor Maine at the Jackson Research Center. I am expected to present my initial research proposal to the faculty there.

In the afternoon, most of the new student's in the seminar were scheduled to take the Fundamentals of Genetics course as well. Joe and I arrived early and found seats in the second row for the fundamentals class. I again succeeded in gaining the respect of Dr. James as well as the other students.

Dr. James continued to use the same Socratic approach during the following days and weeks that followed. The response he received from his students using his different approach was interesting. The students became highly motivated and improved quickly. There were two students who did not do well under this pressure. Once the Professor discovered their weakness, he no longer called upon them to answer any questions. By the end of the second week, they were gone.

I was shocked. I told Joe, "This school is a torture chamber in disguise."

Although our classes had already begun, the Institute was just now announcing the new grad students. I returned to the apartment and dressed for the press conference. I arrived at the BRB five minutes early. The conference was on TV live. Craig Simpkin was not there with us. There was a total of twelve new students, Five women and seven men. Two were from South Korea and one was from Great Britain. It took only fifteen minutes and I was able to return home. It was a very full day. I had "hit the wall." I turned in and quickly fell asleep.

During the fifth week of the New Student Seminar Dr. James began calling on us to present our term papers. He called upon us in random order. My number came up on Friday. As I took the podium, a late arrival entered the room. He takes a seat in the back row. It is my nemesis, Doctor Altschul. Someone certainly is keeping the Doctor apprised of my activities. Perhaps I have him worried, but I'm not. I felt confident as I began my presentation.

"Good morning class. My name is Jacob Cahn. The name of the paper I am presenting today is "The Evidence for Evolution." I am sure most of you have encountered in your philosophy or biology classes the argument presented by William Paley for an external agency responsible for designing and producing all higher life forms. In 1802, he presented the analogy of a watch found in a field. His argument went something like this: Watches display design; therefore, watches are the product of a watchmaker. In biology, we might state it this way: Organisms display design therefore organisms are the product of a Creator, capital C. The sceptics of this line of reasoning long ago pointed to the lack of uniformity between the two things the analogy compares, watches and organisms. However, it is fair to say the discovery of biomolecular motors and machines inside the cell has given new life to this teleological argument. The recent discoveries in nanoscience and nanotechnology have exposed the elegance and sophistication of the component designs within the cell. Whereas the bacterial flagella operate at nearly 100 percent efficiency, the equivalent human designs only function as high as 65 percent. To summarize, at the nano level we can find a much closer analogy reinforcing his reasonable conclusion. Life's chemistry is not the product of a blind natural Darwinian process, but is the end product of an intelligent designer, capital D.

This morning, I am going to challenge the holy grail of naturalistic processes responsible for the myriad forms of life from yet another direction. If one discards the intelligent design line of thinking, i.e. the watchmaker analogy. I am here this morning to present an even more convincing argument to reject the evolutionary theory, namely the problem of the chicken and the egg. Did the egg or the chicken come first? Did DNA first produce proteins, or did proteins first invent DNA? Biologists commonly refer to DNA as consisting of a complex self-replicating molecule because its structural properties make it possible to generate two identical daughter molecules from the original parent. In truth, it is not possible for DNA to replicate itself all by its lonesome. DNA replication requires a myriad of proteins. The bottom line is the fact the synthesis of proteins and the replication of DNA are mutually independent processes. Proteins cannot be produced without DNA and DNA cannot be replicated without proteins. I'm not going to take time to follow the complicated and somewhat convoluted processes involved with protein synthesis and messenger RNA. Let me just summarize it and say the RNA polymerase core is capable of synthesizing mRNA on its own, but cannot recognize the location along the DNA strand where the gene begins. A myriad of proteins making up the ribosomes play an integral part in helping the mRNA bind properly to the ribosome and causing the newly formed protein chain to properly fold. Now you may be thinking, perhaps there is a way of overcoming this problem at some earlier period. Perhaps we just have not gained access to this earlier life form. I maintain this is equivalent to the improper view of theologians when they explain a missing link by reference to the "God of the Gaps" hypothesis. What's good for the goose is good for the gander as well. For the evolutionist to say this vital earlier life form just hasn't been found yet. This is equivalent to the theologian's false assumption that missing pieces of the puzzle can all be attributed to the Almighty. I read in the Bible another important fact for your consideration. Most biologists fail to recognize the difference between past and present. Unlike astronomers who can investigate past events because of the time required for those events to reach our earth at the speed of light, biologists can only observe changes occurring now. Changes of species, or Biblical 'types' have not been discovered taking place. But we should expect to find multiple cases of new life forms at a rate equivalent of what we find in the fossil record. I will agree there are a few exceptions in the case of very simple life forms, amoebas and such. In the book *The Edge of Evolution* by Michael Behe, he certainly demonstrates the limits of favorable random mutations. Behe found in the study of the mosquito that the maximum sequence of beneficial mutations to be no greater than two steps. Hardly enough time to accomplish the diversity of species over the geologic time available.

Given the lack of time available to accommodate favorable mutations, this community of Darwinian enthusiasts has proposed to save their failed hypothesis by proposing a changing rate of mutations in our species, the so-called Human Accelerated Region (HAR) on Chromosome 20. This has yet to be demonstrated in the lab. What about the rate of evolution on the remaining life forms which do not contain this chromosome? I believe it is time to jettison these failed hypotheses and move on.

So, what can be responsible for this diversity of life on our small planet? I submit to you another hypothesis, if you will. The Bible teaches this awesome God, this Designer of life, rested on the seventh day from all of his creative work. We would not encounter new species on this seventh day. Perhaps a holistic scientific approach accepting the existence of a personal God who desires to bless us with life in abundance might present a more plausible, even truthful solution. Thank you for your attention."

I received a lively round of applause. I watched as Dr. Altschul quietly slipped out of the room. Doctor James asked me to remain at the podium for questions. I detected a level of fear hanging over the students. There were only a few questions. Professor James thanked me and I took my seat. The presentation of term papers continued until noon.

The class schedules at JH are intense. In the sixth week, we began the Introduction to Biometrics class. It overlapped the two previous classes by two weeks. The Biometics class is another full eight-week course. The class is offered by the Department of Biometrics, not the Institute. The Biometrics class met in room 725 of the Preclinical Teaching Building located several blocks north of the BRB on North Wolf Street. It was scheduled to start on Monday of the last week of November and would not finish until the middle of January. I began to understand the mystery why these classes were scheduled to overlap one another. Research training begins soon after the student enters the program. The first part of each class teaches the theoretical and informational basis of the field and the final weeks involve actual lab experience.

I slept late and quickly grabbed a sandwich from the food court before heading for the first biometrics class lecture. When I arrived, Joe had saved me a seat in the classroom. It is still fifteen minutes before the class is scheduled to begin. I again hope to get a jump on the other students in this class as well. The classroom is very large. I estimate there are more than a hundred students in the room. At ten minutes after eight the professor shows up.

"Good morning everyone. My name is Dr. John Silvers. I am your door to the exciting field of Biometrics offered by the Department of Biometrics. This course actually consists of two separate parts, Pre-Genomic Biometrics, and Post-Genomic Biometrics. In the first four weeks, we will concentrate on the skills developed in the earlier history of the field. The second four weeks will cover the recently emerging fields of comparative genomics, levels of gene expression, DNA microarrays, and functional and structural genomics."

Dr. Silvers finished his short introduction to the class and then said we were free to leave. I had no opportunity to get a jump on the other students. I am wondering how this professor will be able to cover such a large amount of material in eight short weeks.

On Monday, the first week in October, the Medical School offered a class on Cell Structure and Dynamics. This course examines the cell at the macroscopic level, including metabolic processes as well as cell structure. I spent several days thoroughly preparing for this course. The technical terminology, so much a part of the genetics of the cell, can be overwhelming for most students. I actually revel in its complexity. The human body is an amazing machine. I can see clear evidence of the Creator in its elegant design, its amazing ability to heal itself, to defend itself from virus attacks, and the elaborate mechanisms protecting the cell from random mutation damage. It is simply astounding. The most difficult part of the course for me was interpreting the professor's heavy Spanish accent. Professor Carlos Cortez-Hernandez is a native of San Salvador. The six-week course prepares students for the next step in the learning curve. There was another class offered through the Department of Medicine, Genetic Pathways. It also overlaps other earlier classes. I am especially enjoying the increasing time I now spend in the labs. It is what I have always dreamed of doing. The Institute offered a second seminar beginning the middle of January and extending to the end of May. By now, we all knew what Dr. James expects of us. He had narrowed our cohorts to only ten.

These integrated classes in my first year passed very quickly. There was no opportunity for fun and games. I kept my nose to the proverbial millstone. Near the middle of December, I began to lay out plans for my research project. My first step in approaching the project was to schedule a weekend of intense prayer. My studies are continuing nonstop. I must somehow find time to work on my research proposal. I have studied right through the holiday season without a break. It is now the last week in February. Snow is on the ground and the temperatures have been hovering at freezing or below for the whole month.

In addition to the set schedule of classes, students are given the choice of taking elective courses offered from a wide number of departments. I felt Ha Shem prompting me to enroll in some elective courses. One course on my list is offered by the Robotics Engineering School. Since students begin research in the second year of their four years preparing for the doctorate, I need to add an engineering course in my tool kit for this second year. I decide to contact a Dr. Francis Glick, the head of the School to get his advice on an elective course helpful to my research topic. Also, I hope to share my research plans with him. I had just begun the three-week class in Cell Structure and Dynamics when I placed a call to Dr. Glick's Department to make an appointment. He agreed to meet with me at nine o'clock on the following Friday.

An Outlandish Proposal

The Robotics Engineering School is located in Hackerman Hall on the Homewood Campus. Dr. Glick is waiting for me when I arrived on Friday.

"Good morning, Jacob."

"Good morning, Doctor Glick."

"Well, let's get right down to it. I understand you are considering taking an engineering course."

"Yes, Sir. I have a couple of reasons for asking to meet with you. First, I would like your advice on elective courses pertinent to my research project. Secondly, I would appreciate your opinion as to the feasibility of my research project. Do you by chance remember meeting me several months ago?"

"Yes, I remember meeting you, but I don't recall exactly what topic you were then considering. I believe you were interested in the human fetus."

"Yes, that is my plan. I want to develop the technology to trace the movement of precursor neurons in the brain as they migrate during the first few weeks of life."

"Well, it is an admirable idea, but I doubt the odds are favorable for a successful result. Others have been working on this mystery of precursor neurons for some time."

"I realize it will be very difficult, but I have faith it can be accomplished."

"You say you have faith. What kind of faith, Jacob?"

"I'm referring to God faith, Dr. Glick. If we had more time, I could share some unusual events in my life with you. If I did so, you probably would call security and have me taken to the Department of Psychiatry downtown."

"Are you Jewish like myself, Jacob?"

"There is no way I can deny it is there?"

"Jacob, everyone in our family has accepted Yeshua as Messiah. We attend Beth Israel Congregation in Reisterstown. Our Rabbi is David Epstein."

"I certainly did not expect you to be a completed Jew. This is very encouraging to me, Doctor. Ha Shem has had his hand on me since I was just a young mensch."

"What congregation do you attend, Jacob?"

"None really. I came to the Lord very recently. I attended a small Methodist church in Timonium with my friend Terry Johnson. I believe you met Terry when we visited your open work area."

"Yes, I remember meeting him. Have you ever attended a Messianic congregation, Jake?"

"No, I never have."

"Well, you need to come visit us. I assume you know we completed Jews worship on Shabbat, not on Sundays."

"I'd like very much to pay your congregation a visit. Can I attend tomorrow?"

"Yes, we would love to have you attend tomorrow morning. The service begins at 10:30. Are you living near the Homewood

Campus?"

"Yes, I am in the Bradford Building."

"Have you ever been to the Reisterstown area?"

"No, I haven't. I'm only familiar with a few areas in Baltimore, the JH campuses, Timonium, and the Inner Harbor."

"Reisterstown is in northwest of Baltimore. The easiest route for you to take from Homewood is to use the Interstates. Take I-83 north as if you are going to Timonium, then jog west on I-695 to the I-795. Take I-795 north and turn off at exit 7A. This will put you onto Franklin Blvd going east. When you get to the intersection of Highway 140 and Reisterstown Road, turn left. Follow Reisterstown Road northwest. The road name will change to Main Street. Right after the name changes you will see a sign for Northminster Presbyterian Church on the right. The church sits back off of the highway. Just turn into the long church driveway and park in the back of the church. Enter the door with the red awning over it and climb the stairs to the second floor. Try to arrive a bit early because the church fills up. Do you think you can remember all of this?"

"Yes, I think so."

"Very good. We will be looking for you, Jacob."

"Is a Messianic service much different from the services in a protestant church?"

"Yes, some things are different. Of course, as Christians we all worship Yeshua, our Messiah."

"How else are your services different from a typical protestant church, Doctor?"

"Our worship is different. We primarily sing Jewish songs, often songs out of the Psalms. And we love to dance. But the major difference is at the deeper level of understanding of Scripture. When it comes to interpreting the Tenach, we have the advantage of our Jewish cultural understanding. We have some wonderfully anointed teachers and preachers, true servants of the Most-High God. There are other differences. One example is our great love of the State of Israel. All Jews are excited about the return of the Jewish people to their ancient homeland. Over two thousand years ago the return of the Jews was prophesied. Some gentile churches do not recognize this special relationship with the land God has given to us. We can go into more depth on these things at a later time, Jacob."

"I'm going to take you up on your invitation, Doctor."

"Very good, Jacob."

"You know, Dr. Glick, I believe the Lord might have put us together for a divine purpose."

"Perhaps he has, Jacob. I will believe it when he supplies the necessary funds to accomplish this ambitious project you have come up with. Do you realize the costs involved with genetic research?"

"I really don't. I assume this is where faith comes in. I just will need enough faith to move mountains, mountains of genetic data."

"Can you be a bit more specific? Just what do you have in mind?"

"I can be more specific, but you need to understand what I am going to share is very preliminary. My vision is to remotely mirror what is going on within the fetus without physically entering the womb. You are no doubt aware of the problem in studying the development of the human embryo. I need a way of sensing the replication and movement of neuron cells into the cortex of the cerebrum by these miraculous precursor neurons. The technology needs to record their movement in the brain in three dimensions. I have considered using slices from an improved MRI device. The sequential data would then be stored as imagery by a super computer. Perhaps a quantum computer will be required for this. The researcher would then display the imagery to analyze the emerging cell types and their designated functions."

"Just how many bytes of data are we talking about here, Jacob?"

"I really don't know, Doctor. I think they could run up as high as several trillion."

"A trillion bytes? Jacob, I don't want to discourage you, but your idea is beyond our existing technology."

"Yes, I know. Have you reviewed the capabilities of the new quantum computers?"

"No, I haven't done so in any great detail. I understand they are able to manipulate a whole lot of data very quickly. I'm not very knowledgeable when it comes to quantum physics."

"I believe this is the only machine capable of handling this amount of information within reasonable time constraints. These Qbit machines would greatly expedite our task."

"Our task? When exactly did this project of yours become our task?"

"Doctor, when Terry and I met you several months back, I knew you are going to be able to help me. Let me flesh this out a bit more for you. We will need to find a way of identifying the emerging new cell types. The machine must have the capability of AI. It must be taught to think. The analysis of such a large body of data demands the use of artificial intelligence. For example, once the machine identifies what is going on at the molecular cell level in the fetus, it will identify and store the information in a format allowing the researcher to analyze the interior components of the cells as they undergo division, replication mRNA and resulting cell type of the daughter cells."

"You make it sound so easy, Jacob. I have a question for you. Just how do you plan to remotely access and identify the cell components within an active, growing fetus?"

"I admit you have put your finger on the key question. I will need some help on this. Perhaps DNA, mRNA and other selected cell components may have some unique magnetic resonance or some other radiological signatures I can access remotely. Perhaps the shapes of these components can be identified using sound waves, or some other frequencies not harmful to the ongoing neuron development process. Perhaps a radiologic tracer can be injected into the umbilical cord or placenta."

"I think the problem of identifying these cell components will be by far your greatest challenge, Jacob. You are going to have to either search for someone who has undertaken a similar project, or you will have to do your own research into the response of micro fragments of the cells. No one is going to give you any funding for a research project with such complexity and cost."

"Do you have any suggestions on how I should carry out such a research activity?"

"Yes, I do. I believe the breadth of this project is far too wide. You are going to have to satisfy yourself with developing the technology to identify amino acid fragments and emerging cell types. You will have to leave the mapping of cerebellum development to others, probably a whole host of other researchers. Even narrowed to the development of the technology, it is an outlandish proposal, Jacob."

"What other kind of technology also appears to be needed?"

"The logical choice I suppose would be to alter the functioning of an MRI machine. But you won't be permitted to alter someone else's functioning MRI on our campuses or any other campus. These MRI machines are in high demand and quite costly. We only have three we share them among the entire Baltimore campuses. This is where your ingenuity is going to have to kick in."

"How expensive are MRIs to purchase?"

"I would guess you might find a used machine selling for perhaps a million dollars. Do you have a million tucked away someplace, Jacob?"

"No, I have very few funds. It sounds like I have to find someone who will pursue this idea with me, or someone who already has been experimenting in this direction. If I am able to purchase or rent a machine I would have to secure permission to make changes. Would you consider undertaking the necessary design modifications?"

"I suppose I would consider it, but as it stands now, your project is facing some very daunting issues. It is dead upon arrival, Jacob."

"I've heard there's something called 'Scalar Wave Technology,' but I haven't a clue whether this can be used as a substitute for radio transmissions. What ideas do you have?"

"You are messing with my mind now, Jake. It is plain you are going to have to find someone from the Physics Department to help you with these new technologies."

"Can you suggest someone?"

"There is a third-year physics student of mine who has a talent for thinking outside of the box. His name is Robert Manley. His friends call him 'Nanoman.' I have to warn you Jacob. Robert speaks in the language of mathematics more often than English. Sometimes he mixes them together and ends up getting very frustrated, frustrated at himself and with anyone close by."

"I'm pretty stressed about this problem myself. It sounds like he might be someone who can put me onto a solution. Do you happen to have his phone number?"

"Yes, I can give it to you. Write this down?"

"His number is 663-3431. I'll tell you what. I'll play along with your wild project for now. Let's sit down at the drafting table and take a few minutes to flesh out some kind of approach to put some wheels on this contraption of yours."

I followed Dr. Glick to a large drafting table at the end of the open work space area. He took a piece of graph paper out of the drawer and pulled a pencil out of his shirt pocket. He spent several minutes just staring down at the blank piece of paper in front of him. I searched for some way to encourage him.

"I'll tell you what, Doctor, I'll pray and you just write what comes into your head."

"Do you happen to have the phone number for the psychiatric ward at the hospital with you, Jacob?"

"Doc, they have been trying to diagnose my problem for months now, and haven't made any headway at all."

I began to silently pray, asking the Lord to reveal a clear and concise plan for my research project to this creative engineer. After several minutes, Dr. Glick began to write. He first created a list down one side of the sheet, grouping each entry into major categories. Initially he listed thirteen major categories: Department Research Proposal, Grant Application, Equipment Suppliers, Operating Environment Design, Computer Module Design, Database Design, Remote Sensing Design, Output Formats, Data Processing, Laboratory Procedures, Animal Trials, NIH Research Permissions, and Quality Control. Beneath each category are two and three tiers of sub-categories. Then he began to create a flow chart, transferring the activities from the list onto the chart and connecting the boxes with flow lines and decision points.

After he finished the flow chart he said, "Jacob, this is just a rough outline of your project. I want you to work on improving it, adding additional components, flow lines, and decision points. Fill in as many categories as you can. Once you complete the flow chart, your next step will be to develop a detailed proposal to the Institute based on the chart. I will send you some copies of successful submissions we have presented in the past. Before you submit your proposal and final flow chart, I want you to give me a chance to make some additional suggestions. Also, you are going to have to obtain permission from the Institute before you can apply for any grant funds. The outside entities also will require a very detailed proposal before they give you a single dime. Do you think you can move ahead on this soon?"

"Yes, of course. I will try to have my proposal ready to submit to the Institute's Executive Committee by the time we go to the Jackson Lab at Bar Harbor this Spring. I believe they usually select potential preceptors for research projects during those visits."

"I would advise you to request more than one preceptor for this project. You will need someone already involved in Fetal Development and another person skilled with the research methodology.”

“Would you have time to devote to the latter activity?”

“I can’t give you an answer at this point. Your research idea will be very hard for people to approve. Remember, each of your assigned preceptors must be a person you have good rapport with, someone you can trust. We live in an imperfect world, Jacob. On the positive side, you are working on the cutting edge of new knowledge. There is also a second environment you have to deal with. I’m referring to the political fleshpots of human pride, the desire for fame and fortune, unadulterated competition, and the survival of the fittest. You probably will have to do your early research under the radar. Any conflicts among the members of your team will quickly ‘deep six’ your project. Are you with me on this?”

“Yes, Doctor, I understand the playing field. I believe I have enough chutzpah to carry it through to a successful outcome.”

“Jacob, I have a question for you. Are you fully committed to this topic or are you open to considering another project?”

“I would say I am willing to consider another research project if it lines up with my interest in neurons.”

“I think you should not pursue this project to the extent you become blind to other ones. I wish you success, Jacob. May God be with you.”

We shook hands and I carefully folded the flow chart and stuck it in the folder Dr. Glick gave me. Then I left the Hall. The optimism I felt when I arrived quickly slipped away. I am now very concerned. Just who will the Institute appoint to serve on my research project, Dr. Altschul? If so, he might quickly deep six my proposal. My greatest concern is whether a technology even can be developed to identify the cell types and locations from outside of the womb. I will need to read as many papers on fetal development as I can find over the next few days. Perhaps there is someone I can work with who shares my vision.

I made my way to the food court in the Eisenhower Library located on the west of Beach Circle, only a few yards north of Hackerman Hall. After a bowl of French onion soup, crackers, and a cup of black coffee, I felt a bit better. The library has a completely automated book retrieval system. I sat down in front of a computer and brought up the catalogue program. I did a search for articles or books on fetal development no older than one year. I found only three. I typed my request and waited for my name to be posted on the screen over the reference desk. One paper was not available, but the other two came tumbling down the chute at the desk a few minutes later. I checked them out and headed for the Bradford. I spent the rest of the day researching my chosen field. I didn't finish reading the library materials and the abstracts on the web until half past ten. I still do not have a clear answer on how to approach the study of neuron fetal development from outside of the womb.

Discouraged, I turned on the TV. I found a program called Stranger than Fiction. The topic of this night's show was about an amazing story of a young girl who recently received a heart transplant. The girl frantically told her cardiologist she is remembering something horrible, the murder of a woman in front of a Nature's Best health food store. She tells the doctor she has never visited such a store. The cardiologist was interested enough to track down the husband of the donor and ask if his wife had ever shopped at such a store. The husband says his wife often shopped at a Nature's Best store close to their home. The young girl was able to describe to the cardiologist and police detectives the entire incident, including the perpetrator of the crime. The murderer was apprehended, taken to court, and found guilty. This incredible true story is ringing a bell in my head. Could a transplanted heart somehow contain memories? It must be able to. I went to bed thinking about memories. Where are they stored and how might a person be able to retrieve them using some kind of new technology.

Empowered for Ministry

I awoke Saturday morning in a somewhat better mood. The rollercoaster ride I was on yesterday is now over. I am hoping things will be turning around for me. How I know this I can't really say. After breakfast, I sat down and read a few chapters from my Gideon Bible and prayed. Then I prepared to visit Reisterstown. I haven't any idea what the dress code is for a Messianic congregation, but I'm sure it isn't as casual as the church in Timonium. I wonder if all of the men will show up wearing tallit prayer shawls and kippahs? My kippah is somewhere in the pile of clothes I dumped on the closet floor yesterday. I decide to leave mine at home. I pressed the wrinkles out of my dress shirt, tie, and slacks. Then I grabbed my

Gideon Bible, a sport coat, and left for Reisterstown.

It is a few minutes after ten when I pulled into the parking lot at the rear of the Presbyterian Church. I have no problem spotting the bright red awning over the rear door of the church. People are already arriving. To bolster my confidence, I say a quick prayer before opening the car door. I get out of my car, put on my sport coat, and slip my Bible under my arm. As I'm walking toward the church, several people are greeting me with Shabbat Shalom. I climbed the stairs and entered the sanctuary.

A Rabbi meets me at the door. "Good morning young man. Are you visiting our congregation this morning?"

"Yes, my name is Jacob Cahn."

"Jacob, I am Rabbi Epstein."

"Rabbi, I was invited by Dr. Glick. I am a first-year graduate student studying human genetics at the McKusick-Nathans Institute at Johns Hopkins. Dr. Glick is advising me on my research."

"What is your field of interest at JH, Jacob?"

"I'm just in the beginning stages of my research project to study fetal development."

"I'm impressed! I realize it is difficult to be admitted to the Institute. This is very interesting, Jacob. We don't have time to discuss this further now. Perhaps after the service we can get together for a few minutes."

"Yes, I would like to talk to you after the service."

Just then Dr. Glick and a woman I assume to be his wife comes through the door.

The Rabbi says, "I believe you know this man already, Jacob."

"Yes, I do. Shabbat Shalom Dr. Glick."

"Hello, Jacob. I see you already have gotten acquainted with Rabbi Epstein."

"I have just met the Rabbi. I arrived only a couple minutes before you."

"Jacob, this is my lovely wife Tullah and my two daughters, Amy and Diana. Amy is the oldest. She is now sixteen and Diana is twelve."

"It's so nice to meet all of you."

His wife and the two girls are smiling as I shake each of their hands.

"Jacob, we are going to adopt you into our family this morning.

We usually sit near the front."

"Thank you. I appreciate having someone I know close by. This is a first for me as far as attending a Messianic congregation."

Doctor Glick leads the way to the third row of pews on the right side near the front of the sanctuary. A few minutes later the worship team walks up and prepares to lead worship. A young girl, apparently the worship leader, welcomes everyone and asks us to stand. She first reads a short scripture from the book of Psalms. As the worship team begins to play the first song, a group of six young girls, including Amy and Diana, go forward and begin dancing and waving flags to the music in front of the congregation. The words of the songs are displayed on the screen above the stage. I couldn't pronounce most of them. They are in Hebrew. Everyone is now on their feet in the pews. They are singing, clapping their hands to the beat, or simply raising their arms to praise the Lord. The worship continues a lot longer than the worship service did in Timonium. Then Rabbi Epstein takes the podium and the room grows silent. The next thing I hear is the sound of a shofar from the back of the room. Two young men are walking down the center aisle carrying a large Torah scroll. They place it on a special table in the front of the church designed to hold the scroll. Then the entire room is filled with the sound of people praying. I hear Hebrew, Yiddish, English, and languages I can't identify. Doctor Glick is praying in one of these unknown languages. After perhaps five minutes of prayer, the voices melted away. I can still hear a few people quietly praying. It is as if a holy hush has fallen down on the congregation from heaven above. Rabbi Epstein prays the priestly blessing over the people in Hebrew and in English. Then everyone settles back as the Rabbi asks the ushers to come forward. The ushers pass offering plates down the rows of pews. I slipped ten dollars into the plate, wishing I could do more. Next, the head usher prays over the offering. Following the collection, Rabbi Epstein's wife comes forward and announces the activities planned for the coming weeks.

Rabbi Epstein next introduces the guest speaker. "This morning, we are so blessed to be visited once again by Rabbi Jonathan Weiss who has just returned from Israel. As you know, Rabbi Weiss leads a congregation in Queens. He speaks before many groups worldwide. God has given him a wonderful ministry of the prophetic word. Let us give him a warm welcome."

Rabbi Weiss stands and receives a warm welcome coming from the congregation. He walks to the podium and the two rabbis greet one another with a holy kiss.

"Good morning, mishpachas. It is so nice to be with you again, and to see how the Lord has enlarged your tent. As Rabbi Epstein just said, I have just returned from Eretz Israel. I must take a few minutes to share with you the current situation on the ground there and the thinking of the leadership of God's people. Israel is under great pressure by their neighbors. Although Hamas has been largely subdued of late, they recently have successfully smuggled into Gaza a large number of longer range missiles from the Syrian Hezbollah. These missiles are capable of striking the major population centers of Haifa and Tel Aviv. Terrorist also have been stabbing crossing guards and setting fires in the heart of Haifa. Incredibly, Israel has given an apology to Turkey for attacking the terrorists on the ship which recently arrived from Turkey. The Israelis have even granted Erdogan authority over Gaza. The UN recently passed resolutions attempting to force Israel to relinquish more territory in the West Bank. The UN has voted to exclude any Israeli authority over the city of Jerusalem itself. They have approved a document requiring Israel to hand back the Golan Heights to Syria. Netanyahu has stated Israel will never abide by these UN sanctions. Meanwhile, the Jesuit Pope Francis is working hard to birth a new universal world religion, even convincing many Rabbis to sign on to his new religious agenda. Have we read something like this in Scripture? Yes, we have. Now for the good news. God has his remnant. Many Jews in Israel are waking up. They are now turning to Yeshua. I believe we are witnessing a very significant move of the Holy Spirit there. The dry bones of Ezekiel are now putting on flesh. Can someone say hallelujah?"

The room erupted in praise. The people are clapping and shouting hallelujahs. After his brief update on Israel, Rabbi Weis asked the congregation to turn to Chapter Sixteen in the book of Ezekiel. The text describes Israel's history of rebellion, the corruption brought on by abandoning their God. The Rabbi described the life of harlotry led by God's chosen as they worshipped the evil Gods of the Canaanites, the Amorites, the Hittites, and those merchants of perdition, the Assyrians.

Then Rabbi Weiss raises his right hand and declares, "Ezekiel prophesied God would chastise his chosen people harshly with fire. But the Lord will yet remember his everlasting covenant spoken to Israel in her youth. Scripture states Israel will repent of her ways and be ashamed. In verse sixty-three, Adonoi declares he will never again judge his people harshly. Israel will never again open her mouth against the Almighty, nor will Ha Shem ever again condemn those who are in Yeshua to another holocaust. Rejoice my people, for your God goes before you."

The Rabbi's reference to the holocaust brought tears to the eyes of many in the congregation. They are crying on one another's shoulders, praying against the side walls of the sanctuary, and falling down on their knees before the Torah scroll.

Rabbi Epstein begins shouting out, over and over again, "Let Israel arise. Let Israel arise. let Israel arise and return to her God."

I became completely unglued, standing there crying like a baby. I looked up to see Rabbi Weiss motioning for me to come forward. I walked to the front. He placed his right hand on my head and began to pray in a language I did not understand.

Then he paused and said, "Jacob, Ha Shem has given me a word of knowledge for you. Are you ready to hear it?"

"Yes, of course."

"The Lord says this: 'Your name is no longer Jacob. You are Israel.'"

"Yes, Rabbi, this is true. The Lord told me the same thing in a vision a few weeks ago."

"Truth is confirmed by two witnesses. You are Israel Cahn to your heavenly Father. Also, the Lord says you are not to feel defeated when they say your plan will fail. If you will trust in Yeshua, you will see his mighty arm and the power of the Holy Spirit. Stay in the fight. Do not give up. Does this mean anything to you, Jacob?"

"Yes, it certainly does. I am facing a difficult task with my research project at the University. Some are saying it is an impossible project and I am not sure how to proceed."

"Let's pray."

I watched as the Rabbi Weiss begins bowing his head back and forth in front of me. Finally, he stops.

"Jacob, the Lord says you are going to receive what you require to carry out your research, but the topic of your research will be different from what you currently are planning to pursue. He wants you to empty your check book, except for ten dollars. If you will do this, Yeshua will bless you abundantly."

"If I do what you are suggesting, I will be virtually destitute. Are you sure I am to do this?"

"Yes, if you do this you will gain Ha Shem's blessing."

"Let me understand. I supposed to write out a check? Who do I write it to?"

"I want you do write the check out to this congregation. The Lord has important plans for you, Jacob, very important plans. Through this experience, you are going to learn to wait on Him, and you will be given greater faith."

Somewhat reluctantly, I took out my check book and wrote a check to the church for $832.64, leaving exactly ten dollars in my account.

"How long have you been a believer in Yeshua, Jacob?"

"I've been a completed Jew for just a few days, Rabbi."

"Have you been baptized?"

"Yes, just last week at the Methodist church in Timonium."

"Does this church teach the baptism of the Holy Spirit?"

"I guess so."

"You don't know what I am talking about do you?"

"No, I don't, Rabbi."

"The Holy Spirit actually comes to take up residence in your heart when you are saved, when you have humbled yourself and asked him to forgive your sin, and made him the Lord of your life. This is called the baptism into salvation. When you are baptized in water, you give a public witness to the fact you are now a believer in Yeshua. There is a third baptism Yeshua offers to us. He desires to give us the power to carry out special ministry gifts. This is called the baptism in the Holy Spirit. Jacob, you will need the power of the Holy Spirit to successfully accomplish the Lord's plan for you. I'm sure you will face tasks not yet revealed to you where this anointing will be necessary for you to receive. Are you ready to receive this baptism?"

"Do you mean right now, this morning?"

"Yes, right now."

"Well, I'm not sure. I do know I need Yeshua's Holy Spirit living in me. Access to greater spiritual power could help me contend with difficult situations and people. I already have come up against one person in particular. I suppose I'm open to receiving this third baptism. How do I receive it?"

"So, you are saying you wish to receive it."

"Yes, I want to receive this Holy Spirit baptism."

Rabbi Weiss reaches over and places his right hand upon my head. He begins praying in a strange language. My body is beginning to shake violently. I can't seem to keep my balance. Suddenly, I am falling backward onto the floor. As I am falling, I look up at Rabbi Weiss. He is smiling broadly. Then the Rabbi's face seems to melt away, like there is a frosted glass between us. I am now somewhere else, in some other realm. I feel completely at peace. I don't know how long I am laying there. As I begin to come awake, I try to pray, but the words are coming out funny. The room which before was filled with people now is empty except for the Rabbi Epstein and Dr. Glick and his family. I realize I have missed the rest of the service. Dr. Glick helps me into a chair. He is smiling like a Cheshire cat. He knows something I do not.

"I'm so sorry. I must have passed out on you, Doc."

"No problem, Jacob. No doubt this was your day to come to our congregation. How are you feeling?"

"I'm not sure. I don't understand why I passed out like this."

"What happened to you is good, Jacob. Do you think you can drive?"

"Yes, I think I can."

"We want to take you to lunch with us? The girls like to go to the pizza place a couple of blocks down the street."

"I'd like to. Perhaps you will fill me in on what just happened."

"Yes, I can tell you what happened to you over lunch. Turn left as if you are going back to the Interstate. It's the first pizza place you come to in the town."

"I'll just follow you there."

We walked out of the room and down the stairs together. The girls are asking me all kinds of questions. I didn't have any answers for them. I told them their dad will explain what has happened to me over pizza. I spent another two hours at the pizza place with the Glick family. Dr. Glick pointed me to the passages describing the gifts of the Holy Spirit and how they apply to ministry. I now understand the strange language. My soul was speaking, praising Yeshua and the Holy Spirit. We talked for a long time. Then I made my way back to JH. It was six o'clock before I arrive back in my cave at the Bradford. Having demolished a number of pieces of pizza I certainly wasn't hungry for dinner. I decide not to tell Terry and his family about this morning's experience, at least not for now. I need to skip going to Timonium tomorrow. I spent the evening on the computer continuing to search for research papers dealing with precursor neurons. About eleven I called it a day.

18

Opportunity Knocks

*"For verily I say unto you, if ye have faith as a grain of mustard
seed, Ye shall say unto this mountain, remove hence to yonder
place,
And it shall remove; and nothing shall be impossible unto you."*
(Matthew 17:20)

A Tentative Venture

I woke up Sunday morning early. I ate a quick breakfast of
Raisin Bran and a banana. I have drained my checking account. I'm
not sure how long I can exist on ten dollars. I began to work on my
project by reviewing the schematic Dr. Glick gave me. I cleared off
the kitchen table and unrolled it. The first thing I need to decide is
how to approach the project. Do I begin with a bottom-up approach,
or is there a way to pursue the task from the top-down? Then there
is always the possibility Dr. Glick mentioned to me. I still have the
option of completely changing the topic of my research proposal. I
am very uncertain at this point. I need to keep all three options open
for now.

I began to carefully consider the three options. If I decide to
work from the bottom up, I will need to have the ability to record the
activity of every neuron as they make their way from their origin to

their assigned position in the cortex. This implies I will be working with huge data sets. I have only a vague idea how to investigate the neural development of a fetus with a top-down approach. I still believe the logical choice is to develop the technology necessary to support a bottom-up approach. This also means I will need access to a storage device with tremendous memory and directional ability. There is only one machine I am aware of with the muscle to handle such a large data set. I will need access to a Qbit computer, a computer utilizing quantum physics. Only these new computers are able of storing such vast amounts of data and quickly retrieving it.

I know there are several prominent physics labs working on this new technology, principally in the US, Canada, and England. No one design has been shown to be preferred as yet. Qbit computers have the advantage of using a third state for every bit location. This greatly expands their search capabilities. For example, if a trucking company wants to find the most efficient route to reach their many destinations, a standard two-state computer must test all possible routes, one at a time, until the shortest delivery route is found. A Qbit computer can determine the route with one single pass. What I also need is a way of summarizing into some logical sequence all of this data. This means I will need a programmer familiar with AI. The AI is necessary to control the computer's actions, to teach it how to proceed, making decisions for perhaps a trillion elements. In other words, the computer must know how to think. The capability of quantum computers and AI seems to be the only device capable of handling my particular need. The human brain contains ten billion neural cells, and ten times the number of glial cells. The use of this technology has several problems. First this technology is just now entering its application stage and is still not fully developed. Second, these machines are extremely expensive. A third obstacle to be overcome will be the programming skills to create a dependable and unique design. I certainly do not have the expertise to write programs for AI.

I wanted to get some idea of how these computers are designed. I found two different methods being used to access the capabilities of the quantum world. One method uses lasers to strip away the electrons from hydrogen atoms. Instead of generating heat, this process does exactly the opposite. As the electrons are stripped away the temperature of the nude nuclei drops. Once the temperature of the nude hydrogen atoms reaches less than one degree above absolute zero, the atoms flatten to form a motionless grid of positive nuclei. This grid becomes the coldest place on earth, approaching only 0.3 of a degree above absolute zero.

A second design takes a somewhat different approach. This method places the Qbit grid into a large thermos cylinder filled with liquid helium. Both designs are extremely sensitive to external radiation of any kind. These Qbit grids must be protected by a special metal container called a Faraday Box. Then there is an even more difficult measuring problem inherent in detecting the state of the Qbits themselves. Any attempt by an outside device to measure their state introduces an extraneous variable into the system. This problem when attempting to measure the location of sub-atomic particles is similar to the uncertainty principle known in physics as the observer effect, the well-known Heisenberg Uncertainty Principle. This source of error has been shown to be inherent in the properties of all wave-like systems, and arises in quantum mechanics due to the nature of these objects. According to the sources I reviewed, this measurement effect actually is a property of quantum systems, not just a problem caused by the observer. Any interaction between classical physics and quantum introduces uncertainty. In other words, the task of measuring the state of the Qbits introduces noise into the system. To remedy this problem, very high tech electronic measuring equipment is used. Even so, I could not entirely avoid this problem, unless I could step outside of our universe and run our tests there. After reviewing the problems presented by the use of quantum computers and the difficulty in managing them with AI, I am tempted to discard the whole thing and select another research project.

By lunchtime I am quite discouraged. I took a break. I brewed some coffee and opened a can of vegetable soup. I poured the soup into a bowl and stuck it into the microwave. I found a box of saltine crackers in the cupboard and lunch was served. After lunch, I am not in the mood to continue my search for a solution. I sat back down in my easy chair and turned on the TV. A program on one of the Christian channels is interviewing Robert Matson, the author of a book on UFO phenomena. Robert Matson is a medical doctor. He claims to have taken implants out of people who have been abducted by UFOs. The implants are very small, measuring no greater than 0.5

centimeters in length. He has taken several of these small implants to a lab for analysis. The lab discovered the implants are emitting a very high radio frequency of 370 gigahertz, over a hundred times above the 3.5 gigahertz current limit of radio transmissions. This doctor Matson made one comment I found very interesting. When asked what he thought these implants are doing, he said it was his conviction they are altering a person's DNA. I wondered just how one could extend radio frequencies beyond the current limit? Can a high frequency radio transmission be used in tracing the movement of neurons? My limited knowledge of physics is poorly suited for the task at hand. I wondered, is this Doctor Matson's information correct? Is there a level of frequency transmissions capable of setting up harmonic reactions to alter the elemental parts of the cell? It is plain I will need help on these questions.

After the program, I put on my sweats and took a run. The day was too nice to be confined to the indoor track, so I charted a course circling the entire boundary of the Homewood Campus. Thirty minutes later I reached the east gate of the campus. Stopping for a rest, I poked my head into the Bloomberg Center for Physics and Astronomy. Perhaps Calvin, Jeff, or Robert Manley might be there. I see no one I know when I entered. There is a group of students sitting around a table in the lobby, discussing their astronomy class. I waited for a break in their conversation before asking about Nanoman.

"Hello guys. My name is Jacob Cahn. I wonder, do any of you happen to know a Robert Manley?"

One of the students, evidently their leader, says, "You're looking for Nanoman?"

"Yes, have you seen him?"

"Earlier I saw him working in the lab on the third floor, room 353. He probably is still there."

"Thanks, I'll give it a try."

I took the elevator to the third floor and found room 353. The door was locked, but I could see through the window someone was inside. I knocked and a man came to the door.

"Hello, my name is Jacob Cahn. I'm looking for Robert Manley."

"You found him. Come on in Jacob."

Robert is rather short and overweight. He has eyes you feel are boring right through you. I can tell he is extremely intelligent.

"What brings you to my inner sanctum, Jacob?"

"Dr. Glick recommended I contact you. I was taking a run and stopped for a rest outside the building. I poked my head in to see if you might be here today."

"Really? Why did Dr. Glick refer you to me?"

"I have a physics problem and he said you might be able to help me out. He said you are someone who thinks outside of the box, so to speak."

"I have to plead guilty. It's true, I don't shy away from stepping into forbidden zones on occasion. So, what is your research problem Jacob?"

"I'll try to be brief. I'm a first-year grad student in the McKusick-Nathans Institute, studying human genetics. I'm just beginning my research project. It has taken me into some heavy physics, Robert. My chosen research proposal is to develop a method to observe fetal development at the sub-cellular level."

"Whoa there! I don't know anything about genetics. I'm strictly into black holes and anti-matter, stuff you can describe with mathematics. I couldn't tell the difference between a gene and a genie."

"I desperately need a genie, Robert. I am facing a stone wall when it comes to tracing the paths taken by neurons. I need a method of observing the paths they take from outside of the womb. I believe these precursor neurons travel to their destination following some kind of directive inherent within the developing cortex of the brain. I believe these pathways must be some form of electro-magnetic, or resonance pathway yet to be identified."

"So, you are looking to discover some unknown network present in the fetus using some as yet to be developed external apparatus to sense the movement of these neutrons within the head of a fetus?"

"No, not neutrons, neurons."

"And you will want to chart this network in some sort of a 3D display. Just how many neurons are we talking about in a human brain, Jacob?"

"It is a very large number. In 2009, Dr. Azevedo and his team found the adult male human brain has an average weight of 1.5 kilograms, and is estimated to contain 86 billion neurons, plus many more glial cells. The glia cells account for ninety percent of the cortex, hence the saying we use only ten percent of our brainpower."

"You want my answer? I believe you should change your research topic, Jacob. There is only one machine capable of holding such numbers with their directional locations."

"You are referring to a quantum computer, I assume?"

"Correctamente. Do you know how much the hourly rate is to run a job on a Qbit machine, Jacob?"

"No, I don't, but I suppose it is more than I am making from my student grant."

"Oh yes. The last research project I worked on using the Qbit cost just over a million dollars an hour. Do you think I'm some kind of naïve trekkie? Let me tell you, when I work on a research project for this outfit, I always stay within the bounds of a credible hypothesis. I don't believe in trying to squirm through worm holes. What I will say is this. Gravity is not the only thing making this world go around. We are now finding new facts about this universe. It seems to resemble a very complex electromagnetic system where each and every mass object is in some way tied to all others. Perhaps our perceived reality is a false conception. We may be living in a hologram of vibrating strings, a subset of many other universes known as The Multiverse."

"Yes, I have read a couple of books on string theory. Before you quickly dismiss my proposal, let me tell you about something else I am bringing to the table. I don't know what your position is on the spiritual component of a human being; whether you are a Christian believer as I am, or not. I have experienced some supernatural events convincing me there are dimensions beyond our perceived reality. Miracles do occur in spite of denials by the naturalists. I have experienced some of them."

"So, you are counting on a miracle to support your research? I haven't seen any of those on my watch, Jake."

"Robert, miracles do happen. In fact, I have learned to expect them."

"Nice tale, Jacob, but I can't buy into this neuron thing of yours. It does not appear to be viable. It's not worth my time. Sorry."

"I may be able to furnish more information soon. Can we talk again?"

"Okay, I will give you the benefit of the doubt. Give me a call when you have something more credible to share and I will hear you out. Don't get your hopes to high. I would hate to see you get thrown out from the McKusick-Nathans Institute on grounds of psychological instability."

"Thank you, Robert for hearing me out. I appreciate your input."

I left the physics building, having very serious doubts about my research project's chances of success. Perhaps Nanoman is right. I should just set this precursor neuron topic aside and switch to another topic, perhaps something on human memory?

When I finished my run, I came up with another home-cooked meal: I brewed some coffee, boiled up some spaghetti, warmed up a jar of tomato sauce in the microwave, opened a jar of aged parmesan cheese, and sat down to dinner. After dinner, I reviewed Dr. Glick's research plan. I came to the conclusion it is not going to work for me. Tomorrow I will attempt to draw up a better one. If I can successfully demonstrate the existence of a network in the brain existing prior to neuron placement, or possibly some external force controlling the radiation of neurons, I can yet rescue my research project.

As I am considering what approach to adopt, I remember Nanoman's comment that the universe is really a complex electrical system where every mass object is in some way electrically tied to all others. I also have been thinking about the news report I heard the other day about the young girl who had received a heart transplant and also received the memories of the donor. I spent the remainder of the evening researching the effect of electromagnetism on animals, the field known as biomagnetism. I also searched the literature on the topic of human memory. Although I found several papers on the topic, I was surprised to learn the brain's ability to store memories remains a great mystery. It was half-past eleven when I finally turned in.

Monday morning arrived early. I was up at five. I am convinced I need to know more about the capabilities of the diagnostic instruments used to sense the activity of the neurons in various parts of the brain. After breakfast, I went online to find information about the interesting field of Magnetoencephalography. These machines commonly are referred to as MEG machines. I am encouraged by what I found. The Superconducting Quantum Interference Devices (SQUIDs), are currently the most commonly used magnetometer device. They have recently been improved to simultaneously cover the entire brain area. Another new device, the Spin Exchange Relaxation-Free

(SERF) has recently been designed to support future MEG research. These MEG machines are used in several research applications such as: the study of perceptual and cognitive brain processes, localizing regions affected by pathology before surgical removal, the function of various parts of the brain, and neurofeedback experiments. The latter two study areas may be well suited to my research project. I searched course offerings for Biotechnology at JH and found the Krieger School of Art and Science offers advanced classes in Biotechnology. I made a mental note to contact them later in the day.

I finished my on-line queries by half-past eight, leaving me just enough time to get to my morning class. This class covers genetic pathways, cell structure, and cell dynamics. The course is offered by the Department of Medicine and taught by Doctor Jason Baker. I arrived ten minutes early. The lab assistant at the door handed me the course syllabus. The lecture hall holds about sixty and it is nearly full already. I found a vacant seat in the second row. At exactly nine o'clock Dr. Baker enters and walks to the white board at the front of the room. He writes the name of the course, his name, and the names of the three lab assistants and their email addresses. Only one assistant is in the classroom. I have never met Dr. Baker before. It does not take long for me to discern this professor is all business. He briefly welcomes us to the class and then quickly goes over the information on the syllabus.

"Students, my lectures are scheduled for Monday and Wednesday mornings at nine o'clock. The labs are on the same days in the afternoon. I have asked one of the lab assistants to start the projector. Please take careful notes. I will see you again on Wednesday."

Professor Baker leaves the room. There is no chance for me to make a good impression on this man today. At the end of the hour-long tape the lab assistant asked if we had any questions. I raised my hand and asked if the powerpoint lessons are available in the library. He said they are. He also demonstrated how to access the powerpoint lessons. Then the assistant goes to the white board and lists the chapters in the text we are required to study before next Wednesday's lecture. Finally, he takes roll by reading off our names and assigning each of us to a lab session. My lab meets at three o'clock on Mondays and Wednesdays. Fortunately, my lab schedule does not conflict with any of my other classes. The assistant then excused the class.

I returned to my apartment to eat lunch. There is a letter in my mail box. It is from the School of Medicine. In the envelope is a check in the amount of $832.64, the very same amount of the check I wrote to the Beth Israel Congregation in Reisterstown. There can be no doubt. The Lord has orchestrated this amazing coincidence. Tonight, I will dine out.

The first thing after lunch, I contacted the Krieger School of Art and Science. I wanted to make an appointment to talk to one of their counsellors, perhaps someone involved in MEG research. The person on the phone transfers me to a Ms. Wilcox.

"Krieger Counseling Department, Ms. Wilcox. Can I help you?"

"Good afternoon Ms. Wilcox. My name is Jacob Cahn. I am a first-year grad student in the McKusick-Nathans Institute. I am interested in taking an elective course from your school. This is in relation to my research project."

"I see. What particular field are you interested in, Jacob?"

"I am considering a research project to study the development of neurons in the cerebrum."

"A very interesting research project, Jacob. It so happens we have someone on staff with an interest in the same subject area. His name is Dr. John Majors. I will contact him and have him give you a call. You both can take it from there. Can I have your cell phone number please?"

"Yes of course. My cell number is area code 402 661-2336."

"Is it 402-661-2336?"

"Correct. Thank you, Ms. Wilcox. I look forward to hearing from Dr. Majors."

"He is on campus this week. I'm sure he will be in contact with you soon. It has been nice speaking with you, Jacob. I wish you every success in your research project."

"Thanks again, Ms. Wilcox."

After the phone call, I reviewed Dr. Baker's class assignment. I had just opened the textbook when my cell phone rings. It is Dr. Majors.

"Hello, Mr. Cahn?"

"Yes, this is Jacob Cahn."

"Jacob, I understand you are interested in taking an elective course in Biotechnology from our school to support your research project."

"Yes Sir, I am interested in expanding my training to include a class in biotechnology. I am enrolled as a grad student in the McKusick-Nathans Institute. My undergrad degree is from Emory University. My major there was pre-med."

"Do you have a particular research project in mind, Jacob?"

"Yes, I tentatively plan to explore the factors determining the pathways followed by neurons during the early developmental stage of pregnancy."

"I assume you desire to study this from outside of the womb?"

"Yes, from outside. I do not want to risk damaging an unborn fetus. It is my understanding your school has been provided with the latest biometric equipment for studying the brain."

"Yes, our lab has procured the very latest equipment. In the lab, we use the latest MEG machines. We have recently acquired two SQUID machines. These machines are a significant improvement as they simultaneously cover the entire brain area. Also, we now have a new machine known as a Spin Exchange Relaxation-free Device. The acronym used is SERF. This improved machine will support our future MEG research efforts. We use these MEG machines for several different research applications, such as the study of perceptual and cognitive brain processes, localizing regions affected by pathology before surgical removal, determining the function of various parts of the brain, and neurofeedback experiments. Jacob, I wonder if you are aware of just how complicated the human brain is?"

"Yes, I am. I have read articles from the major neuroscience organizations such as the Society for Neuroscience (SNF), The International Brain Research Organization (IBRO), and The British Neuroscience Association (BNA). The SNF held its yearly meeting in Atlanta last year. I was attending Emory University at the time. Doctor Michael Stein told me about the meeting and I was able to attend one session."

"Jacob, I'm going to be very frank with you. I believe you have fallen into a familiar error, often committed by first year graduate students at Johns Hopkins."

"Oh? How have I errored, Dr. Majors?"

"Jacob, the topic you have selected far exceeds your ability to adequately research in the time allotted for you to complete your research project necessary to achieve your higher degree. I have witnessed this over and over again. A young bright graduate student dreams of the fame, self-respect, and financial rewards he will gain by making some major discovery, hoping to achieve a giant leap forward in his particular field of interest. Failing to achieve the goal, the student becomes greatly disappointed. The aftermath often results in the student leaving the university for greener pastures, like one student that started a landscaping business. A few students are humbled enough to narrow their research project to some manageable task. They may even be content to join in a team effort, entering the trenches of day-to-day research as part of a dedicated team. In other words, they are patient enough to become, so to speak, a cog in a larger gear. The research topic you have selected has been the life's work of many fine researchers over several decades, not just years. None have gained any grand achievement award as yet. What I am advising you to do is to narrow your scope, perhaps assisting in an on-going research project here at the University. Am I making myself clear to you?"

"Yes, you are Dr. Majors. I must confess I have been greatly humbled already. I am concerned whether my research topic has any high likelihood of success. In fact, I just shelved my latest flow chart because I could not visualize how to proceed."

"Would you be willing to join a team of researchers, content to contribute your genetic expertise toward achieving the overall goals of such a research group?"

"I certainly am willing to seriously consider such an option. I would have to take some time to mull over such an approach, and to consult with my advisors whether, as a member of a larger research team, such a project would satisfy my degree requirements."

"Fair enough. I want you to take some needed time to assess this approach. I am willing to pursue this further with you, to explore the possibilities. I suggest we meet together to consider the matter in greater depth. How does this sound to you?"

"It is like a breath of fresh air, Doctor. I very much want to discuss my research goals with you again. My class schedule occupies my time on Mondays, Wednesdays, and Fridays."

"Why don't we meet next Thursday then, say at nine o'clock here in our department?"

"That works for me. Is there anything you would like me to bring to the meeting?"

"Yes. I would like you to bring your records from Emory and

your grades from the first semester here at JH."

"I can bring them. I also worked as a lab assistant this summer for Dr. Adelman from the Biology Department."

"Very good. Do you have any problem with me contacting Dr. Adelman and Director Hobbs at the Institute?"

"No, not at all. Doctor this conversation has really encouraged me. I look forward to meeting with you then on Thursday at 9:00."

I hung up the phone. I am feeling greatly relieved of this burden I have been carrying. As a member of a research team, I would not be faced with the many required organizational tasks required for an individual research project. Even more importantly, the required specialized equipment and testing procedures would come along without cost. I would be aided by ideas and advice from others on the team. I do need to check with Dr. Hobbs whether participation on a team research project would satisfy my research requirements for the degree.

I immediately placed a call to Dr. Hobbs. He was not in the office. I asked Robbie to have him call me when he had time. Then I continued my review of the assignment for Dr. Baker's class. The material is pretty straight forward. I finished the chapters before it was time to attend the lab. After the lab, I stopped by the bank and deposited my miracle check. I am flush with cash. No gravy over biscuits tonight. I'm having dinner at the restaurant in the inner harbor. As I was pulling out of the parking lot, my cell phone rang. It is Dr. Hobbs.

"Hello, Jacob. How are you?"

"I'm just fine. I called you to ask a question. My question is in regard to the requirements for my research project."

"Oh? What is your question Jacob?"

"Sir, I have been in contact with Dr. Majors at the Krieger School of Art and Science about taking an elect course in biotechnology from his school. I described my interest area. In our discussion, he suggested I might consider participating as a member of his team of researchers in a project his department is now beginning which closely parallels my interest areas. Such an approach would provide several advantages to me over an individual research project. I have not committed to such a plan as yet. I first need to find out whether such a team research project can fulfill my degree research requirement."

"Jacob, I believe we have approved a couple of team projects as fulfilling the research requirement in the past. A final determination will have to be made after the specific research project is known, evaluated, and your participation has a clear contribution to the research. Knowing your abilities at this stage, I am of the opinion you should pursue this approach with Dr. Majors. Please keep me informed on the details you and Dr. Majors come up with."

"Thank you, Doctor. Also, you will be getting a phone call from Dr. Majors concerning my aptitude for such a project."

"Fine, I certainly will talk the matter over with him. Thank you for keeping me informed of you plans, Jacob."

The Director hung up the phone and I continued on my way to the Inner Harbor. I'm looking forward to a seafood dinner. As I walked in, I asked the hostess whether Judy was working. She tells me Judy has taken the night off for some reason. I am disappointed. After dinner, I returned to the apartment and just relaxed.

I set aside the assignment from Dr. James and pulled out one of my favorite DVDs, *The Terminal* with Tom Hanks. I sat down in my easy chair to watch the movie for the umpteenth time.

Tom's character, Viktor Navorski, just finished collecting quarters from the baggage cart machine when the phone rang. It was Ramona.

"Hello Jake."

"Hi Ramona, What's up?"

"I just want to keep you up-to-date. Terry and I have decided to go ahead and get married right away. We want you to be the best man. We will let you know what day the wedding will be."

"I'm not surprised at all. Tell Terry I am looking forward to helping you two get married. I will wait to hear from you. I'm really working hard right now, but I will definitely squeeze it into my busy schedule. Thanks for letting me know ahead of time, Ramona."

"We will get back to you, Jake. Bye."

I hung up the phone and returned to my movie. After the movie, I took some time to consider this new option suggested by Dr. Majors. His phone call has convinced me to put my old proposal on the shelf. I literally put my proposal materials in a box and set it on the top shelf of the closet. Then I stuck in another Tom Hanks movie into my DVD player, *The Castaway*.

After the sad ending of the movie I began to consider the possible outcomes of my research proposal. Should I take Dr. Majors' advice? If I continue to press on with my individual approach, will I fail to achieve my research goal? Will I become a castaway like Tom Hanks, lost in an endless sea of an uncertain future? It's time to bring such thoughts to a close. I took a shower, said my prayers, and turned in.

Sleep failed to show up. I lay awake, staring at the alarm clock for several hours. I can't stop my mind from dwelling on the issue raised by Dr. Majors. At three o'clock, I gave up and got out of bed. I made a peanut butter sandwich and brewed a pot of coffee. Then I sat down in my easy chair and slowly ate the sandwich, sipping my coffee. Out of my window I can see the Homewood Campus, illuminated by hundreds of street lights. Everything is very calm. An occasional passing car in the street below is the only noise I hear. Gradually my brain settles down and turns to something I read recently. I came across a blog discussing something called the Schumann Resonance. This naturally occurring resonance was first identified by Winfried Otto Schumann. He briefly worked in the US after WWII. The blog caught my interest at the time, and I took time to look into the history of this phenomenon. Schumann claimed the earth's atmosphere produces a low frequency electromagnetic resonance. He named it what it was, Electromagnetic Low Frequency (ELF). Scientists recently have confirmed this low frequency does exist and resonates at 7.83 hertz. The interplay among lightning bolts, the denser areas of the ionosphere, and the size of the planet are the three factors believed to produce this phenomenon. Scientists also have determined animals are attuned to electrical fields, allowing them to sense the occurrence of earthquakes and violent storms. Higher multiples of the ELF frequency of 7.83 also exist, but are much weaker in their effect upon the brain. Perhaps this Schumann Resonance might yield a clue to the earliest phases of neurons in the unborn. According to recent research, magnetic fields can and do alter the behavior of all living systems, plants as well as animals. Charged proton ions flowing from the sun penetrating our ionosphere are sensed by animals, warning them of earthquakes and storms. Some frequencies can harm our neural cells. Our brains attempt to filter out disruptive frequencies by using a complex system involving cell membranes containing small magnetite liquid crystals, Favorable frequencies are allowed to penetrate into the brain. It is believed by some researchers these specific frequencies help maintain a rational state of mind and a more peaceful spirit. According to several studies, the heart responds strongly to electromagnetic frequencies, but it is the brain that is the most sensitive. Monitoring centers located at several locations around the globe keep track of these electrical resonances.

New agers have seized on this resonance frequency to make wild claims of the benefits of the Schumann resonance. They have often confused this electromagnetic resonance with the frequency of sound waves. The two are different animals. The frequency is often referred to as the God frequency, perhaps because there is a key musical tone corresponding to the Schuman Resonance frequency. It is the pitch of 'A', the key most every stringed instrument is tuned to. Although the sound frequencies are not related to the electromagnetic resonance, one can certainly make a case for the beneficial effect of music upon meditation and prayer. Music certainly influences our human feelings.

Another electromagnetic current known as the Hall Current travels along the surface of the earth. This current actually links everyone and everything on earth to one another. Some believe the Hall linkage may provide a means to control the collective behavior of the population, either in positive or negative ways. The technology could be used to create social unrest, incite riots, or even cause individuals to commit crimes.

I began to ponder what the connection might be between the earth's electromagnetic currents and the brain. Is there a connection with the precursor neurons and their glia? I will have to pursue this avenue further.

I realize I need to draw a clear line of distinction between what is science and what is new age gobbledygook when it comes to applying these still mysterious resonances and electrical waves to my research if I am going to retain any credibility. Only after I had exhausted these somewhat wild speculations could I go back to bed and get to sleep.

Resolution

On Tuesday morning, the phone woke me up early. I struggled to reach for it on the bedside table. The call was from Ramona.

"Good morning, Jacob. Time to get up."

"What time is it anyway?"

"I'm not sure. I think it is about six."

"Six? What day is it?"

"This is Tuesday, Jake. I called to tell you Terry and I are going to be married at the farm at ten o'clock on Saturday."

"On Saturday, you say?"

"Yes, this Saturday. It will be a country wedding. Nothing fancy. You might want to buy yourself some new jeans, a cowboy shirt with snaps, not buttons, and a suitable hat."

"Do I need boots as well?"

"No. New boots will hurt your feet. Just wear some nice shoes. Also, while you are at it, Terry asked me to have you to pick up some Champaign, maybe four or five bottles."

"Where am I going to find this cowpoke getup?"

"There is a western store on Pimlico Road just south of the race track. Do you know how to find you way there?"

"Yes, I know where the track is located."

"Are you now fully awake, Jake."

"Yes, I'm awake. I'll see you at the farm a bit before ten o'clock on Saturday."

"Okay. By the way, you will be escorting the maid of honor, of course."

"And who is the maid of honor?"

"I believe you met her once before, Rhonda Sutherland."

"You aren't adding my name to her list are you Ramona?"

"Now why would I add your name to her list? It's already on it."

"Yes, very true. Okay, I'll see you on Saturday. Have a good day."

"See you, Jake."

I rolled out of bed and made breakfast. I was just clearing the table when my cell phone rang again.

"Good morning."

"Good morning to you, Jacob. This is Robbie. I hope my call is not too early for you."

"No, not at all. I just finished breakfast."

"I have been asked to set up an appointment with you by Director Hobbs. He and Dr. Altschul want to discuss a few matters with you concerning your research proposal."

"I see. When do they want to meet with me?"

"I wonder, could you meet with them later today?"

"Yes, I'm available anytime. I can catch up on my lectures on audio tape."

"Fine. Let's make it for nine o'clock, Jacob. The meeting will be held in Dr. Hobbs' office."

"Fine, I'll be there. Is there anything they want me to bring to the meeting?"

"No, Dr. Hobbs did not mention anything for you to bring. You don't have to dress up. We will see you at nine o'clock."

"Thank you, Robbie. Goodbye."

I am wondering what Dr. Altschul is up to now? Perhaps Director Hobbs may be attempting to bring some peace into the camp. I spent the time before the meeting reviewing Dr. Baker's last lecture and taking time to pray over the nine o'clock meeting. I decided to dress up anyway. I put on my suit, a dress shirt and tie, and took the shuttle downtown. I arrived at the Director's office at five minutes after nine o'clock. Robbie welcomed me into the anti-room.

"Good morning, Jacob. Have a seat."

"I'm sorry for my late arrival. I took the shuttle, but we ran into a traffic jam."

"Yes, Dr. Hobbs just called to say he also is running late. The traffic is horrible this morning for some reason. I just arrived a few minutes ago, myself."

I sat down and tried to remain calm, but my insides are churning. "How are your classes going this semester, Jacob?"

"Everything is going well. I am enjoying the lab work we are doing. I'm learning a lot."

"How is your social life?"

"Not much time for a social life, Robbie. I am standing in as Terry Johnson's best man on Saturday."

"Oh? Is he a student here?"

"Yes, he wants to be a veterinarian. His father has a thoroughbred farm in Timonium. You probably can't guess who he is marrying."

"Do I know her?"

"Yes, you do. He is marrying Ramona Bergman."

"Really, our Ramona?"

"Yes. Ramona changed her major and is now enrolled in the vet school."

"Oh, my goodness. Jacob, I seem to remember Ramona has some physical problems. Is she capable of holding her own as a vet?"

"Yes, for two reasons. First of all, Terry will be treating the large animals and Ramona will be taking care of the domestic pets. The second reason is Ramona has been miraculously healed of her crippled leg and her back problems."

"Really! How wonderful."

"Yes, she was healed in a little church in Timonium a few weeks ago. She is now riding race horses and looking forward to the Pimlico meet this Spring."

"My goodness, what an incredible story Jacob."

Just then the door opens and Director Hobbs and Dr. Altschul enter the office. I stood up to shake hands with both of them. Dr. Hobbs is smiling as I shake his hand. Good morning, Jacob. I believe you know Dr. Altschul, so no introductions are necessary. Robbie, we will meet in the conference room. Please bring your notebook with you."

Robbie made a quick call to ask a secretary in the nearby office to answer the telephone during our meeting.

I followed the two men into the conference room. There is no doubt who is in charge. Dr. Hobbs sits down at the head of the table and Dr. Altschul finds a chair across the table from me.

"Jacob, I have called this meeting to discuss a serious issue with you both. I believe you and Dr. Altschul are aware of the matter we are here to discuss this morning. As I have watched both of you, I soon came to realize your relationship has been, let's just say, strained. I have no problem with disagreements between professors and/or students just as long as it does not reflect upon the reputation of the Institute. Unfortunately, your intellectual and philosophical differences have been the subject of conversations well beyond our Department. It is called gossip, gentlemen. So, I have called you together to see if it might be possible to arrive at a reconciliation. As you know Jacob, Dr. Altschul's contribution to the advancement of genetic knowledge is well known, and I don't need to detail them here. In your brief tenure here at the Institute, Jacob, you have certainly excelled. I have seldom seen a student with as much potential as you have. I want to stress to you both, this is not a legal tribunal. I need you to present your opinions about what we can do to bring a solution and an end to the problem we are addressing this morning."

I gathered my courage and began with my apology and defense direct to Dr. Altschul.

"Dr. Altschul, let me start by saying I recognize there has been no shortage of differences between you and myself. I am willing to openly discuss this matter with you and Director Hobbs. I confess I have been a thorn in your side. I must plead *Mea culpa*. I deeply regret my actions at the debate. The question I raised at the debate was motivated by an immature desire to subject you to open criticism. I want today to apologize to you for my unseemly disruptive behavior. I truly am sorry I have created this gulf between us, especially in view of the public notice it has gained. As you know, Doctor, we have very different world views. I am committed to a deep spiritual conviction which is in diect conflict with the Theory of Macroevolution. Very recently this conviction has blossomed into an even deeper faith. I have accepted Yeshua as my Lord and Savior. With this new transformation, my life has totally been changed. I have come to love and respect all men. Jesus teaches us to love every person, for each has been created in the image of God Almighty. We all are His children. Dr. Altschul, I promise never to repeat such a foolish action again."

Dr. Hobbs quite plainly was pleased by this new attitude of mine. "Jacob, this belief in the children of God you are referring to does sound quite different from your earlier views of your fellow man. I do believe you have experienced a change of heart."

"Yes, Sir. I have truly had a change of heart. I do not want to sound like a holy Joe, but let me take time to tell you about two events which led to my new-found faith. My secular Jewish upbringing has always caused me to resist the acceptance of 'that man' Jesus, as being the Son of God. Even so, Ha Shem has been calling me through the years to seek a deeper spiritual life. A few weeks ago, I needed to come to grips with this struggle, much like my namesake of the past. A pastor in Atlanta encouraged me to begin studying the Scriptures, both old and new covenant passages. Through this reading of his selected biblical passages, I came to the intellectual conviction of the validity of these Scriptures. This Jesus, is the divine Son of God. Still, I found it difficult to accept him as my personal Savior and Lord. Only recently did I succeed in overcoming my Jewish cultural resistance to accepting this truth, in effect moving it from my brain cells to my heart. Following this salvation experience, my friend Terry Johnson invited me to attended his little church in Timonium where I asked the pastor to baptize me. Incidentally, our friend Ramona was saved the week before my baptism. Dr. Glick invited me to attend his Messianic Congregation in Reisterstown where I received the Baptism in the Holy Spirit. My life has been totally transformed by these two recent events."

After I finish my story, Dr. Hobbs remains quiet for a few minutes. Then he turned to Dr. Altschul. "Well, Fred, do you think the two of you can mend some fences?"

"Yes, Jonathan, I am willing to set our differences aside."

"Very well, I have a prescription ready for both of you. We must find the way to turn around this public conflict into a visibly evident reconciliation in the minds of our scholarly community. We need to show them we are willing to bring peace where there has been conflict within the Institute, real peace. Gentlemen, this is what we need to do. Fred, I am now assigning you as Jacob's thesis and research advisor. Are you willing to serve your fellow man in this way?"

I watch as Dr. Altschul painfully considers the Director's request. He looks shocked. It is clear he had no previous expectation of such a request by the Director. He swallows several times, and clears his throat before answering.

In a slow and measured response, he replies, "I understand your wishes Jonathan. I deeply respect your keen wisdom. You always know how to protect the reputation of the Institute. I am willing to cooperate fully with Jacob as his thesis advisor and preceptor. I will seek to help him accomplish the completion of his research project, and not in any way oppose the successful conferring to him the doctoral degree."

"In return, Jacob, I will request you communicate the progress of your schooling and your research proposal status with Dr. Altschul."

"Yes Sir, I agree to communicate fully with the Doctor, and will not in any way voice criticism of the man himself, but reserving the right to present my full position on all topics under discussion."

A broad smile breaks out on the Director's face. "Well now, I think this has been a very useful meeting. As Jacob's Advisor, Fred, I want the two of you to meet on a regular basis to facilitate the progress of his proposal. No doubt you will want to involve Dr.

Majors in your meetings as well."

With those words, Dr. Hobbs rose from his seat and shook our hands. Then a very unexpected thing happened. Dr. Altschul walked around the table and gave me a hug, an actual hug. I wondered, perhaps this man is searching for answers in his life? Time will tell. When I returned to my apartment, I gave Dr. Majors a call to let him know the results of the meeting. He sounded very relieved, and said he is looking forward to our meeting on Thursday.

I rose up early on Thursday morning. It is raining. I decide to take a few laps on the indoor track at the Rec Center. When I arrive, the lacrosse team is there working out. I spotted Joe running near the front of the pack. I ran onto the track as he is coming past the entry gate.

"Good morning Joe."

"Hi, Jake. Where have you been lately? I haven't seen you except during the class lectures."

"I've been trying to decide on how to approach my research project. Have you been doing the same?"

"Yeah, but I'm still in the very early stages, trying to decide on a topic I can reasonably expect will end in success."

"You are moving in the right direction, Joe. I have rejected my original topic. It is just too difficult. I do not have any assurance it is going to end well. I am considering joining a research team. The advantages are several. The team director would be responsible for the overall organization of the research. Secondly, the equipment required would of course be furnished by the department undertaking the research. Finally, I would likely benefit by interaction with others working on the team as well."

"Jake, do you think such an approach will satisfy the degree requirements?"

"Dr. Hobbs defined the Institute's requirement for me. My work on the research project needs to make a significant and recognizable genetic contribution to the research project."

"Sounds interesting, Jake. Keep me informed. I like the sound of this team effort approach."

"Okay Joe, I will."

Just then the team coach takes the team to the weight room. I said goodbye to Joe and continued my run. After running four more miles, I returned to my apartment, showered, and dressed. I ate a bowl of cereal and brewed a cup of coffee. Then I pulled my grade transcripts at Emory along with my honors certificate out of my desk drawer. I even put on my Legacy Medallion. I checked my old umbrella in the closet to see if it is still functional. With my transcripts wrapped in a plastic bag I left to meet with Dr. John Majors.

I arrived at the Wyman Park Building on the north side of the campus at five minutes before nine. Dr. Majors is waiting for me downstairs in the lobby. We said hello and then we took the elevator to the Krieger School of Art and Science offices on the fifth floor. Dr. Majors ushers me into a small conference room adjacent to his office.

"Take a seat Jacob. Do you want a cup of coffee?"

"No Sir, I just finished breakfast."

"Let me begin by saying both Dr. Adelman and Director Hobbs think very highly of you. Director Hobbs shared your fine grades here at JH with me. Dr. Hobbs also told me about your ongoing tiff with Dr. Altschul which has been resolved. Not only are you a very promising student, you apparently are a man of strong convictions. I like someone with the courage to speak his mind."

"We call it chutzpah Doctor. My disagreement with Dr. Altschul has been fully resolved."

"Very good. We don't want to have to deal with any skeletons in the closet. Jacob, let me get right down to business. We have a variety of research projects just now being implemented. You have expressed an interest in the development of the neural network of the fetus, specifically the pathways taken by precursor cells."

"Yes, this is, or was, my chosen topic. I may decide to pursue a second, but related research topic instead. I recently have been looking into the brain's ability to store memory. I have read of cases where people are able to recall events of long ago in great detail. Some researchers are suggesting we even have the ability to pass on memories to progeny."

"Well, I must say you are ambitious. We are not planning research specifically on precursor neurons, but we are interested in identifying the processes involved with human memory. Our research on the glial lattice structure may reveal something about your second interest area. We want to determine the activities of these glial cells. Are you familiar with glial cells?"

"I am somewhat familiar with them. I know they compose ninety percent of the mass of the cerebellum, with neurons accounting for the remaining ten percent. I know some glial cells compose the lattice structure used by the migrating neurons."

"You are correct. In addition to this lattice structure, we are finding there are other very interesting functions performed by these glial cells. Some recent research has suggested the glia may be where our thinking occurs. We may find they are the answer to your question about memory. They could be the depository of memory, the source of dreams, and possibly the vehicle for abstract creative thinking. Does this research topic interest you Jacob?"

"Absolutely, Doctor. The glial cells are very interesting to me."

"Now, assuming you are interested in participating with this research project, what do you think your contribution would be?"

"I believe I can investigate the activities going on within the glia, their nuclei and the activity of the genes of the DNA. I don't wish to sound conceited, but I do bring to the project a photographic memory and a creative bent."

"A creative bent, you say?"

"Yes, Sir. I seem to be gifted with the ability to conceive of new hypotheses to explain the phenomena being observed. I have received insights at times from dreams and visions."

"Really. Perhaps we will have to put your head into the SERF."

"I am willing to do so, Doctor."

"You are not afraid of getting water on the brain, I take it. Alright, Jacob. I will present your name to the planning committee at our next meeting and see what they decide. The committee will be meeting next Monday. I will get back to you as soon as I have something to report. Do you have any questions, misgivings, or other concerns?"

"No, Sir. I feel very comfortable with what you have outlined this morning."

"Very well, Jacob. Why don't we take a look at our new SERF machine? Do you have time for a tour of the lab?"

"Yes, I would like to see the SERF as well as the rest of your lab."

"Fine, let me pick up some paperwork in the office and we will leave."

I waited outside the office door while Dr. Majors retrieved his paperwork. Then we took the elevator to the basement. Dr. Majors inserted his card key into the door and we entered the lab. The room is huge, perhaps two hundred feet by three hundred feet. Dr. Majors asked me to wait for him as he walked to an office located at the far end of the room. Next to the office is a conference room. Along the near side of the lab are work stations. On the other side of the room I see a series of strange box-like containers resembling cold storage boxes. Each of these containers houses an MRI-type machine. The machines are fitted with chair-like structures, allowing a person to be seated beneath large hood-like devices. The hoods can be lowered over the head of the person being examined. No doubt these hood-like structures contain the electromagnetic coils used to scan the brain. In the middle of the room there is a large number of smaller-size devices. Two computer work stations and two electron microscopes line the far wall.

Dr. Majors walked out of the office and motioned me to join him. He begins explaining the contents of the laboratory as we walk through the room. He describes the development history of the field of biotechnology in great detail. He explains the functioning of the MEG machines, the SQUID and SERF models. He explains their use when used in clinical settings. Dr. Majors then gives me a history of the development of these machines.

"Jacob, the first MEG machine was developed by the physicist David Cohen at the University of Illinois in 1968. These early machines were very simple. The major problem encountered by these machines was the elimination of the magnetic background noise. Cohen later built a machine with a better shield at the Massachusetts Institute of Technology (MIT). Meanwhile, James E. Strahlerman, a researcher at the Ford Motor Company in Detroit, developed a machine which more effectively measured MEG signals. He developed the first SQUID detectors used to measure product quality at Ford.

This machine's design achieved signals almost as clear as those of an electrical encephalograph, or EEG. The successful development of the more accurate machine stimulated the interest of physicists who were looking for a machine with the capabilities of the SQUID. Cohen decided to use one of these first SQUIDs to take advantage of the vastly improved signals. Then in the 1980s, MEG manufacturers began to arrange multiple sensors into arrays to cover a larger area of the head. Present-day MEG arrays are set in a helmet-shaped vacuum flask typically containing 300 sensors, covering most of the head. In this way, information for the patient can be accumulated more rapidly and efficiently."

Dr. Majors pointed across the room. "As you can see Jacob, we have several of these shielded containers. In 2012, we demonstrated MEGs could work with a chip-scale atomic magnetometer (CSAM). This CSAM vastly improved the capability to measure very weak magnetic fields down to 10 femtotesla (fT) of cortical activity and 10 cubed fT for the human alpha rhythm. We still have major problems. I assume you have noticed the heavy doors and walls housing the MEG machines. We apply AC currents to the protecting materials of the doors to ensure against eddy currents. Also, precious silver and gold plating is used to improve the conductivity of the aluminum layers. Built into the systems are wires to degauss the surfaces of the inner protective layers. We make use of music composition software algorithms to provide noise cancellation, reducing both low and high frequency noise. With our latest systems, we have achieved a noise floor slightly above 1 Hz. We are hopeful this new shielding may allow us to access new portions of the brain now emitting very weak magnetic signals. As I say, the strength of these emissions only amounts to a few fTs. We even have to protect the machines from the Earth's weaker magnetic signals, especially the Hall Current. This Hall Current travels along the surface of our planet."

"Yes, the Hall Current is an interesting phenomenon. We apparently are all connected to one another by this current."

"Jacob, even with the isolating components of these boxes, the signatures are still significantly smaller than the remaining ambient background noise. Since this difference is so small it is hard to separate the signatures from the remaining low residual noise. In addition, we have a more serious problem. The currents received can be thought of as current dipoles, pointing to position, orientation, and magnitude. But these currents do not point to their spatial extent. In order to generate stronger signatures, the dipoles must have similar orientations, thereby reinforcing one another. Otherwise the magnetic fields are too weak to be detected and separated from the background residual noise. Only where the layer of these cells is situated orthogonally, perpendicular to the cortical surface, can we receive stronger magnetic signatures. We must access the parallel bundles of neurons located in the sulci, the spaces between the gyri, the folds in the brain

These bundles in the sulci are tangentially located to the surface of the scalp. Only these neuron bundles project measurable magnetic fields to the outside of the head. The action potentials from deeper in the brain do not produce an observable field because the currents associated with them flow in opposite directions and the magnetic fields cancel one another out. We have only been able to measure the peripheral neurons. We are experimenting with various signal processing methods to detect deeper brain signals. So far, we have not found a clinically effective method to do this."

I asked, "Sir, perhaps the neurons and glia located on the lattice structure could furnish a useful magnetic signature? They are grouped in a common orientation."

"No. A good try, Jacob, but no cigar. The lattice structures of the brain are too deep. No one has successfully recorded magnetic fields from the lattices."

"Doctor, I assumed these machines would furnish access to the internal portions of the brain. These two problems seem to be very limiting. How will I be ever trace complete pathways of neurons. How will I be able to determine where memory is actually stored in the glia unless memory happens to exists in the sulci?"

"You have put your finger on your major problem. You will have to solve it for us, Jacob."

"Doctor, I have another question. Are the higher frequency radiations, such as Xray, and gamma rays, capable of penetrating your shielding materials?"

"Xray is not a problem, since the booths are protected with three nested layers of aluminum, iron, and molybdenum. Each layer is insulated from the adjacent layers. These insulated layers offer protection from radio frequencies also. You are correct in your concern about gamma rays, If the earth was to receive an intense gamma ray pulse penetrating the ionosphere and reaching our detectors, our machines would be cooked. We face other obstacles, such as something called the 'inverse problem.' More later on the inverse problem. No doubt you have many more questions. We will have to save them for another day. I have some paperwork waiting for me here in the lab office, you are free to leave."

"Thank you for your thorough introduction to this amazing laboratory Doctor."

"Yes, it is truly amazing isn't it. I will keep you informed as to the committee's decision."

"Thank you. I'll wait for your call."

I returned to my apartment very encouraged. I believe I made a good impression on Dr. Majors. What's more, I realize I may have an open door, or at least a foot in the door to this MEG research team.

The tour has given me some valuable insight into the limitations of the research tools. There still remains my own specific problem. How can I materially contribute my knowledge of genetics to the ongoing research by the Doctor's research group? Not only are they limited to orthogonal bundles of glia and neurons, I cannot visualize a method of piercing the inner workings of the cells themselves. I must come up with some way to do so from outside of the brain. I must find some avenue to explore the interior of these glial cells short of performing a lobotomy.

Many thoughts are coming to my mind, some of them very weird. Could the Schulmann resonance play a role in some way? What about the story about the connection between a transplanted heart containing the memory of the young girl's death? I know the Lord will have to play a key role, revealing answers to my questions. I continued to ponder these thoughts over the rest of the morning. Perhaps I need a trip to the farm. I picked up my cell phone and called Terry.

"Hello Terry."

"Hi, Jake. Nice to hear from you. What's been going on with you? Have you been hiding from us?"

"No, I have been busy with classes and planning my research activities. I did talk to Ramona and she said you two are getting married soon."

"Yes, we are going to get married right away, before she changes her mind."

"Not any chance of her changing her mind, Terry. She has you securely hooked and you are a keeper. So, Saturday is the big day?"

"Yep. It will be here at the farm. No need to pay a visit to Yudi, your Jewish friend. It's going to be a cowboy wedding in our hay barn."

"Yes, Ramona said you are planning to be married in the hay barn. Do you intend to spend your first night in the loft?"

"We are going to take a honeymoon trip to sunny California, Jake."

"I have a question. Should I buy a Stetson hat?"

"Sure, spend your last dime on a Stetson, Jake."

"I take it Stetson hats are expensive?"

"Yes. Old man Stetson died a couple of decades back. By the way, you need to know when to take your hat off. Cowboys only take off their hats for three reasons."

"I'll play along. What are the three reasons, Terry?"

"Only when they say the pledge of allegiance, when they hear the Star-Spangled Banner, and when they kiss their girlfriends."

"Okay, Terry, I'll keep my hat on except for those situations. Ramona asked me to bring some Champaign. Is there anything else I can get for you? I assume you already have her ring."

"Yes, and Ramona has one for me. You might want to bring a camera. I'm not sure if Ramona has scheduled a photographer or not."

"Okay, I will bring along my camera. I guess I know who Ramona has pared me up with?"

"Yes, you will be escorting Rhonda Sutherland all day, Jake."

"She's the one who wrote her name and phone number on the paper plate covering her apple pie."

"She had her hooks into me for a while. Really, she is a nice gal, but not flashy, if you know what I mean."

"Yes, I know what you are saying. You mean ugly."

"Are you serious? She is not ugly, Jake. She's just, as I say, not flashy."

"I'm not serious. Forget what I said. Well, I better let you go. I have to find the western store Ramona told me about."

"Rawlings is located just south of the race track on Pimlico Road. Do you know where the track is?"

"I plan to take University Parkway North to the Northern Parkway and then west to Pimlico Road."

"That will get you there."

"I'll see you Saturday morning, Terry."

"Okay. See you then."

I realize I did not have time to go to the farm today, and I won't be able to go to Reisterstown tomorrow. I need to call Dr. Glick. I picked up the phone and called his cell.

"Hello Doctor."

"Hello Jacob. How are you?"

"I'm well. I have called you to let you know I won't be able to go to the congregation on Saturday. My friend Terry Johnson is getting married. He has asked me to stand in as his best man. I'm very sorry. I will have to miss out on your service."

"No problem, Jacob. I'm glad you let me know. I will tell Rabbi Epstein for you. How is your research proposal coming along?"

"I have made a major decision. After considering the advice from you and my meeting with Nanoman, I have abandoned my plan to investigate precursor neurons. Also, I took your advice to look into some elective courses. I contacted the Krieger School of Art and Science. They referred me to a Dr. Majors who is in charge of their MEG research program. After meeting with him this morning, he has agreed to put my name forward for a possible position on one of their research teams. I contacted Director Hobbs for advice as to whether such would fulfill the graduate research requirement. He said he believed it would providing I am able to make a meaningful contribution to their research project. It sounds like I have a good chance of being accepted."

"Jacob, I think you have made an excellent decision. I will pray for you. When do you expect to hear back on their decision?"

"I believe Dr. Majors intends to put my name forward at their meeting this coming Monday."

"I work with Dr. Majors quite often. Have you seen his lab?"

"Yes, he gave me a tour. The machines are awesome. He took time to describe their functions, and also some of the problems they have to contend with, such as the ambient noise and the orthogonal requirement of neuron bundles."

"You will face some problems of your own as well. As a geneticist, you will need to find some way to penetrate the cells themselves, even into the nucleus. Do you have any ideas about how you could achieve such a goal?"

"No, I don't. I realize I am up against a tough problem. I do not have any method in mind how it can be accomplished. I do have some ideas of avenues to explore. They are too outlandish to share with anyone at this point, Doctor. At least I will have access to the full range of lab equipment, including MEGs, EEGS, electron microscopes, sets of slide staining equipment, centrifuges, and a host of other devices."

"Yes, and at no cost to you. I can share with you the MEG lab has deep pockets. They obtain funding from several governmental agencies. You will be asked to obtain a security clearance. I hope you have stayed clear of any protest groups while at Emory."

"Dr. Majors didn't mention the need for a clearance, but my history is clean. The only protest I have ever made was to the dorm manager about the loud noise from the radios in the dorm."

"They certainly won't worry about your need for peace and quiet, Jacob. The lab is well secured. Everyone uses a card key to enter. Did you realize you were scanned when you entered?"

"No, I did not. Is there a camera?"

"There's more than one, Jacob. Welcome to the black economy."

"Well, I guess I better let you get back to your work. I will let you know what I hear, one way or the other."

"Very good, Jacob. God bless you."

I glanced at my watch. It is now three o'clock. Tomorrow I will be working in the genetic pathways lab all day. I need to go shopping for my cowboy gear, and locate some bottles of Champaign right away. As I walked to my car, I am pondering Dr. Glick's warning about the black economy. Am I about to enter a forbidden zone? The Lord previously warned me to be aware of wolves in sheep's clothing. Is there a dark side to Dr. Majors? Only time will tell. I drove north along the University Parkway, then west along the Northern Parkway. I reached the racetrack and turned south. About a mile from the track I spotted Rawlings Western Wear store. A friendly salesman helped me select a western shirt and showed me their selection of cowboy hats.

I wanted to test the salesman's knowledge of cowboy gear, so I asked him, "Sir, do you carry Stetson hats?"

"You want to buy an original Stetson?"

The man tried to hold it in, then he broke out with a fit of laughter. I had achieved my objective.

"Are you serious?"

"No, I was just testing you. Just a bit of Jewish humor."

"Well, for your information city boy, Olé Man Stetson died about fifty years ago, in Death Valley, California. The company went bust a few years afterward. Stetson hats are now found only in museums."

I took a look at the various styles. "I believe I'll take one of those black Montana hats. I have always wanted to visit the Big Sky Country."

"Are you going to Montana?"

"Nope, just to Timonium for my friend's wedding on Saturday. Do you know the Johnson Family?"

"Yes, matter of fact I do. Who's getting married, Terry?"

"Yes, he is marrying a girl named Ramona Bergman."

"Don't think I know her. My name is Craig Grady."

"Hi, my name is Jacob Cahn. I need a couple of other things, Craig. I want to buy a nice cowboy shirt with snaps, not buttons, a pair of Wrangler jeans size 32 by 30, and a nice string tie."

Craig picked out my cowboy duds, and then he asked, "Do you need anything else Jacob? Perhaps a pair of boots?"

I heeded Terry's advice. "No, I do not want to torcher my feet at this time. I'll get along without the boots."

At the cash register, my eyes fell upon a cabinet containing a collection of shiny gold and silver belt buckles. I was tempted to buy one until I looked at the price tags. I settled up with Craig and thanked him for his help.

I have no trouble finding a liquor store on Pimlico Road. There were several. I purchased several large bottles of Champaign and then returned to my apartment. I put the wine in the refrigerator and made dinner. After dinner, I studied the lab manual, said some prayers, and then went to bed.

A short time later I am awakened by the Lord. I feel the love of Yeshua's anointing all over me. I break out in quiet praise, praise in words I do not understand. Then Yeshua gives me a brief but stern word. He tells me I am now a Nazarite and I cannot drink any wine at the wedding, or any alcohol from now on.

I quickly asked, "Lord, I won't touch the alcohol, but what about this MEG opportunity? Should I accept an offer to work with them?"

There is no answer from the Lord. Surely the Lord must know about this decision I am facing, whether to join or not to join Dr. Major's research team. There is no further word. As I ponder the decision I will have to make, I drifted back to sleep.

Friday passes quickly enough. I am enjoying Dr. Baker's lab exercises, learning the correct methods of processing electron scope slides. I arrived back at the apartment after six o'clock. I took a pizza out of the frig and then watched another old movie from my collection. This time I picked out the movie *Dave*, starring Kevin Kline and Sigourney Weaver. Some movies never get old, and this is one of them. It is the humorous tale where Dave Kovic agrees to impersonate the President. Dave assumes it is a matter of security, but it is really to cover up the President's extramarital affair. He has succumbed to a severe stroke. The movie has a happy ending when Dave and the deceased president's widow are reunited. After the movie, I finished off the remains of the pizza and went to bed.

I awoke Saturday morning at eight o'clock, Ramona's wedding day. I opened the door to the refrigerator to retrieve a bottle of milk. The bottles of Champaign are calling my name. I am being tempted by Satan or one of his minions. I closed the door of the refrigerator and said a quick prayer. I turned on my coffee maker, scrambled two eggs, and burned two pieces of toast. The two cups of coffee got me fully awake.

After breakfast, I opened the door of refrigerator and boxed up the Champaign. Then I dressed in my new cowboy duds. I grabbed my camera. As far as I can tell, I am ready to leave for Timonium. I arrived at the Johnson farm, wine in hand, at nine-thirty. The crowd already is gathering. I took the wine to the kitchen and said good morning to Eldora. Then I walked out to the hay barn. Terry has hired a disc jockey for the wedding. The guy is playing a country song I recognize. It's *Get Up John* by Ricky Skaggs and the Kentucky Thunder. Everyone is in a happy mood. The young kids are playing tag in and out of the barn. Pastor Grafton gathers the wedding party together for some last-minute instructions. Ramona gives Terry's wedding ring to me. I put it into the pocket of my jeans. Then Ramona introduces her bride's maid of honor to me, Rhonda Sutherland.

Pastor Grafton walks over to me and shakes my hand. "Hello Jacob. Nice to see you again. Terry tells me you are an experienced best man and you don't need any special instructions."

"Yes, Pastor, I have done this before."

Rhonda is smiling as I take her hand. "Well, we meet again, Jacob. Where have you been hiding?"

"I have been keeping my nose to the grindstone, Rhonda."

"I see. We missed seeing you at church."

"I am attending a Messianic congregation in Reisterstown now."

"Are you enjoying the services there?"

"Yes, very much. Last week I received the baptism in the Holy Spirit."

"Really! You will have to tell me about it."

I am saved by the PA system asking everyone to be seated for the service. Pastor Grafton goes through the service quickly. Terry and Ramona are a beautiful couple. True to Terry's word, the cowboys kept their hats on throughout the service, except when kissing their girlfriends, the bride, and any other opportunity presenting itself. I forgot to use my camera. Fortunately, Ramona hired a photographer. After the photographer finished taking the photos, we all sat down in the barn to another wonderful meal prepared by Eldora. The toasts are made and the wine and spirits flow. Remembering the message from the Lord, I carefully manage to pour out my glass of Champaign onto the straw floor of the barn. Rhonda keeps me company the whole time. Terry was right. She isn't flashy, but she certainly is persistent. Shortly after, Terry and Ramona prepared to leave on their short honeymoon trip, I said goodbye to everyone and made my way back to town.

I spent Sunday morning reading the textbooks and pondering how a geneticist might make a valuable contribution to the field of Biology using MEG. The limitations placed on my decision are worrying. The technology is severely limited by the lack of a strong signal from the deeper regions of the brain. Could the Schumann Resonance somehow be used to trigger a memory response by the neurons during sleep or meditation? What about the cases where those receiving organ transplants are recalling memories communicated by the organ donor? I wondered if the research team would agree to an investigation along such unorthodox lines. Such a proposal could meet strong criticism. At the minimum, I will need verified accounts of memory episodes during sleep. My research proposal must be based upon solid evidence, evidence I as yet do not have. I have the better part of a week to come up with some answers to these concerns.

The Collaboration

My next week was uneventful. Dr. Glick did call me on Tuesday to say he heard on the grape vine I am seriously being considered for the research project. I did not hear back from Dr. Majors until

Thursday. I just finished lunch when the phone call finally came.

"Hello, Jacob. This is Dr. Majors."

"Hello, Doctor."

"Jacob, I have some good news for you. We have agreed to take advantage of your services for one of our research projects. The project is entitled 'The Electromagnetic Potential of the Neural Glial Network.' We are ready to begin right away. This project will be what we like to call a pure research program. In other words, it will be basic research into the electromagnetic system of the soma. This will give you an excellent opportunity to develop your specific research topic investigating the genetic properties of the neural and glial system. How does this sound to you?"

"This sounds daunting, but very exciting Doctor. When I was talking to Dr. Glick he advised me I will be undergoing a background check. Is this correct?"

"Yes, you will be required to furnish the names of ten references. Once you are approved you will be given a security clearance. You can come to the office and pick up the forms, or I can send you the computer site where you can download the forms on-line."

"I would like to take care of the clearance requirement as soon as possible."

"Good. Let me send the site to you now."

I could hear the keys clicking on his computer.

"Okay, Jacob. You should have the address in your inbox now. Do you have any other questions?"

"Yes, Sir. I am curious as to when I can begin working in the lab?"

"I discussed this with the management team. We are going to give you a special pass allowing you to begin training on the equipment right away. You won't have to wait for your security clearance. You should be able to begin by next week."

"Also, do you want to see my tentative research proposal being prepared for the Institute?"

"I don't really need to approve your graduate research proposal, but I will need to approve your specific lab procedures and your overall hypothesis, testing formats, and method of evaluation. Is this clear?"

"Yes. I understand you will want to review my team research proposal to assess my methodology and hypotheses."

"We will want to review this with you and possibly suggest some changes to you. The research director is responsible for the approval of all research proposals. By the way, I am the research director for this research effort. Is there anything else?"

"I assume I will be able to schedule my lab activities to avoid conflicts with my classes."

"Yes, of course. Perhaps you may want to know you have been approved to receive a salary of twenty thousand dollars."

"Twenty thousand dollars for my part in the research project is fine with me."

"I don't think you heard me correctly, Jacob. You will receive a salary of twenty thousand dollars per month for the duration of the research project. I take it you are in agreement with the salary amount. The only condition for the salary is the committee's approval of our research contributions to the project. They will be assessing your activities from time to time."

"Yes, of course. I am very surprised. I really did not expect a salary. I already receive grant money from the Institute and my family has supplied me with some funds."

"You are a valuable commodity, Jacob. Don't sell yourself short. If this goes well for you, the salary you are now receiving will appear miniscule compared to your future salaries. Congratulations. I look forward to a very interesting year together."

"Sir. I certainly will put my utmost effort into the research. I look forward to working with your team."

"Very well. Have a good day, Jacob."

I sat there with my phone still in my hand trying to get my mind around what has just happened. It is an awesome day. I quickly put in a call to Dad.

"Hello, Dad. This is your wayward son."

"Hi Jacob. What have you gone and done now?"

"I have just landed a research position here as a geneticist with the Krieger School of Art and Science. The School is involved in MEG research."

"I see. Does this mean you have changed your major yet again?"

"No, Dad. My work will be involved with the genetic aspects of their research. The overall project is entitled The Electromagnetic Potential of the Neural Glia Network. They are planning to begin right away. This project will be a pure research program. In other words, it will be basic research into the electromagnetic system of the soma. This research effort will be acceptable to the Institute, fulfilling my thesis requirement. Dr. John Majors is the director in charge of the program."

"Will you be receiving any pay for your work, or are they just taking advantage of your graduate research requirement?"

"Dad, I will be receiving a salary of twenty thousand ducats a month."

"Really? Not bad gelt."

"Yes. Dr. Majors said if I prove myself, I stand to earn a lot more than twenty thousand working on other projects."

"Well, I'm happy to hear this is working out for you, Son. Keep us informed. I'll tell your mother the good news."

"Is she there now?"

"No, she is playing bridge. Take care."

I hung up the phone. Dad seemed pleased about my information, but I sensed a certain coolness in his voice.

I called Ms. Wilcox to see if she happened to have a card key for the lab.

"Hello, Ms. Wilcox. This is Jacob Cahn."

"Yes, what can I do for you Jacob?"

"Dr. Majors said I would be given a card key to the lab. I wonder if you have received the card yet?"

"No, I don't think so. I will check on it and give you a call right back."

"Thank you."

I turned on my computer and began filling out the security request form, listing my ten references. A few minutes later my cell phone rang again. It was Dr. Majors.

"Hello, Jacob. I have arranged for a student to drop off your card key later today. I said you would be at your apartment by five or so. Does this work for you?"

"Yes, I will be here at five. I don't remember if I told you, but Dr. Altschul has been appointed as my research director."

"Yes, I believe you did share it with me."

"Dr. Altschul has promised to give me his full support. Also, Dr. Hobbs said you might want to sit in on our meetings, so I will let you know as we schedule them."

"Is this a firm request for me to attend the meetings? I have a lot of meetings to attend already."

"No, it was just a suggestion. I will however let you know when we are meeting. You can decide whether to attend or not to attend."

"I can live with it. Have you applied for your security clearance yet?"

"I am just now submitting the form along with my ten references. I assume I am permitted to tell them they will be contacted soon?"

"Yes, you are free to let your references know you have applied for a clearance and they will be contacted."

"I plan to submit my proposal to my research committee soon."

"Yes, I want you to do so. Send a copy to me as well."

"Fine. Is it possible for me to begin training in the lab this weekend? I would like to go to the lab to get acquainted with the technicians."

"You can come in the evenings if you want to. The lab is open from seven in the morning until eleven o'clock in the evening weekdays. It is open on Saturday from eight in the morning until four in the afternoon. If you decide to come in the evenings, David Strahlermann will meet with you. He is our lab supervisor on the night shift."

"Okay, I will probably be working in the evenings. I'm not sure yet just what night I will be coming this week."

"I understand. I have to go now, Jacob. Keep in touch."

After the phone call, I took time to contact each of my references by phone to let them know they will be contacted. I told each of them not to worry. I'm not being sent to jail. The student dropped off the key card at five o'clock. This whole change of direction is moving ahead very quickly.

19

The Search Begins

"Surely the Lord God will do nothing but he revealeth his secret unto his servants, the prophets."
(Amos 3:7)

Rachel Weeping Over Her Children, Because They Are Not

I am still trying to get my head around the fact I am free to pursue my very own research project. What an incredible opportunity. It is Thursday evening already. My life is unfolding quickly now. I set aside a specific time to work in the lab on my research project. I need to contact the lab supervisor, David Strahlermann. I want to ask him some general questions about the capabilities of the machines. These night sessions in the lab will put a heavy strain on my already busy schedule, but there's no other possible solution. I will visit the lab tomorrow after dinner.

I had been studying for about an hour when my cell phone rang. "Hello, Jacob Cahn."

I didn't recognize the voice at first. "Hello. Is this possibly Dr. Mike?"

"Yes, I need to talk to you."

"Oh, it's so nice to hear from you Dr. Mike. I suppose you want my comments on your translation?"

"Perhaps later. No, I called to tell you about something else. It's about Rachel. You know she has agreed to marry the artist from New York, right?"

"Yes, I received an invitation, but I cannot go."

"Have you heard from her recently?"

"No, I haven't. Is there something I should know?"

"Yes. I happened to have a conversation with Sam Kaufman at synagogue last week. Sam is very upset. I don't like to tell tales, but under the circumstances I feel you need to know the situation. It seems Rachel's husband is not anxious to raise a family. He is very involved in his art gallery and doesn't want the complication of children in his life. Sam and Hannah are completely disappointed and distraught about this."

"I understand. Rachel always looked forward to having children. Is she going to divorce him?"

"No, she can't divorce him. There is a knotty financial problem. Apparently, this fellow borrowed a lot of money from Sam to build his art gallery in New York. He built it on very expensive property, right downtown at the Battery. As you know, Sam is Orthodox. According to the Torah, he must forgive all debts at the end of the seventh year. The property in New York is held in joint tenancy with Rachel. This Leavitt guy has no way of coming up with funds to purchase Rachel's interest in the gallery. I doubt the gallery will ever pay its way. The art gallery will probably end up in foreclosure. If Rachel were to divorce Leavitt, the Kaufman family would lose at least half and probably more of their investment."

"You are saying his gallery sales would probably not even cover the interest on a buyout loan."

"Yes, I'm saying the business will never be able to carry the debt. Besides, he has no intention of helping the Kaufmanns out of the situation. So right now, Rachel is living in a big house on Long Island, estranged from her future husband who spends his waking hours working at the gallery. Sam says he recently encouraged Rachel to get out of the house and join a dig in the Middle East someplace. This would give her something to look forward to. He said she has agreed to go to Egypt on an archaeological dig soon. Jake, Sam feels really bad about his decision to oppose your engagement to marry Rachel."

"Dr. Mike, please tell him his decision not to approve our

engagement was not a mistake. Perhaps he made a bad decision to hastily choose Jeremiah Leavitt, but I can't blame him for not accepting me. His reasons were valid Dr. Mike."

"You have grown in wisdom since our last conversation, Jacob. I will relay your story to Sam. I'm sure he will be pleased to hear of your forgiving attitude. If you have a pencil and paper, I'll give you Rachel's new address."

"Yes, I'm ready."

"It is Rachel Leavitt, 105 Elmwood Dr., Northport, NY 11768. Her phone is 631-261-4415. I told Sam I would contact you to explain the situation. He was pleased when I said I would give you a call."

"Thank you so much for letting me know about her situation Dr. Mike."

"How are things going for you at JH Jacob."

"Things are going unbelievably well. I have been given a spot on a Biotechnology research team under a Dr. Majors. I will receive a nice salary, plus access to the MEG lab at the Krieger School of Art and Sciences. Dr. Hobbs has tentatively approved this work as fulfilling my research requirement for the doctoral degree."

"Sounds awesome, Jacob. Keep in touch. I want to know what you come up with in your research. Also, if you need any materials from the library, just give me a call."

"I certainly will. Thanks again for the update. I will contact Rachel soon. I think a letter will be more fitting than a phone call. Do you agree?"

"Yes, no need to open old wounds, Jacob."

"I may include an allegory in the letter I once heard from an old librarian in Atlanta. I believe it will minister to her pain. Take care, Dr. Mike."

I closed my cell phone and took a few minutes to reminisce. Yes, Rachel and I have taken different roads, but perhaps we will end up at the same destination, in love with Yeshua.

I have the urge to have dinner at the Inner Harbor tonight. I really want to have a talk with my friend Judy. I want to give her my testimony. I arrived at the Harbor at six. The restaurant is very crowded. I should have waited, or come earlier. Judy is very busy working tables on another section and she does not notice me. As I was preparing to leave. I wrote her a brief note on my napkin describing my salvation experience, and how the Lord is blessing me with my research project. I paid my bill and slipped the note into Judy's hand. She stopped and quickly read it. Then turned and gave me a big hug.

"Jacob, I am so blessed to read this, how the Lord Yeshua is blessing you. I'm so sorry you picked this busy night to come. Please come again soon when the restaurant isn't so busy. We can talk then. God bless you."

I shook her hand and said goodbye. As I drove back to the apartment, I kept turning over in my mind what I should write to Rachel. I didn't want to postpone the letter. It would be too easy to end up not writing her.

When I got back to my cave, I sat down at my desk, took out pen and paper, and began to write:

"Dear Rachel, I have just finished talking with Dr. Mike. He has brought me up to date with your unfortunate situation. I am so very heart-broken about it. I told Dr. Mike it was too painful for me to contact you by phone, but I said I would write you a letter instead. So, here it is. I hope this allegory may help to heal your pain.

There was young woman who lived alone in a very big house. She was so very lonely, for she had no one to keep her company and to share her life with. In desperation, she decided to leave her large lonely house and take a trip to a far-away land. So, she went to the library to find a map to follow on her trip. An old librarian showed her a map with a city he said was just wonderful. The people living in the city lived in harmony with one another. There were no wars or problems such as she had faced where she was now living. So, she decided to go to this wonderful city located a very long distance away. There were two roads that left the town where she lived. One road was a well paved road. It would be easy to travel on, but she was not sure whether the road would lead to her destination. There was a second road. It was a rough road filled with may potholes, stones, and dangerous detours. She asked the old librarian, "Do both of these roads go to the same city?" "Yes, said the old librarian. It doesn't matter which road you take. They both end up at your destination. You will have to decide." So, she decided to take well-paved road. She made good time the first day, but on the second day the paved road suddenly became a gravel road filled with large potholes. That is when she discovered the two roads had become the same road. She had no choice but to continue on toward her ultimate destination or just quit and giving up her search for the wonderful city. She decided not to stop. She struggled on. Finally, after many difficult days she arrived at her destination. It was so wonderful. She was so thankful she had decided to continue her quest. It was then she realized what the old librarian really meant when he said it did not matter which road she followed. She learned it is the destination which is important.

Rachel, whatever type of road life takes us on makes little difference. Our eyes must be fixed upon our ultimate destination. Sometimes in life, we are given wonderfully paved roads that degrade into very difficult roads. Then we must decide to press on to our ultimate destination. We all must complete our journey, and we must complete it well. My recent journey here at John's Hopkins has gone well. In fact, I am sure I am on the road to my ultimate destination. My ultimate destination is not a doctoral degree in genetics, or fame, or fortune. The destination I have found is a spiritual one. I have accepted Yeshua as my Savior and the Lord of my life. I am married to him. He is my *Khosn*, my bridegroom. I am one part of many in the Bride of Christ. Two thousand years ago, Yeshua wrote his signature on my ketuba in his own blood. As I now encounter rough roads, I can count it all joy. Rachel, you have taken a rough road, but I want you to seriously consider choosing to arrive at the same destination where I will be, safely in the house of the Lord. Please write and let me know how your journey is progressing. You know I will always cherish the wonderful times we have experienced together. Why not spend eternity with me as well?

Love always, Jacob."

I took time to address the letter and slip it in the outbound mail slot in the apartment lobby. Then I watched the news for an hour and went to bed.

The Testing

I spent Friday working in my class labs, anxious for the day to end. I arrived back in my apartment at five. I grabbed something to eat at the café next to the apartments. The café is run by this Irish woman and Friday's are still fish nights. I ordered fish and chips and a Pepsi. The cod is good, but the Pepsi had lost its fizz. I ordered a chocolate Sunday for dessert, then I made my way across the campus to the lab located in the basement of the Wyman Park Building. I pushed the basement button in the elevator. As I approached the door to the lab I now noticed the cameras Dr. Glick mentioned. There are four cameras designed to capture an intruder's image from all sides. I slipped my new card key into the reader at the side of the door and it swung open. As I entered I hear a low hum emanating from one of the machines. The door to the MEG booth is open and there is a person working at the machine. He is the only person in the lab. I assume he must be David Strahlermann. I called, "David, David Strahlermann?" The man left what he was working on and came to meet me.

"David, my name is Jacob Cahn. I'll be working here in the lab for the next few months."

"Oh yes. Dr. Majors said you might show up here on my shift. Are you looking to begin training this evening?"

"No, I just want to ask you a few questions if you have time."

"Sure, I have time. Let me just turn the MEG off."

"No, please. I would like to take a look at the machine."

"Okay, come on over. We'll take a look."

"David, are you able to generate a number of frequencies, or is the frequency set in stone?"

"We can generate a force field of frequencies from a little below 1 Hz to a thousand Hz. Do you have a particular frequency in mind?"

"Yes, I do. I want to experiment with the resonance of 7.83, and multiples of the frequency to perhaps 33.8 Hz."

"Do you want me to set a frequency now?"

"Yes. Can you generate 7.83 Hz now?"

"Yes, of course. Give me a couple of minutes. By the way, do you have any metal on you, either in your pockets or pins in your bones?"

"No pins in my bones, but I will have to empty my pockets."

"There are some trays on the table near the door. We use them to store metal and any material possibly causing problems for the machines or people. Even though we usually are working with very weak fields, we have strict rules to abide by."

"Let me go empty my pockets, and I will be back."

"Don't forget any rings, ear piercings, or metal shoe arch

supports."

"How about my dental fillings?"

"Do you have metal fillings?"

"Yes, I have a couple."

"Well, you may want to replace them with composite fillings at some point, depending on the type of testing you are doing. We don't want to have to scrape your fillings off of the ceiling."

"Automatic tooth extractions? What an interesting idea."

"Yes, we could put an ad in the school paper and make a fortune. Since you are planning to work at very low frequency levels and probably low gains, I would not worry about it."

I walked over and emptied my pockets, watch, ring, and belt buckle into the tray and returned to the machine.

David hit the start button.

"Do you feel or hear anything, Jacob?"

"I hear only a low hum, but I think I am sensing a force field. I felt a certain pulsing of energy on my skin."

"Good. You have passed the test. Most people who experience the MEG say they hear the machine, but the sound is not where the action is. It is only the mechanical sound coming from the generator motor. The sound, we are interested in is far below your olfactory threshold. The sensation of a pulse on the skin is the business end of this machine. Sound frequency is the equivalent of blowing over the top of a beer bottle. We are dealing here with electrical field frequencies, not sound. Take a look at the oscilloscope and compare it to the pulse you are sensing on your skin."

"They both seem to be agreeing with one another."

"Yes, unlike the sound you are hearing, the scope and your skin sensation are in resonance. Shall I turn the machine to a higher frequency?"

"No, David, you can turn it off. I would like you to show me around the lab."

David gave me several demonstrations of other equipment, such as the two electron microscopes, centrifuges, optical microscopes, and a number of spectrometers.

"I have another question. If the operator wishes to test someone in a sleeping state, how do you put them to sleep?"

"We have several ways to achieve various sleep states. The chairs in the booths are very comfortable. We can play music, or give them a drug, or just hit them over the head with a hammer. No, actually we usually administer a mild sedative and it works most of the time within a few minutes. If the person is a bit up-tight, we give them a shot."

"I see the chairs are nicely padded, but who do you draft to sit in them?"

"Undergrad students needing some financial help are eager to volunteer. We just hired a foreign student from Kabul, Afghanistan, Walid Shinar."

"Are these volunteers afraid of the machines, what they might do to their brains?"

"Many are very fearful, but when you are a struggling undergrad student, you take chances. We subjected Walid to a test yesterday. He was completely relaxed. He actually went to sleep during the test without any sedative."

"Does he speak English?"

"Yes, he speaks excellent English. He attended Kabul University for two semesters before managing to immigrate to the US."

"He is just the kind of person I need for my research. Can you assign him to my project?"

"Yes, I can. Do you want to schedule a meeting with him?"

"Yes, and I would like very much to have a talk with him."

"Okay. I actually have him scheduled for a second test on Monday at four o'clock. Does this fit your schedule?"

"I'll be here."

"Fine. Will you be using the MEG, because I need to give him one more test with a sedative?"

"Yes. I want to put him to sleep for an hour or so. Is an hour enough time for your test?"

"Yes, an hour is fine. I will have to give him either a pill or a shot. We can make this decision after he arrives."

"Very good. I believe I should let you get back to your regular routine. See you Monday at four."

"Okay, Jacob, I'll see you then."

"By the way, David, you can call me Jake."

"Sure, I'm fine with Dave as well. See you Monday Jake."

Dave spent over an hour with me. I accomplished everything I wanted to. I am looking forward to my visit with the student Walid from Afghanistan.

I slept in a bit on Saturday. I got up just in time to get dressed and make my way to Reisterstown. I did not want to miss another Saturday with the Messianic Congregation. When I walked into the room, Rabbi Epstein gave me a warm greeting. I made my way to the front of the room where Dr. Glick and his family were seated. Amy has a place saved for me.

"Good morning Amy."

"God's morning to you, Jacob. I'm so glad you are able to come this week."

"So am I, Amy, so am I. Are you going to dance this morning?"

"Oh yes. We are going to use our banners and flags."

I was so taken by the innocence of this young girl. I love both of them, but Amy is something special.

The service ministered to me greatly. I made it official. This is my spiritual home. I filled in the membership application and handed it to Rabbi Epstein. He was very pleased and prayed over me before I left. Dr. Glick invited me to go to lunch with them, but I declined. I face a lot of homework to complete before next week. I did bring him up-to-date about my visit to the lab and my appointment on Monday with the Afghan student Walid Shinar. Back in my apartment, I cracked the books all day Sunday.

On Monday I am up early, looking forward to my return to the lab. I took some time to pull up the ancient history of Afghanistan off of the internet. Throughout centuries this nation has never really been conquered. It is composed of a large number of tribal groups, many of them nomadic. The ruling elite has always been the Pathan. The British were defeated in the Khyber Pass in the late 1830s. The Russians were defeated with the help of our CIA in the 1970's. The US has been embroiled in Afghanistan ever since. So, the story goes on and on.

I read an interesting theory on the possible link between the ancient Jewish diaspora and these Pathan people. It seems their origin may trace back to the fifth century BC. It is claimed by some historians the Hebrews of the early diaspora arrived in Afghanistan very early and were assimilated into the local Persian culture. In turn, this Persian culture was then overlaid by Islamic culture in the seventh century AD. I copied the material. I want to give the information to Walid.

When I arrived at the lab, Dave was just about to administer a sedative to Walid. I took Dave aside.

"Dave, I don't want Walid to hear us. I am quite interested in Walid's reactions to the experience he has during the transition from wakefulness into sleep."

"Yes, I understand your request."

"This state of consciousness is called 'the hypnagogic state'. Have you heard this term before?"

"Yes, I have. Dr. Majors is also very interested in it. Sometimes the patient experiences hallucinations, very intense dreams, or flashbacks."

"Exactly. When his breathing becomes regular and deep we will know he is fully asleep."

"Dr. Majors also records eye movements. I believe they indicate a state called Rem Sleep."

"Dave, I want you to place Walid in the chair before you give him the sedative. I want to hook him up to an EKG also."

"Okay. I will wheel an EKG over next to the MEG and wire him up."

I walked over to where Walid is standing. "Hello, Walid. I'm Jacob Cahn. How are you?"

"Allah be praised. I am very well, thank you."

"Walid, you speak very good English. Did you learn to speak English in Afghanistan?"

"Yes, I learned at Kabul University."

"I suppose Dave told you we will be working together."

"Yes, certainly."

"The first thing we will do is to seat you in the MEG. Have you done this before?"

"Yes. I have done this, three times before."

"I am also asking Dave to hook you up to an electrocardiogram (EKG) machine. Have they done this before with you?"

"Yes, certainly."

"Fine. So, let's go over to the MEG machine now. Do you know what MEG stands for?"

"Yes, I know. You do not have to say the long word."

"Walid, do you know what an EKG machine is?"

"Yes, I know it is used to listen to my heart, but I don't know the whole words."

"The full name is an electrocardiogram machine and yes, it is used to monitor your heart activity."

David set up the DVD camera on a tripod directly facing Walid. He started the camera rolling. Then he ran tests to make sure the MEG machine was putting out a good magnetic signal. Dave seated Walid and hooked him up to the EKG machine. While this is going on, I am preparing the form I will use to record Walid's reactions during each minute of the test. I will be keeping track of eye, facial, and body movements. The EKG also will monitor his blood pressure, heartbeat, and nervous reactions. I then began my little planned interview.

"Walid, before we put you to sleep and start the machine, I want to tell you something about myself. This talk may touch a sensitive issue for you, okay?"

"Yes, I am listening."

"You are Islamic. Did you know I am Jewish?"

"No, I did not know you are a person from the Book."

"Can we still be friends given this fact of life?"

"Of course, we can be friends as long as you do not criticize my religion. Allah, be praised."

"Very good. Then I will call you my khaver. This word means friend in Hebrew."

"Friend is 'sadiq' in Arabic."

"I also must tell you I am a completed Jew. You probably do not know what the term completed Jew means?"

"No, Sadiq, I do not know what it means."

"A completed Jew is a Jewish person who believes in Yeshua, Jesus."

"Yes. We also believe in the Prophet Jesus."

"Walid, Jesus is more than a prophet in our religion."

"The Koran says Jesus is a prophet who will bring in the Twelfth Maudie."

"I know Jesus personally, Walid. Jesus is not just a prophet, he is more than a prophet."

"Does he predict things and they happen?"

"Yes, he does prophecy."

"Then he must be prophet."

"Jesus is the Son of God, Walid. Christians and completed Jews believe God exists in three persons, Father, Son, and Holy Spirit."

Walid is looking at me and shaking his head no. He doesn't

answer. He is looking at me as if I just committed an unforgiveable sin.

"Walid, are you feeling upset by what I just said to you?"

"Yes. I am, as you say, upset. I believe there is one god Allah, and his prophet is Muhammed. But I will be your 'sadiq' when in this place. I am ready to receive the medicine."

"Fine. Dave, I want you to administer the sedative now."

Dave assured Walid the shot would not harm him. It takes only a couple of seconds for the shot to go into effect. Dave turns on the machine and sets the frequency to the Schumann resonance of 7.83 Hz. He presses the red start button. I watch Walid as the MEG kicks in. His eye movements reveal he has arrived in Rem sleep quickly. He takes a sudden deep breath, and then begins breathing slowly. His eyes are inattentive now. He is completely passive. We kept him under the frequency for ten minutes, then I asked Dave to raise the frequencies ten minutes for each of the five higher harmonic steps up to 33.8 Hz. After the fifty minutes are up, we shut down the machine and waited for Walid to recover. He remained very stable through the whole experiment until the end. With a sudden jerk, he came fully awake.

I told Dave, "This reaction upon waking is called a hypnic, or involuntary twitch. The involuntary twitch marks the division between sleep and wakefulness."

I turn my attention to Walid. "Hello again, Walid. Did you have a nice sleep?"

"Yes, Allah be praised."

"You are back with the living again. What a nice way to earn a living."

Walid did not show any reaction to my humor. "Walid, did you hear me?"

"Yes, I here you."

"Did you have any dreams?"

"Yes, I have several dreams. One very bad."

"Can you tell me about the bad dream, Walid?"

"No, I don't want to tell you about the bad dream."

"Walid, we are doing research here. I need you to tell me about everything, even your bad dream."

"No! I will not tell you about this one. I will tell you about the others, but I will not tell you about the bad one."

"Very well. Tell me about another dream."

Walid spoke slowly, describing the dream to me. "I am back at home in Afghanistan, in our garden picking squash. The squash, they are very large and I make several trips from the garden to our house. Then the dream ended."

"What color was the squash, Walid?"

"Most are yellow. The ones looking like a hat are green."

"You mean they look like a turban?"

"Yes, like a turban."

"Was this the first dream you had?"

"No, it was the last one."

"How many dreams did you have, Walid."

"I had six dreams."

"Was the first dream the one you don't want to tell me about, Walid?"

"Yes, I will not tell you about the first one. Allah will protect me."

I now know Walid has experienced six different dreams for each resonance frequency, and the dream upsetting him most is the first one, the test using the lowest frequency pulse.

"Walid, can you describe for me the other dreams?"

"Yes. I am in the bazaar in Kandahar, looking at goats. The next dream I am riding my horse Kashi, playing Buzkashi, our national game. Then I am in the street near my house, playing soccer with my brothers. The last dream I am, how do you say, picking the opium poppy juice. Those are the other dreams."

"You mean, in the last dream you are harvesting the sap from opium poppy boles?"

"Yes, I scrape the juice into the clay pots."

"You are collecting pot in pots."

"No, not hashish. It is opium juice."

"I understand. I just made a bad joke, Walid. We need to fix a problem we are having If I am going to be able to do my research well, I need you to tell me about your first dream. I'll tell you what. I have a present to give to you. I copied something from the internet this morning. It talks about all of the tribes in Afghanistan. It tells the story of your people, the Pathan Tribe. The Pathan tribal has been the dominant rulers in the country for many years, in fact centuries. Some, like the Kuchi are nomadic. They move up and down the mountains, even into neighboring nations like Tibet and Mongolia."

"Yes, Kuchi means dog. I am Pathan. We do not like Kuchi. They have goats and sheep, too many. They eat all of the grass in Springtime. I am a loyal Pathan. My father is leader of our Pushtuk Village near Kandahar."

"Walid, your tribe was not always Islamic. Before your tribe became a follower of Muhammed, your tribe was Persian. Before they were Persian, your people may even have been Hebrew. According to ethnologists, the Pathan people are distantly related to the Jews. They have traced the origin of the Pathan Tribe back to the great dispersion brought about by the conquering Babylonians. I have underlined where they say this. It seems you and I may be distant relatives. How does this make you feel?"

I handed the printout to him and watched as he slowly read it. He then folded it neatly and put it into his shirt pocket.

"This is nice Sadiq. Thank you, Sir."

"Walid, these historical facts about the Pathan genealogy has given me another idea. I will ask the School to run a complete DNA analysis on both of us. Just perhaps we will find we are actually brothers. What do you think?"

"I don't know about my forefathers. All I know is I am a Pathan and Muhammed is my prophet and teacher."

I found an empty test tube in a storage cabinet. "Walid, let's give it a try. I want you to spit into this test tube for me please . . . Thank you.

Dave, this is enough for tonight."

I thanked Walid and Dave. Then we scheduled another meeting for the following Thursday. Walid was not willing to return until Dave gave him his check. He reluctantly agreed to continue the experiments with us. I am glad Dave had his pay ready because I certainly need to give Walid more tests. After Walid left, Dave and I replayed the video to observe the outward actions of Walid's body through the sequence of dreams.

"Dave, his reactions are most extreme during the first dream. If I am going to be able to continue working on this research topic, I need to find a way to strengthen the neural-glial signals we are receiving from the sulci. I want to try administering some brain supplements on next Thursday's test. I am hoping to improve the neural transmission using these supplements. Perhaps they will strengthen the signals from the sulci."

Dave agreed it was a good idea.

Even if I manage to gain new information from these MEG experiments, a major task remains. I can't simply stick a probe into Walid's brain. If I am going to qualify my research under the heading of genetics, I must somehow find a way to penetrate and observe the interior of the cells themselves.

There was little time to pursue my research during the week. I did schedule the DNA tests for myself and Walid, but the results will not be available for several weeks. I took time to research the various brain cognitive enhancing supplements. The market is overwhelmed with these products. Some of them must be reconstituted by the liver in order to be make them available to the neurons. None can be directly administered through the blood system, although they can act to aid neurons by increasing the localized circulation.

I went ahead and ordered a supply of the supplements containing vitamins, various plant abstracts, and specific compounds, including Dopamine (DA), Dimethylaminoethanol (DMAE), Centrophenoxine, and Docosahexaenoic acid (DHA).

DHA is a well-known neurotransmitter. DMAE seems to work by elevating the synthesis of acetylcholine, and also is a neurotransmitter. Centrophenoxine is synthesized from DMAE, but may be a more effective option due to its higher bioavailability for acetylcholine conversion. DHA acts to improve the membrane of older cells, reducing the 'canibalization' of the muscle tissue, and generally stabilizing the neural system, reducing levels of stress. Dave cut a DVD disc of the test and I took the completed form and disc back to my apartment. I just arrived back to my apartment when Dr. Majors called.

"Hello Jacob. I just wanted to check in with you. How is the lab work going?"

"Hello, Doctor. Walid, the student we are using, is working out quite well. We obtained a fairly strong signal from his neuron bundle. I am going to run another test on Thursday. I plan to give Walid some brain supplements this week. I want to see if they might improve the signal strength. The ads for these supplements always claim to strengthen the neuron connectivity. We will see whether these claims are valid, or just false wishes."

"Yes, go ahead and give them a try. Did you include some DA?"

"Matter of fact, I did. I also am planning to give him some DMAE and Centrophenoxine."

"Be sure to check for possible reactions."

"Thank you for the reminder. I'm still not sure how I am going to gain access to the innards of these cells, Doctor."

"Yes, you do have a problem there. I may be able to help you out once you have your security clearance. I have access to a new scanning machine. It is located at the National Security Agency (NSA) at Fort Meade. Would you be willing to drive down to the facility?"

"Absolutely."

"Okay. I will try to line up some time on the machine. We will have to use one of their Guinea Pigs."

"Guinea Pigs?"

"Not actual rodents. Volunteers with security clearances."

"Oh, yes of course."

"Just keep me posted on how it is going, Jacob."

"Yes Sir, I will."

Dr. Majors' request for results from my tests, plus the cryptic message about a new machine really is encouraging me. After my conversation with Dr. Majors, I called Walid and asked him to meet me in the lobby of the Wyman Park Building on Monday afternoon at one o'clock. Walid showed up right on time. I gave him four doses of each of the supplements and told him to take one of each pill every morning. I asked him to contact me if he experiences any bad reactions to the supplements. He said he will meet with me for another test at six o'clock on Thursday.

The Spring Semester is rapidly drawing to a close. I soon will be travelling to Bar Harbor, Maine to present my research proposal. I began searching the internet for clues as to the human brain's ability to reason and to memorize. I did find a post by a Messianic Rabbi who suggests a person's memory can be stored in the DNA. He cited some scriptures and a passage from the apocryphal book, "The Assumption of Moses." He suggested the total history of a person's lineage perhaps is hidden within the brain. Could this possibly be true? Could the glial cells within the brain contain the total history of a person's lineage? If so, glial DNA might be awakened to reveal the total store of memories back to the beginning of mankind, even to the Garden of Eden. It is an intriguing thought, but I have to dismiss it this time around. I cannot give much credence to such an outlandish speculation based on one Rabbi's internet posting. The idea went in one of my ears, but did not entirely come out of the other.

Over the next few weeks David and I ran a series of tests with Walid. The cognitive supplements did improve the magnetic response from the neurons by a little over ten percent, permitting us to dig deeper into the sulci. I again tried to convince Walid to divulge the contents of his first dream. He continues to refuse and is becoming more hostile to my requests. Then I received some very interesting information. The technician at the DNA lab called on Tuesday to tell me they had finished Walid's and my own DNA profiles. They said they posted the results on my email. I lost no time in retrieving the data. It is nothing short of amazing. We both are blood type B. Our DNA markers are similar, some identical. Pathan Walid is a Hebrew and he doesn't even know it. Perhaps this will finally convince him to share his dream with me. I decide to call Walid on my cell phone to give him the results.

"Hello, Walid. This is Jacob Cahn. Do you have a few minutes? I have some interesting news for you."

"Yes, of course. I am having some tea in my dorm room."

"Walid, I just received the results of our DNA and blood tests. We both have the same blood type. More remarkable than the blood test, we have matching markers in our DNA. Your people, the Pathans, actually are the remnant of the Hebrews who migrated from Babylon over twenty-five centuries ago. Isn't it amazing?"

"Yes, I know."

"You know this already? How do you know this?"

"It was in my dream, the first dream I had in the laboratory."

"This is awesome. Are you willing to tell me about the dream now?

"No, I won't tell you about the dream."

"But why do you continue to refuse to tell me about this first dream? It is important for my research, Walid."

"Allah will not be pleased. I know this. I will not go to paradise to enjoy seven beautiful virgins if I tell you the dream."

"Walid, I have news for you. There is only one way you are going to paradise. Jesus is the Messiah sent to Earth as a sacrifice for our sins. He loves us. He gives us forgiveness of all our sins if we will confess them and repent. His offer of forgiveness is free."

"Nothing of value is free. Allah is the one and only god, and Muhammed is his messenger."

"Walid, does your Allah love you?"

"I don't know if he loves me, but I must be obedient to the teachings of the Koran and the teachings of the Imani. I will not tell you about my dream. Goodbye Jacob."

I was disappointed once again Not so much about the lack of information for my research as by Walid's refusal to accept Yeshua's grace and forgiveness. I took time to pray for him.

20

Divine Assistance

"I saw in the visions of my head upon my bed,
And, behold, a watcher and a holy one came down from heaven."
(Daniel 4:13)

Hand of God

The rest of the week I worked to complete my class lab assignments. My evenings are spent preparing for the trip to Fort Meade. I am completely frantic. I will need to present a description of my research hypothesis and methodology to NSA. I need to frame this in a rigorous scientific method format, namely, I must clearly spell out my hypothesis, a review of the pertinent scientific literature, a proposed methodology, and my preliminary test results. If I do have some preliminary test results, they may ask me to present a revised hypothesis. I want to give this research some kind of a catchy title. The only thing I am coming up with is "The Genetic Investigation of Glial Memory." All this must be accomplished before meeting with the managers at Fort Meade.

On Thursday morning Dr. Majors gave me a call. "Hello, Jacob."
"Hello Doctor."

"Jacob, I am in possession of your top-secret security clearance. We are free to schedule a meeting with the people at Fort Meade.

How well are you prepared?"

"I believe I am prepared. I have completed a double-spaced, twenty-page proposal. I also have included attachments containing the results of my lab tests, pertinent research papers, as well as some background information from non-scientific sources. I am still in the dark concerning the capabilities of this new machine. Until I know the operation of the machine, it is impossible to accurately define my final approach to the investigation."

"Yes, I understand. I believe they also know the situation regarding the new machine. Can you deliver a copy of your proposal to me today, Jacob?"

"Yes, I will send a copy to Ms. Wilcox this morning."

"Fine. I will give it a look. If I find you need additional preparation, I will get back to you. Otherwise we will meet with them tomorrow morning."

"Assuming they agree to meet tomorrow, will you be picking me up here at my apartment?"

"Yes, of course. We will take a van from the motor pool. There will be several other of my researchers coming along with us. I'll let you know later today, Jacob."

"Thank you. I'll wait to hear from you."

I turned on my computer and sent an email to the attention of Ms. Wilcox, asking her to print the attachment for Dr. Majors to read. A queasy feeling is rising up in my stomach. I need to find a way to relax. I decide to take a nap. I prayed, but I am still too keyed up to fall asleep. I took a couple of aspirins. Then I got up again and made a peanut butter and marmalade sandwich. For some reason, eating something seems to quiet my nerves.

Perhaps it was the sandwich, but more likely the prayer which allowed me to fall asleep, to fall into hypnagogia. I began to dream at once. I saw a bright light emanating from above. It was Yeshua. He is smiling at me. He said nothing. He stretches out His hand and places a scroll in my hand. Then he disappeared. I awoke feeling completely rested and revitalized. I looked at my hand. I am not holding anything. In my spirit, I know the Lord has given me a gift. It is a gift of understanding. I now know beyond doubt my research efforts at NSA will be successful.

It is three o'clock in the afternoon when Dr. Majors calls me back.

"Well, Jacob, you have done a very thorough preparation for our meeting. I have scheduled our visit for nine o'clock tomorrow morning. We will pick you up at half past seven. Be sure to dress up a bit. Casual is fine. Do you have any questions?"

"No, Sir. I will be ready when you arrive."

"I have arranged with Ms. Wilcox to print seven copies of your proposal. I will give you your security badge, security paperwork, and the copies of your proposal when I pick you up. We will be having lunch there at the NSA commissary. They may ask you to give a short presentation to them. I'm not sure. They may just be satisfied with your written proposal. When we arrive, they will take us on a tour of their lab. We will get a first look at the new machine. Then you will be given a security interview by a military officer on their staff. During your interview, they may ask you to sign a non-disclosure statement."

"What kind of a non-disclosure statement, Doctor?"

"Well, they will ask you not to reveal the capabilities of this machine. It is top secret."

"How can I conduct research on a secret machine and not reveal to the public the methods I am using?"

"There are ways to take care of such situations. You don't have to be concerned at this point, Jacob."

"Okay. Thank you for the information, Doctor."

"See you tomorrow morning, Jacob."

He hung up his phone. I am hopeful this machine will be the answer to my dilemma. Although I am excited about the prospects, I do have one major concern. I remember Dr. Glick's warning when I told him about the offer Dr. Major's made. His reply was "Welcome to the black economy." Am I stepping through a dark and dangerous doorway?

The rest of Thursday I gave my mind a rest. There is really nothing I need to do before tomorrow's meeting.

I remembered Terry had said they would be back from their honeymoon by this week. I called Terry. "Hello, Terry, this is Jake. How are you people doing up there?"

"Very fine, Jake. It's good to hear from you. Is everything going well?"

"Yes, tomorrow I am going to Fort Meade to be interviewed. Dr. Majors has arranged for me to meet with the NSA to gain permission to use a new machine. He says he believes it will solve my problem of remotely accessing the cells within the brain."

"Very good news, Jake. If you have some time, why don't you have dinner with us. Eldora is cooking a brisket. She has been marinating it for hours."

"This sounds too good to pass up. I'll put on my cowboy duds and be on my way."

"Can you leave right away?"

"Yes, I don't have anything holding me back."

"Great, we will look forward to seeing you."

I put on my cowboy shirt with the snaps, my jeans, grabbed my Montana hat and I was on my way.

When I arrived, everyone greeted me with handshakes and slaps on the back. Rhonda was present. I kept my hat firmly in place. Terry took me into the game room and we played a game of eight ball. Terry won when I knocked the eight ball into the wrong pocket.

"Jake, are you going to leave soon for Bar Harbor?"

"Yes, I will be leaving on the tenth of June. I hope to have some results by then to show my advisors. I may be able to present some amazing things to them."

"Your attitude is really positive, Jake."

"Terry, I know I am in the Lord's will on this project."

"Have you heard anything from Rachel?"

"I heard about her situation from Dr. Mike. She is very lonely and living in a big house on Long Island. Her husband is not interested in anything except his art gallery. He does not want the inconvenience of having children."

"Oh, my goodness. You shared how much she looked forward to motherhood. Will she be divorcing him?"

"No. She cannot divorce him because of the family's financial situation. Her dad loaned the husband a lot of money when the guy purchased an art gallery in New York. It is jointly owned by Rachel and himself. Rachel must protect her dad's investment. If they divorce, the court no doubt would give the art gallery to the husband. This would saddle the Kaufman family with at least half of the unpaid debt."

"Ouch. Have you called to talk to her?"

"No, I just recently sent her a letter. She hasn't had time to respond."

"This is a sad story, Jake. Let's pray about it right now."

"Thanks Bro."

We joined hands and Terry and I prayed for Rachel. Then Terry gave me a big hug. His hug really cemented our relationship. We truly are brothers in the Lord.

Eldora begins ringing the triangle. Everyone gathers in the dining room for dinner. Howie prays. Terry prays again for Rachel. Following dinner, we all sat around the fireplace telling stories. I told them how the Lord blessed me by preparing a way for my research project to proceed. Terry shared how Ramona has blessed his life.

Then Terry says, "Everyone, I have an important announcement to make. I know how all of you are looking forward to Ramona's first start at Pimlico. Dad and I met with the stewards last week. We were able to get approval for her to ride. Also, Black Gold passed the starting gate test, so we are all set to enter him in a maiden race very soon. Ramona will be up when he starts."

Everyone lets out a cheer. This was the capstone of the evening. It was getting late. I told everyone how much I enjoyed seeing them again. Then I made my way back to a lonely and dark apartment.

Into the Lion's Den

I rose early on Friday. I showered and dressed in my suit. I ate a quick meal of scrambled eggs, toast, and two cups of coffee. I took some time to pray over the breakfast. Then I took a few minutes to reread my proposal. When I finished, it was time to leave. I walked down to the lobby and waited for the van. The van arrives right on time. Dr. Majors is driving. As I enter the van, Dr. Majors hands me my documents and my security pass. Somehow the NSA has managed to include my picture on the pass.

There are three others passengers in the van besides myself. Dr. Majors takes time to introduce everyone.

"Gentlemen, Jacob here will be our resident geneticist. He is in the process of completing his first year and looking forward to receiving his doctoral degree. Jacob, this man in the front seat here is Charlie Simpson. He is a computer whiz and will be handling our data. Helmut Vogel on your left is an expert in physical anatomy. Helmut will be analyzing patient treatment responses. Finally, the guy asleep on the backseat is Joe Hernandez. Joe is responsible for keeping our financial books in order. Fasten your seat belts. This is going to be quite a ride."

We all greeted one another, except for Joe who continued sleeping, perhaps dreaming of balancing his ledger accounts. We journeyed South on the BW Parkway past Jessup and turned into the NSA off ramp. There are security cars parked along the exit road. The guard house was expecting us. They waved us on through the gate and a guard on a golf cart escorted us to a large imposing building. The sign in front of the building reads Site M - High Performance Computing Center Two (HPCC2). We parked at the visitor's parking lot in front of the building. The building exterior is covered completely in black glass. Dr. Majors points to the building and tells us, "This is one of the new computing centers. It is perfect for working on MEGs as the building is protected with copper shielding beneath the glass exterior, ostensibly to protect against electronic eavesdropping."

The plethora of black-glassed buildings gives me the creeps. One building in particular is an immense five or six story building covered entirely by black glass.

Helmut tells me, "Jake, everyone calls the large black building you are looking at the Rubik's Cube. The Cube measures three million square feet and is the home of the NSA Directorate."

As we all enter building number two, a Master Sargent asks to see our security badges and then we are electronically frisked. The guards thoroughly inspect my briefcase containing my papers. Then the first guard escorts us into the lobby and asks us to be seated. A few minutes later we are joined by a young female First Lieutenant dressed in tailored fatigues.

"Good morning, gentlemen. My name is Lt. Mary Swift. I will be your tour guide today. We will first take you to meet General John Weaver. He is the Commandant of the HPCC2. Right this way please."

We follow the Lieutenant down the hallway to the office of the Commander. The Commander is waiting for us and asks us to follow him to an adjoining conference room. The Lieutenant serves us coffee and doughnuts.

The three-star general shakes Dr. Major's hand and says, "Nice to see you again John. Please introduce me to your friends."

Dr. Majors introduces us and gives the General a brief description of our areas of expertise. The General then asks the Lieutenant to contact a Major Conners. Then he tells us, "Gentlemen, Major Conners will be giving you a tour of our Qbit computer installation as well as the new machine you are interested in. After the tour, we will be going to lunch at the Cube. Following lunch, the security interviews will begin. I believe Mr. Hernandez and Mr. Vogel have already been interviewed previously, so we will be interviewing only Mr. Cahn and Mr. Simpson. I'm sure Dr. Majors has briefed you somewhat on our activities here. If not, we can take some time at lunch to fill you in. Also, you will find a lot of information about the NSA on the internet, actually more than we are comfortable with. The WikiLeaks fiasco has given us a lot of unwanted exposure."

Just then, Major Conners enters the room and asks us to follow him. He takes us down to a lower floor of the building and into a large laboratory.

"Gentlemen, this room is our MEG lab. I take it you are already familiar with most of the equipment here. Our new machine is housed in a separate room, adjacent to our Qbit computer. Do you have any questions?"

Joe asks, "Should we have brought along heavy coats today?"

"No. The Qbit computer is housed in a deeply insulated casing. We keep the room at a fairly cool temperature of 15.6 degrees Celsius, or sixty degrees Fahrenheit, just above the shivering point, Joe."

The Major opens the door to this inner room. It is small, perhaps twenty feet by thirty feet. The computer takes up most of the room. At one side is the new MEG machine.

I ask the Major, "Sir, this machine looks quite different from a MEG. I don't see any large coils?"

"You are very observant, Jacob. This machine does not use large coils, or even magnetic fields as you are used to."

"Does it have a name?"

"Yes, it has a very long name. We call it the Integral Analyzing and Magnifying Grouped Optical Device (IAMGOD). I need a volunteer who would like to be today's guinea pig, to take a seat in the IAMGOD machine?"

Oh, my goodness. I can't believe the acronym they have assigned to this machine. Still, I didn't wait to volunteer. "Yes, I will volunteer. Will I have to be sedated or can I remain awake?"

"You can remain awake, or asleep. It's your choice."

"Perhaps I can experience both states?"

"Certainly. We can start with you awake and then give you a sedative. If we have a second volunteer this will allow everyone to observe the operation of the machine."

Joe also was willing to be a guinea pig. I asked Joe to go first. It is no surprise when Joe agrees to take the sedative. The Major seats Joe on the device chair and lowers the helmet over his head. Before giving Joe the shot, Conners fastens him firmly into the chair and turns on the camera to record Joe's reactions. Conners then walks over to the work station and turns on the Qbit computer. A large screen comes to life. On the screen is a three-dimensional image of Joe's skull. Then Major Conners gives Joe the sedative. I watch as Joe enters the hypnagogia state.

Major Conners asks, "Gentlemen, please watch the large screen. I am going to demonstrate the abilities of this machine to you. Let me begin by saying everything electrical going on in Joe's brain is now being instantly recorded as to the location in the four dimensions of length, width, depth and time Also the amplitude and frequency of the neuron signals are recorded. As you can see on the screen, the machine is programmed to displaying a still frame of the visual holographic image of Joe's head for ten minutes. After ten minutes, the computer stores the first data and cycles to a new set of sequences. As an analogy, think of these cycles as sequences on a roll of camera film. The film consists of a series of frames. The holographic images recorded by the Qbit computer are similar, except the actual data being recorded includes all information down to the maximum resolution of the images. This resolution is incredibly miniscule, slightly less than the resolution of an electron microscope. This machine is penetrating everything down to the strands of DNA. I'm sure you must have some idea of just how much data is involved here. The electron microscope is able to enlarge a target up to about ten million times, limited only by the size of the electron. As you approach this degree of magnification, the object becomes less distinct. When you approach the minute scale of the electrons, the machine shows us they are actually not clear particles, but misty rings around the positive atom. As I say, the machine we are demonstrating today is limited by the size of the electron. Since it records multiple layers of one electron width instead of a single layer like the electron scope, we can greatly sharpen the images by using enhancing algorithms. Right now, you are viewing Joe's brain at minimum amplification. I am now going to zoom in and show you the internal makeup of a cell in real time."

I watch as the Major selects a portion of the brain near the eye. He enters the xyz coordinates of the mouse into the computer. Suddenly the screen clears and we see the outline of a nucleus within a cell wall. It is pulsating.

I have to ask, "Major, this pulsation, are we seeing on this screen the movements of a living cell?"

"Yes, Jacob. You are seeing it in real time, and it is being recorded for future examination."

"Is there any way of interacting with this cell, or is it untouchable, so to speak?"

"Ah, you have asked an important question. I believe it may be possible to interact with the cell and its nucleus sometime in the future. To date, we have not examined how this might be accomplished."

I want to press him further. "I can detect what appears to be a cell wall and possibly a nerve attachment?"

"Yes, it may be possible to trace neuron networks down to the cell level. I suppose if you could precisely induce a low hertz current onto the connecting neuron, you might be able to observe its effect on the cell in real time."

"Could the guinea pig possibly react in a violent manner to these lower level stimulations."

"We believe the machine is safe. There may be some degree of response. We are not sure how strong this would be."

I am overwhelmed by this awesome machine. I have one more question for the Major.

"Sir, I am curious as to who developed this machine. Was this developed in-house, possibly by yourself?"

"I'm not at liberty to share the source of this technology, Jacob. The answer to your question must come from someone above my pay grade."

"I see. Well, thank you for demonstrating the capabilities of this miraculous machine to us, Major Connors."

"Yes, it truly is miraculous. I'm now going to give the computer and the machine a rest. We will terminate Joe's nap as well."

Joe gradually came back to life. There was no expected jerk. As soon as he was fully awake, I asked him if he had dreamed. Joe said he dreamed about a trip to Bermuda with his wife and kids. He said he had a wonderful trip and is looking forward to returning there again.

I asked, "Joe, have you ever visited Bermuda before?"

"Why, yes. We went to Bermuda, three years ago."

"And, did you see the same things in your dream you saw during your trip to Bermuda?"

"Yeah, I did. My dream was like an exact replay of our trip three years ago, the same hot dog stand, the same boardwalk, the same everything."

"Thank you, Joe. You have given me some very interesting and tantalizing information."

I asked Major Conners, "Sir, will we be able to get copies of these demonstrations?"

"Yes, we can give you all a copy."

Then it was my turn to play guinea pig. Major Conners straps me into the chair and sends me to dreamland. My dream is the repeat of my scroll vision. When I wake up I realize my dream is an exact replay, just as Joe experienced with the exception my dream is of another dream or vision.

Everyone is looking at me strangely. I finally asked, "Why are all of you staring at me? Did I say something I shouldn't have while sleeping?"

Major Conners answered, "Jacob, we witnessed something very strange at one point in the session. The machine was in minimum resolution mode. Suddenly, a very bright light floods the computer screen. It lasts only a minute or so, and then is gone. I have never seen this happen before. I don't know if there was a short circuit or what the cause could be. Can you tell us what you were dreaming when this happened?"

"Yes, I can, but you may not believe me."

Major Conners is pressing me for more details. "Jacob, tell us exactly what happened to you?"

"Okay. My dream was a dream of another dream, or vision I had last week. I received a visit from the Lord. He just smiled and put a scroll into my right hand. Then he left me."

"What was written on the scroll, Jacob?"

"Major Conners, I don't know what was written on the scroll. When I awoke from my vision, there was nothing in my hand. I believe the Lord intends to reveal the message on the scroll at some future time."

My comments left them scratching their heads. About this time General Weaver came through the door and said it was time for lunch. None of them wanted to discuss my dream in front of the General. My experience from the spiritual dimension was beyond their ability to get their heads around. We all walked out of the building and drove over to the Cube.

When we finished lunch at the Cube, the General took us to an office upstairs. The sign on the door read "Director of Compliance."

We entered the director's office. The director, Mr. Ronald Logan, rose from his desk and introduced himself. He asked how we enjoyed our visit to NSA. Everyone agreed it has been very interesting. What an understatement. The Director then asks General Weaver to take everyone except Charlie and myself on a tour of the Cryptographic Museum across the highway from Fort Meade while Charlie and I are being interviewed. The Director says he will call the General when our interviews are completed. General Weaver escorts everyone except the two of us out of the building. The Director's secretary contacts two staff interrogators on the intercom and tells them it is time for the interviews. Shortly afterward a Master Sergeant appears and escorts Charlie and I down a hallway to separate rooms. The sign on my door reads "Interrogation Room 204." No one is in the room when I enter. The room is quite small and only contains a small desk and two chairs. The room is lit with recessed lighting fixtures. There are no outside windows. I notice two large mirrors along one wall. I know these are not ordinary mirrors. I can barely make out the walls of a room behind the one-way windows. There will be others observing my interview. I also notice the room has a camera mounted on the ceiling facing where I am sitting. I feel very vulnerable. My mind harkens back to the interrogation rooms described by Aleksandr Solzhenitsyn in *The Gulag Archipelago*. Am I being subjected to a Gulag-type interrogation? The door opens and a bird colonel walks in carrying a briefcase. He is a muscular black man about forty years old and dressed in tailored fatigues.

"Hello, Jacob. My name is Colonel Ben Jones. I have been assigned to give you an interview today. I see you have noticed the one-way mirrors and the camera. Don't let this room get you up tight. This interview is just a normal part of our vetting process here at NSA. We have to be very careful when it comes to allowing new people such as yourself access to this facility. I hope you understand."

"Certainly, Sir. I understand. You have your procedures."

"May I take a look at the materials you brought with you?"

"Gee, I thought you guys already knew what is in my briefcase."

The Colonel isn't smiling; I can see my humor will not change his approach. He means business. I hand him the contents of my briefcase. He opens his briefcase and pulls out a large folder. I can make out the title of the folder. It reads. "Dossier: Cahn, Jacob C45926" He slowly begins reading each and every page of the contents of my briefcase, making a few notes in his file. I'm just sitting there, trying to remain calm. Occasionally I hear a quiet "h'mm." It takes him about a half hour to finish examining the contents of my briefcase.

"Well, Jacob I am impressed. Your scholarly achievements are quite remarkable. Now I need to ask you a few questions. Are you still a single man with no current plans to get engaged or married?"

"Sir. I have no current plans to be married."

"Let's forget the 'Sir' thing, Jacob. Just answer my questions. From your dossier, I believe you recently were dating a Miss Rachel Kaufman?"

"I am no longer in contact with Rachel. Her father refused to agree to our engagement. She recently married another man."

"I suppose this explains her living now in New York. So, your relationship with Rachel is now completely over?"

"Yes. As far as marriage is concerned, it is kaput. I still hold her in high regard."

"I see you often are in contact with several people from Atlanta. I assume these are people you met while attending Emory University."

"Yes, they are friends I met while at the University."

"I also see you are acquainted with a Mr. Terry Johnson living here in Timonium?"

"Terry and his wife Ramona are good friends of mine. Terry's father is in the thoroughbred breeding business."

"Do you like to go to the track, Jacob?"

"Yes, I have attended the races on occasion, but not lately. I do enjoy watching these animals run. The last time I attended a race track was when I was in Omaha, Nebraska. The track is now closed."

"Would you consider yourself a professional gambler?"

"Hardly. I don't enjoy seeing my hard-earned money go to feed horses."

"I see your father is a general practitioner in Omaha. Is your relationship with your father, a good one? Are the two of you close?"

"I love my father very much and he feels the same about me. We do have our differences. Dad did not really want me to come to Johns Hopkins. He asked me to stay at Emory, become a general practitioner, and take over his practice in Omaha when he retires."

"So, you have some disagreements about this geneticist choice of yours. I take it you are still on speaking terms with your dad?"

"Absolutely."

"You are of Jewish descent, but I understand you have recently became a Christian. Can you explain why you made this choice?"

"Jewishness is in our DNA. I'm still a Jew. I believe in a Jewish Jesus, Yeshua Ha Mashiach. Messianic Jews like myself prefer to be known as completed Jews, completed in Yeshua."

"So, you prefer to be called a completed Jew? Interesting. We don't encounter this brand of faith often. Let me ask you a faith question. If you had in mind to help a poor person, would you ask Yeshua for his blessing beforehand, or would you simply follow your conscience and not seek the Lord's permission before helping this person?"

"You have asked an interesting question. I believe most believers would simply follow their conscience and their good hearts to help the man, like the Good Samaritan."

"What would you do, Jacob?"

"I hope I would first ask whether it is the Lord's will beforehand. I can't claim to have always done this in the past."

"You have a strong faith, Jacob. I need to ask if you know what the word transhumance means?"

"Yes, I know the word. It has two meanings, and their meanings are quite different."

"Can you give me both definitions?"

"Yes, I can define each of them. The original use of the word has a pastoral meaning. It is the seasonal movement of grazing animals up and down slopes according to the seasonal availability of forage. The second is a more recent use of the term. It literally means to alter the human DNA, producing a hybrid being with more desirable physical and/or mental capabilities. I personally have a problem with this, especially if the alteration occurs at the germ level. Altered DNA at the germ level is passed down to the individual's progeny, forever altering the original DNA."

"This next question is important to the agency. Do you consider the alteration of a persons' DNA at the germ level a good thing, Jacob?"

"No, I do not. I am convinced the altering of DNA at the germ level is extremely dangerous, and even evil. God has destroyed whole people groups because of altered DNA in order to preserve the purity of the genome."

"Jacob, I know you are a person with strong opinions. I have another critical question to ask you. What if you were asked by your government to do something which conflicts with your belief in the sanctity of a human's DNA, even where the altered genome appears to have great merit and is done above the germ level? Would you acquiesce to your government's wishes and agree to carry out the procedure?"

"Depending on the situation, I probably would agree to perform a DNA change to therapeutically correct an abnormality if it was above the germ level. I would not agree to such a change if it was to create a being with qualities exceeding a normal person's abilities."

"So, you would not agree to alter the DNA to produce a hybrid being capable of superhuman abilities, even though your government requested you to do so?"

"I can only give you the answer Jesus gave us. I would render unto Caesar what is Caesars and unto God what is God's."

"But you really haven't answered my question, Jacob?"

"When they asked Jesus whether he was the Messiah, he refused to answer because he knew they wanted to crucify him. Do you want to crucify me, Sir?"

The Colonel paused and let out a great sigh. "No, Jacob. I will not crucify you."

I watch the Colonel do something rather strange. He leans over the desktop with his body, turns over one of the pages of my proposal. He then quickly writes something on the page and sticks the page beneath several other pages. Then he places my proposal back into my briefcase and hands it back to me.

"Jacob. This concludes our interview. Our staff will discuss your responses to my questions and let you know the results in a few days. Thank you for your help and have a good day."

The Colonel stands up, shakes my hand, and leaves the room. The Master Sergeant enters and escorts me back to Director Logan's office. I see Charlie has already returned from his interview. The Director puts in a call to General Weaver, telling them the interviews have been completed. He says we are ready to be pick up at the Cube. Then the Director shakes our hands says we are free to leave. We thank him for the opportunity to view the new machine. The Director smiles and directs his secretary to escort Joe and I down to the lobby. The van is waiting for us in front.

When I am safely in the van, I opened my briefcase and retrieved the paper the interviewer had written on. It reads, "I understand your predicament, Jacob. I share this problem with you. I am a believing Pentecostal Christian from Atlanta, Georgia. God bless you."

Once again, I have been taken care of by the Lord. I believe I will gain access to the IAMGOD machine.

The traffic on the Interstate was very heavy. I did not arrive back in my apartment until half-past six, too tired to cook dinner. I decided to give the nearby café another chance to redeem themselves. I ordered fish and chips with coffee. The coffee was terrible and the cod fish was dripping with oil. Somehow, I managed to finish the meal. I returned to my room and was about to start another old movie when the telephone rings. It is Dr. Glick.

"Hello, Jacob. Do you know what tomorrow is?"

"Yes, it is Saturday, Shabbat."

"And what else?"

"I don't know. What else?"

"It is Passover. We want you to have Passover with our family. After service tomorrow, you will be coming home with us."

"Thanks, Dr. Glick. I will look forward to it."

I hung up the phone and sat down to watch the *Flight of the Phoenix*, with James Stewart as Frank Towns, the pilot of a twin-engine cargo plane flying from Jaghbub to Benghazi. Richard Attenborough was Lew Moran the navigator, Hardy Krüger was the German aeronautical engineer Heinrich Dorfmann, and Earnest Borgnine played a great supporting role. It was a great caste. I took some soda water to help my digestion and then went to bed.

21

The Discovery

"I have written unto you, young men, because ye are strong, And the word of God abideth in you, and ye have overcome the wicked one."
(I John 2:14–15)

Part Way Home

I woke up on Saturday, anxious to share my visit at Fort Meade with Dr. Glick. I ate a quick breakfast and made my way to Reisterstown. The sun is shining and the recent rain has left a clean sweet smell in the air. I arrive very early. When I enter the room, the band is still practicing their set. Rabbi Epstein is busy straightening chairs. I put my Bible down and went over to help him.

"Good morning, Rabbi."

"Good morning Jacob. You are early today."

"Yes, I got a quick start. Can I help you with the chairs?"

"Yes. I'll take the right side and you can take the left."

"I am a registered Republican I'd rather take the right side."

"What?"

"No, I'm just kidding."

"If I were to guess, I would have thought you would be a Democrat."

"No, not any more. The Democratic Party has been taken over by these progressives."

"Yes, I agree. Perhaps Trump will pull the party back to the center again."

"Perhaps, but I doubt it. The leadership is filled with hate. They are irretrievably changing our Nation."

"So, how is the good Lord treating you Jacob?"

"I am being very blessed. I actually experienced another vision last week."

"Do you want to share it with me?"

"Yes, I do. On Thursday, I took a short nap. I had just fallen asleep when I saw a very bright cloud descending from a blue sky. I knew it is Yeshua. He didn't say anything. His hand reaches out from the bright cloud and hands me a scroll. I woke up, expecting to find a scroll, but there is nothing in my hand. I felt completely rested and revitalized after the vision. I know the Lord has given me a gift of some kind. Perhaps the scroll means my research proposal will be successful. The vision was partly confirmed yesterday. Dr. Majors, the director of our research group, myself, and a few others went to the NSA at Fort Meade. They interviewed me and one other person. The interview was very intensive. It was more like an interrogation than an interview. In the course of my interview, this bird colonel asks my position concerning possible conflicts between the government's policies and my Judeo-Christian belief. I'm sure he was instructed to ask me this question by higher-ups. I noticed three one-way mirrors at one side of the room. I was being watched. I believe the interview also was video-recorded as well. At the end of the interview, something unusual happened. The colonel shields my paperwork from the windows by bending over the table. Then he quickly writes a note on the back of one page of my proposal. Next, he put my documents back into my briefcase and hands it back to me. After I arrive back in the van, I opened my briefcase to read what he wrote. His note read, 'Do not worry. I am a believer raised in a Pentecostal church in Atlanta.' He knew from my information I am familiar with the city of Atlanta. I know his report will be favorable. How he will handle the tough question about the conflict between governmental authority and my moral belief is hard to say."

"Very courageous, Jacob. I feel we are one step away from

slavery to this powerful government. I would have you share your experience with the congregation except it might endanger the colonel. Also, it might harm your chances to gain access to this new technology."

"Yes, I can't take the chance of such a disaster taking place."

Rabbi Epstein left me to finish organizing the chairs to welcome the first arrivals. Dr. Glick and his family came in shortly after I finished the chairs. I sat down with the family. I briefly told Dr. Glick about my trip to NSA. I did not tell him about my vision or the secret note from the Colonel. After the service, Dr. Glick asked me to follow him to his home to celebrate the Passover Seder with his family. It was a wonderful experience. How perfectly this festival foreshadows the Lord's ministry, his crucifixion, and also his second coming. I will cherish it for many days to come.

When I arrived back in my cave, I took the time, to write a summary of my interrogation. I emailed it to Ms. Wilcox, asking her to give it to Dr. Majors. The rest of the day I spent getting some exercise. I took a run across the campus, then stopped at the Rec Center and worked out on the weights in the exercise room. The physical effort felt good. I cooked up some pasta for dinner and settled down to read about the town of Bar Harbor posted on Wikipedia. Bar Harbor is located on Mount Desert Island in Hancock County, Maine. The community was first settled by Europeans in 1763 by Israel Higgins and John Thomas. It was incorporated on February 23, 1796 under the name of Eden, after Sir Richard Eden, an English statesman. Today, the population of the town is only five thousand or so, but it expands during summer and fall as it is a popular tourist destination. The town is home to the College of the Atlantic and several research labs; one is the Jackson Laboratory. This lab has close historical ties to Johns Hopkins and the Institute. The Jackson Lab will host the Institute's thesis evaluation meetings. The town is served by Cape Air and Pen Air with connections to Boston and a few other cities in the Northeast, but I am considering taking Old Blue instead of flying. I have never visited New York or Boston, plus having wheels will give me the freedom to take a look at the Bar Harbor area.

The town of Bar Harbor has a rich history. Early industries included fishing, lumber, and shipbuilding. Where the soil was good, agriculture developed in the 1840s. The rugged maritime scenery attracted the art community. The Hudson River School in the town was the familiar territory for quite a few famous artists such as Thomas Cole, Frederic Church, William Hart, and others. The area also attracted journalists, sportsmen, and most recently, movie and TV personalities. Bar Harbor has become synonymous with the elite wealth. Prominent politicos and industrialists, including President Howard Taft. John D. and Nelson Rockefeller, Cornelius Vanderbilt, and J. P. Morgan, made the town their playground. From the climate data, I found summers in the town can be quite warm, averaging in the high 90's from May through September. The fishing industry caught my eye. I will be treated to lobster, crab, clams, and several other delicacies from the sea.

I want to make this trip a vacation experience. I took a look at the mileage. It is a total of six hundred seventy-six miles from Baltimore to Bar Harbor. From Baltimore to New York is just under two hundred miles, and from there to Boston is another 210 miles. This leaves a final leg from Boston to Bar Harbor of 280 miles. If Joe and I get an early start on the first two legs, we may have some time to see a few sites in the Big Apple and Boston, provided the traffic doesn't delay us.

I took a look at my calendar. Finals are scheduled to begin on Monday. I already have completed my course work in the labs. This gives me about four weeks to continue my research in preparation for the conference and interviews. Hopefully I will be given some lab time on the new machine. If so, I may be able to achieve a breakthrough and have new results to bring to the conference. To say I am excited is an understatement. I am very hyped.

I spent the rest of Saturday reviewing the data from the lab tests with Walid, plus beginning to cram for my finals. I am still feeling frustrated by Walid's refusal to tell me what he experienced in the first dream. I ran the video of the first test several times. I did note one small piece of evidence slightly different from the other dreams, a marked movement of Walid's right eye. It seemed to roll backward twice as he fell into the sleep state. Perhaps this is a clue, but a clue to what? Without knowing about his dream, I cannot associate it with any specific memory. I ended the day by watching another film from my collection, *The Hunt for Red October*. The film stars Sean Connery as the Russian Captain Marko Ramius, and Alex Baldwin as Jack Ryan of the CIA. The film is a real high-tech thriller. The Soviets create a nuclear submarine with a revolutionary propulsion system, allowing the sub to run silently. The Russian Captain realizes this weapon will result in nuclear World War III. He decides to defect with the boat to the United States. Some films never grow old, and this is one of them. As I view the current state of affairs in our world, I fear we now are slipping into a Second Cold War. Unlike the Russians of the First Cold War two decades ago, the foes we now face are not adhering to past norms of rational thought. This new reality troubles me greatly. As a result, I did not fall asleep until 2:00.

I have not given my class work my full attention for several weeks since gaining access to the biology lab. On Sunday, I continued to cram for my final exams. I went to the student help office in the Student Union. The office stores the past exams of all instructors except those who are new to the campus. The past final exams of my professors gave me some perspective on how each of them structure their exams. All of the tests were standard formats, a combination of multiple choice and short essays, except for Dr. Baker's exam. A note in his folder says there are no past exams available because of the format Dr. Baker uses. He no doubt will use his Socratic approach, verbally testing each individual student, just as he has done throughout the year. I am comfortable with it, although I do not understand how he will be able to individually test us within the allotted time. His final exam is scheduled for this Monday morning at eight o'clock.

I slept well and woke up ready for Dr. Baker's final exam. When I arrived, Dr. Baker is just opening the classroom. He marks time until everyone has been seated and then addresses the class. Good Morning class. Today is your day of reckoning. I believe you are aware I have come to know each of you quite well over this semester. I am familiar with your individual talents and capabilities, your weaknesses and your strengths. I have posted your lab grades on the bulletin board in the hallway. Let me explain how we are going to conduct this final exam. I will ask each of you one or more questions. I will then consider each of your answers to arrive at your overall grade for the course. I will then record your grades in my ledger and on a slip of paper. My proctor will hand you the slip of paper with your grade for the course. After you receive your grade, and if you agree with the grade, you are free to leave. Of course, if you want to remain to test yourself on the other questions, feel free to do so. If you disagree with your grade, please remain until I have finished my questions. Then we will have a private discussion. Do any of you have a question?"

The classroom is as silent as a tomb. You could cut the tension in the air with a knife.

"Very well then. Let us begin. Jacob Cahn, you are the first person on my list."

"I am here Doctor."

"Jacob, you have done exceptionally good work in my class and the Seminar as well. I am going to ask you a single two-park question. Please spell and repeat the term. Then I want you to clearly define it. Jacob, can you spell and describe a prokaryotic cell?"

"Yes, Sir. The term prokaryote is spelled 'p r o k a r y o t e. Prokaryote cells are typically about one micron in diameter. They appear to be much simpler than eukaryotic cells. Apart from a cell boundary, prokaryotes lack any visible defining features. They don't have a nucleus, organelles, or internal membranes. Their genetic material consists of bare and highly coiled DNA residing in the cytoplasm. Bacteria and archaea are prokaryotic organisms. Although the archaea are similar to bacteria in appearance, they differ fundamentally in their unique biochemical makeup. Do you wish me to describe their chemical constituents Doctor?"

"No, Jacob. Your description is sufficient. I may ask someone else to tell me about their chemical properties. Thank you, Jacob. You are excused unless you wish me to test you further."

"No, Sir. I sure don't wish to run the risk of trying to answer possibly a more difficult question. I will leave the next question to someone else thank you."

My response cuts a bit of the tension in the room. The assistant hands me my slip of paper. Dr. Baker has written a large 'A' on the paper and a short note. The note says, "Jacob, you have an awesome career in genetics in front of you." He signed his name at the bottom.

I quickly folded the slip of paper and put it into my shirt pocket. Then I decided to stay in the classroom to listen to the questions until Joe received his grade. After Joe's test, we went to lunch together.

Dr. Majors called me first thing on Friday morning.

"Hello, Jacob."

"Good morning Doctor Majors."

"Jacob, I want to congratulate you on the results of your first year. You have done well."

"Thank you, Sir."

"I have received word from NSA about the use of their machine. They say you can use the machine when it is not being used, but there's a problem. NSA has agreed to provide an operator for the machine, but they will not supply you with guinea pigs for the research. You will not be able to get a security clearance on Walid, since he is not a citizen. We don't have any other individuals with NSA clearance. This pretty much puts the kibosh on your use of the machine for now. I'm sorry."

"I have a suggestion, Doctor. What if I volunteer to be the guinea pig? This way, I will be able to directly experience the dream state. The external observations will be available to me on tape. I can give the operator instructions as to the parameters to be used."

"Are you sure you want to subject your brain to this new rather untried machine?"

"Yes, I am ready to do this. I really can't go to my conference in Bar Harbor with this thin a stream of data. I believe a few sessions on the machine is essential to part the waters in front of me."

"Very well. I'll tell them to go ahead and schedule you some time. I will get back to you."

"Thank you, Doctor."

I did not have long to wait. Dr. Majors sent me an email about an hour later, telling me I am scheduled from two to four o'clock on Mondays, Wednesdays, and Fridays. I am free to begin this coming Monday.

I considered taking some of the supplements, but decide the initial tests should be done without them, giving me a better base to evaluate the results against. I realize I now need to revise my own research proposal as well as the research plan for my contribution to Dr. Majors' program. I spent all weekend working on them, neglecting to go to Reisterstown on Saturday. By the end of the week, I received an envelope with all of my grades. My overall grade point is a 4.0. The envelope also contained a letter from Director Hobbs. He congratulated me on my first year and enclosed a certificate stating I have received the distinction of my name on the Director's list of honor students. I am a quarter of the way home.

The Awakening

I awoke on Monday morning, ready to begin my research at NSA. I felt the need for intense prayer. The results of my work at NSA will either support or defeat my hypothesis. I cannot seem to get in touch with the Lord. In my mind, I am telling him, "Are you working in China today, Holy Spirit?" I thought, Seriously, perhaps he is. I took time to pray for the people of the Inner Kingdom, China. Only after my prayer of intercession for China did I feel the presence of the Holy Spirit. I mentally told him, "Welcome back Holy Spirit. How was your trip to China?"

I did not get away with my Jewish arrogant humor. The Holy Spirit corrected me by quoting a Scripture from Hebrews, Chapter Thirteen and verse five. *"I will never leave thee, nor forsake thee."* What an encouragement this is to me at this pivotal point in my life.

I thought of one of my favorite movies, *Chariots of Fire*. The plot revolves around two runners competing in the 1924 Olympics, Eric Liddell, a Scott portrayed by Ian Charleson and Harold Abrahams, a British Jew played by actor Ben Cross. The title of the film is drawn from Second Kings, Chapter Two in the Bible. This great film is ranked as nineteenth in the British Film Institute's all-time top one-hundred films. It is at the very top of my list. I decide to take a serious run across the campus. There is a special energy flowing through my body today. I will put my heart into this run. Perhaps I will succeed at NSA today like Harold Abrahams triumphed. I did not get back to the apartment until twelve noon. I ate a quick lunch, then prepared for my trip to NSA.

I pulled up to the NSA guard house at half past one. Since I am not expected to pay them a visit today, the guard directs me to park my car at one side of the guardhouse and follow him inside. Another guard is examining my car from the hood to the trunk while the first guard is examining my security pass. He compares my face to the photo on the pass. He wants to know where I am headed, what I am going to do there, and when I will be leaving. I tell him this is only the first of several trips I will be making to NSA. When he hears this, he also wants to know my schedule, when I will be working in the lab and for how long. He finally makes a call to the lab to check me out. Apparently, Old Blue does not fit their frequent guest profile. After giving me the okay, they place a sticker on my bumper.

The officer in charge tells me I will be able to enter without any delay in the future. I thanked them and drove to Site M, the "High-Performance Computing Center Number Two" (HPCC2). A young warrant officer is waiting for me when I arrive.

"Mr. Cahn?"

"Yes, I'm Jacob Cahn."

"Jacob, my name is Officer Dan Jenkins. I will be helping you with your research on the IAMGOD machine."

"It's nice to meet you. Would you mind if we just refer to the machine as 'The Machine?' I really don't want to call it by its acronym."

"Yes, I agree with you. I don't like the acronym either. Let's get started."

I followed Officer Jenkins into the lab. I know Qbit computers must be kept at very low temperature and I forgot to bring a jacket along.

"Dan, do you keep the lab room pretty cold."

"Yes, Mr. Cahn, we do. Did you bring a coat along with you?"

"No I didn't, Dan."

"Not to worry, the computer is kept in a special container. The room temperature is kept at sixty-five degrees Fahrenheit."

"Dan. Let's dismiss the formality. You can call me Jake."

"Okay. It's a habit, but I will try to dispense with the military vernacular."

I gave him a copy of the instructions I previously prepared and he took time to carefully read them before we took the elevator down to the lab.

I gave Dan a verbal summary. "Dan, we will be examining the glial cells close to the right eye. I want you to direct the low Schumann resonance to my glial cells connecting to the neurons in the pupil."

Dan strapped me in and began his set-up procedures. He gave me a tranquilizer shot. Suddenly I was elsewhere. I see a young boy on a hospital operating table. The doctor is removing his tonsils. After the operation is complete, the doctor takes off his mask. It is my father! I came slowly back to the real world. The boy on the operating table was a young frightened Jacob Cahn. This restored memory of my dad taking my tonsils out, took me back to when I was in the third grade. I was having throat and ear infections, one after another.

Then I asked Jenkins, "Dan, did you see anything unusual?"

"Yes, Jacob, I certainly did. Chromosome 22 in the glia cell attached to your right pupil lit up the screen. It was a brief flash, so I didn't have time to zoom in to examine the DNA inside of the chromosome. Of course, all of the layers stored in the Qbit computer can be output to tape. You will be able to take a look at each individual frame to see what was going on. What did you experience?"

"I relived the time my father removed my tonsils. I was only about eight years old. I saw myself on the operating table and recognized my father when he took off his mask after the operation."

"Wow. This is very interesting."

"Yes, it is. We will be able to review the data from the computer so I can better understand what has just occurred."

"Yes, of course. You will be able to zero in and stop the recording on the frame of interest to see what is going on in the interior of the cell. We then can advance the recording and zoom in to view the internal workings of the cell."

Dan started the machine and stopped it at the point of the bright flash on the screen. He began to carefully view the frames just proceeding the flash, tracing the exact location of the origin of the flash on chromosome 22. Then he pointed to the location of the origin of the flash on the DNA chain. We couldn't believe what we were seeing. The Qbit recorded the precise location of my memory when I was eight years old.

"Do you realize what we have just accomplished, Dan?"

"Yes, I do. We are seeing the exact location on the DNA chain of a memory in the human brain for the first time."

"That is correct. And we have the evidence. We will put these images on a DVD and store it for future posterity. Does this mean memory is stored within the entire genome, not just in the glia?"

"Are you going public with this discovery now, Jake?"

"No. I will save this news until the summer convention in Bar Harbor, Maine. Can you agree to not spill the beans on this amazing experience?"

"Yes, I can agree not to make it public, but I am required to report everything occurring on the Qbit and IAMGOD machines to General Weaver. He then will report it to the Director of NSA."

"I was afraid this would be the case. I suppose this is the price I will pay for access to this beautiful machine."

Yes. It is sad, but true."

"Will you be the person setting the priority on this research?"

"Yes, I have already classified your experiment project as top secret. What title shall I give your research, Jake?"

"Let's title it Project Cepher."

"Is this a long anagram? What does the word Cepher mean, Jake?"

"It is the Hebrew word meaning book. I am Jewish, but I only learned about this word Cepher a few days ago."

"Does it mean perhaps the Bible?"

"Not exactly. Bible is the English word we use for God's Word. It is derived from the name of the ancient city of Biblos. This Hebrew word simply means The Book."

"Very interesting, Jake. I'm sure we will have an opportunity to talk about this word Cepher in the future. You should expect to get a call from General Weaver soon. When he calls, you need to be up-front with him on everything. Today's results are going to be examined closely by NSA and possibly other agencies."

"Yes, I'm sure it will be. No doubt they will want to explore the possible military applications."

"Yes. I imagine they will be looking especially at the potential for interrogation, mind control, and other applications. This is the world we live in now, Jacob."

"I understand what you are saying. I assume I can take the DVD along with me to review, as long as I keep it secure? Also, I will need to share these results with Dr. Majors?"

"You can take the disc with you, but unless you have access to a Qbit machine, you won't be able to access it. You can verbally share our results freely with him. Majors has a very high security clearance. Be sure to tell him this is going to be sheltered under Project Cepher. Keep the disc under lock and key, and don't mention it to any other living soul."

"Very well, I will carefully protect it. Dan, I am now realizing there may be serious ramifications flowing from this discovery as related to my graduate proposal."

"Yes. No doubt you are going to be confronted with some serious conflicts of interest, Jake. You will have to work this out with NSA and your Institute."

"I am sure I will."

I watched as Warrant Officer Dan Jeffers extracted the DVD disc from the machine and labeled it Cepher Project with the date and time. I gathered up the disc and my papers and left the building. Driving back to Homewood, I continued to turn over in my mind how to manage this now sequestered information while still making it available to the general scientific community. I will have to play hard ball with the NSA.

I am too keyed up to go out to dinner and no desire to cook a meal. My concern about this looming conflict has taken away my appetite for food. I need to get my mind off of this. My solution to finish the day is to watch another old movie. I picked out the spoof called *The Russians Are Coming*. The 1966 film depicts the chaos following the grounding of a Soviet submarine off a small New England island during the Cold War. The captain decides to send a landing party to try to remedy the situation. The results of his decision are hilarious. Carl Reiner, Eva Marie Saint, Allen Arkin, Jonathan Winters and a number of other great actors make up the cast of characters. Not far into the movie, I fell asleep in my easy chair. I awoke about two o'clock and finished the rest of the night in bed. Tuesday is a day off for me. I really don't know what to do with myself. I gave Joe Harmon a call.

"Hello, Joe. This is Jacob. What are you up to today?"

"Nothing important, Jake. What do you have in mind?"

"I wondered if you have planned your trip to Bar Harbor yet?"

"No, not really. I believe there are a couple of airlines serving the town. Why?"

"Well, I thought, why not make it a vacation? I am going to drive up there. I want to take a look at New York, and Boston. Perhaps spend a little time in each to see the sights. Would you want to come along?"

"Well, I don't know. I suppose we could split the expenses."

"Yes, we could. Why don't you think about it and let me know? I plan on leaving in three or four weeks from now."

"Okay, Jake. I will get back to you. I have a relative living in Jersey. We probably could stay with him, saving the cost of a motel in New York."

"It sounds like a good idea. I was a little concerned how expensive staying even one night in New York would be."

"Yes. I'll let you know later today. Take care."

I hung up my cell and put on my running gear. Just before leaving the apartment the phone rang.

"Hello, Jacob. This is General Weaver. Is this a good time to talk?"

"Yes, Sir. I was just about to take a jog around the campus."

"Jacob, I believe Warrant Officer Dan Jeffers told you to expect a phone call from me."

"Yes, he did Doctor."

"Jacob, we need to talk about these results you have come across in the lab. If possible, I would like to meet with you later this morning."

"Yes, Sir. I can meet with you this morning. I don't have anything important on my schedule."

"Fine. Why don't you drive down here this morning? When you get to the gate, tell them you are meeting with me. I will tell the guardhouse to expect you."

"Do you want me to bring along any paperwork?"

"No, I want to review your results of yesterday with you."

"Will you be inviting Dr. Majors as well?"

"Yes, certainly. I don't want to mislead you. If you feel you need someone else to come with you, please invite them."

"Are you thinking someone like an attorney?"

"Yes, if you think you will need legal advice, bring your attorney."

"I will have to think about this. Can I call you back?"

"Yes, of course. If necessary, I can postpone the meeting to another day."

"I will get back to you as soon as I can."

"I will wait for your call, Jacob."

I hung up the phone and put in a call for Dr. Glick.

"Hello, Doctor, this is Jacob."

"Hello, Jacob."

"I need some advice, Doctor. Do you have a few minutes?"

"Yes, I just arrived in the office. What's on your mind?"

"I just finished talking to General Weaver at NSA. Because of what occurred yesterday in their lab, he has asked me to attend a meeting with him and Dr. Majors. The General said if I wished to bring an attorney, I could do so. I believe I do need to be represented by one. The problem is, I don't know any attorneys here in Baltimore. Do you think it is necessary to have an attorney present?"

"Absolutely, Jacob. You need to protect yourself. Do not sign any agreement with the NSA without the advice of an attorney. I have a suggestion. Isaac Flowers in our congregation is an excellent corporate lawyer. I have used him in the past. Let me give him a call. Do they want to meet with you right away?"

"Yes, they want me to meet with them today."

"I will have Isaac give you a call. I will ask him whether he can meet with you and NSA today, but he probably will need to have them reschedule the meeting. Stay close to your phone."

"Yes, I will Doctor, and thank you."

I did not have long to wait. My phone rang only a few minutes later.

"Hello, Jacob. This is Isaac Flowers. Doctor Glick asked me to contact you. I understand the NSA wants to meet with you concerning your research project."

"Yes, Mr. Flowers. They are very anxious to meet with me. I believe they are going to present a legal agreement concerning the project. I really don't have the skill set to analyze what they will be presenting."

"I understand completely. You need to have legal advice. They want to meet with you later today?"

"Yes, but of course I can ask them to reschedule."

"And in the meantime, you will be on pins and needles?"

"Yes, I will be coming unglued."

"Jacob, I'm going to cancel my appointments and join you at the meeting. Can you arrange for me to get past the gate?"

"Yes, I can arrange it. Do you know there is a special off ramp on the BW Parkway to the NSA?"

"Yes, I've seen it."

"To save time, why don't I keep you on the phone while I call General Weaver on my cell?"

"Fine. Go ahead."

I picked up my cell and called the General back. "Hello, General. This is Jacob Cahn."

"Yes, Jacob. Are you able to meet with us today?"

"Yes, it turns out I am. My attorney is on the other line. His name is Isaac Flowers. Can you arrange to give him permission to enter the gate?"

"Yes, I will arrange it. Tell him a security guard will escort him to the office. We will expect to see you in an hour or so?"

"Yes, I will be leaving the campus in a few minutes."

"Very well, Jacob. Thank you for expediting this. We'll see you and Mr. Flowers soon."

The General hung up the phone. I told Isaac he will be escorted from the guard gate to General Weaver's office. I grabbed paper and pencil and left for the meeting, still dressed in my running clothes. I arrived at the NSA at ten o'clock. The guard passed me through without delay. When I entered General Weaver's office, my attorney, Isaac Flowers, and Dr. Majors already were there. The General welcomed us and showed us into the conference room. He introduced two other persons, a Mr. Daniel Switzer and a Ms. Chris Donovan. They are not in uniform. The General did not tell us whether they are NSA or from some other agency.

The General began the meeting. "Jacob, Warrant Officer Dan Jeffers furnished us the results of your first experiment using the IAMGOD machine. I must compliment you on your ability using the machine. The results of your experiment are simply breathtaking."

"Yes, Sir. The machine has furnished a major breakthrough in our understanding of human memory."

"We are very proud of our new machine as well as the capabilities of the Qbit computer. Many other people will be very interested in using these machines. This is why we have secured it from access by most outside users. Our approval of Dr. Majors' request for your use of the machine may have been premature. You have not been asked to enter a firm agreement with NSA. This has been careless on our part. Nevertheless, what is done is done. After reviewing the situation, we have decided to present a proposal to you. The proposal allows you to further your research using these two machines. In essence, we are proposing to adopt you. We want to offer you a research position on our staff at NSA, Fort Meade. I realize this is not what you have planned for your career. The problem is we cannot allow the results of your research to enter the public domain. I believe the solution to this conflict of interest can be resolved if all concerned are willing to work together. We have just prepared a document outlining our offer of employment. It includes a generous salary and other benefits. We have outlined the details of our offer here in a document we want to present to you today. I would like you and your attorney to review it. If necessary, you may do so in private."

"Sir, I don't think it is necessary to meet in private at this point."

"Very good. I have made a second copy for your attorney. We will have some coffee and donuts while you confer with one another."

Mr. Flowers and I read the document. Then we conferred quietly together.

"Jacob, what do you think of their proposal? The remuneration seems quite generous."

"Mr. Flowers, I do not want to work for the NSA. I must decline the offer."

"You sound convinced. Is it a matter of the money, or something else?"

"It is something else. We can discuss my reasons later if necessary. I am adamant about my decision."

"I believe I understand your concerns. Jacob. Do you want me to respond to the General, or do you wish to tell him directly?"

"I will explain the reasons to the General myself."

"General Weaver, before I give you my decision, let me try to summarize the provisions of the document you have presented to us this morning. In Section One, you are taking full control over the results of my research. I cannot publicize or reveal the results of this research. The penalty for breaking this agreement is a prison sentence of one-hundred years in a private prison facility administered by NSA. This would seem to eliminate my ability to complete the Institute's requirements to receive my degree. This seems quite one-sided to me. You are asking me to, in effect, be in servitude to the NSA. Is this correct?"

"Well, this is our initial proposal. We are open to suggestions how you can successfully receive your doctoral degree. Would you agree the financial benefits we are offering are generous?"

"Yes, they are very generous. I found Section 2 to be quite unusual. You are offering me protection in case of a major conflict or disruption of our economy and environment. Are you saying you are offering to place me in some kind of underground bunker?"

"Perhaps you have been watching too much alternative news programs on YouTube, Jacob. I cannot comment on the existence or lack of any so-called underground bunkers. Let me just say we believe you are a valuable asset to our agency and we feel you should be protected against possible disruptions."

"General, let me first say I am a patriotic citizen of this Nation. I realize this discovery can easily be used by your agency as an interrogation tool. I am sure you will find it very useful in ferreting out dangerous individuals. Let me clarify my concerns to you. First let me say I am not motivated by the desire for fame and fortune. My interest is to pursue the search for new knowledge and the application of such knowledge to the benefit of society as a whole, American, Russian, German, or otherwise. Setting aside my research interest, I have an even greater calling on my life. My ultimate calling is to serve my Lord Yeshua as a minister of reconciliation. I'm willing to play ball with the NSA, but I cannot relinquish my God-given freedom to pursue my destiny."

The General and his two observers are plainly suffering pain. The general pauses a few moments before he responds to my statement. "So, what is your proposal, Jacob?"

"Sir, I want to finish my course work here at the University. I want to be able to complete my research using your lab. I can agree to relinquish my access to your lab as soon as I complete my research. I can limit my time in your lab to end before my trip this spring to Bar Harbor. Once the locations of memory are determined, I believe I will be able to access them using a conventional or modified MEG machine. You will be rid of me once I depart from the convention. The problem I see is your prohibition on my freedom to present the results of my research to my thesis advisors at Bar Harbor. How can I obtain my degree without publicizing my findings?"

"I take it you are firm in your rejection of our offer to work with us as a researcher?"

"Yes, I am. I don't want to be part of the black economy. I will build my own bunker, if necessary."

"So, let me review your position. You are willing to step out of the picture once you have completed your research in our lab. You are also willing to surrender the results of your research to the NSA. You will not reveal the results of your research to the public arena, but desire to be allowed to complete your degree requirements, including the approval of your research by the Institute's faculty advisors. Have I covered all of the bases, Jacob?"

"Yes, you have General. I believe I can apply this new knowledge without access to your lab facilities, thereby avoiding any reference to my experiments at NSA. I must add another requirement. I want assurance the NSA will protect my rights to publish new knowledge gained through the use of machines in common use by the general public. When I say protection, I mean I do not want my father reading my name in the obituary column having died of mysterious causes."

"I take it you would not reveal to the public the method you first used to determine where memory exists in the DNA chains, or any reference to the existence of the IAMGOD machine?"

"Yes, I agree to those two requirements. There seems to be one problem remaining. What are you willing to do about my presentation at the Bar Harbor conference?"

General Weaver quietly consulted with his two cohorts for several minutes. It's plain, he is having a tough time selling our tentative agreement.

Finally, he said, "Jacob, you pitch a hard fast-ball, down the middle of the plate. We will have to get back to you on our decision and how to handle the Bar Harbor issue. I do believe we can arrange something agreeable to all the parties involved. Can we take you two to lunch?"

"Sir, I believe Mr. Flowers said he would not be able to stay for lunch. I also have some work to do. Thanks for the offer, but we will be leaving now."

"Very well, Jacob. Thank you both for coming on such short notice and for your willingness to work this issue out with us."

Isaac Flowers whispered, "You don't need my help, Jacob. You have them in your mitt."

Doctor Majors, who has been silently listening to the meeting, gives me a thumbs-up. We all shake hands and part company. When I arrive back at my apartment there is a message on the phone from Joe Harmon. He says he wants to take me up on my offer. His brother has agreed to put us up for as long as we want to say there. I sent Joe an email, telling him I am looking forward to the trip. Then I went on line and downloaded the baseball schedules for the Yanks and the Red Socks.

22

Spring in Bar Harbor

"But there is a God in heaven that revealeth secrets, and maketh known to the king Nebuchadnezzar what shall be in the latter days Thy dream, and the visions of thy head upon thy bed are these."
(Daniel 2:28)

The Gift

The days are flying by now. I realized I would have to spend more time at NSA if I am going to finish the research using their lab. I need to put in a call to General Weaver. "Hello, General. This is Jacob, Jacob Cahn."

"Good morning Jacob. We haven't finished considering your proposed solution yet."

"Yes, I understand. I do have one additional request. I believe it is to both our advantages. I need more time in your lab if I am going to complete my research. I have only four weeks before my trip to Bar Harbor and the conference. I wonder if you can scare up additional time on the machine?"

"Would you be willing to do your work at night, because the machine is extensively used during normal working hours?"

"Yes, I am willing to take a night shift. I am even willing to surrender my day slots if I could work both swing and graveyard."

"Very well. We will trade your daytime slots for the two nighttime shifts. I believe I can get someone trained on the computer and machine to volunteer to help you, but you would have to pay them overtime."

"General, I do not have deep pockets. What would the approximate cost per hour be?"

"It would cost approximately seventy-five dollars an hour."

"That's not workable for me. What if I get trained on the computer and the machine and I find a guinea pig?"

"Yes, we could give you a quick training on the machines. Your volunteer would have to have top security clearance."

"No, I don't have time to get someone through the classification process. What if I can furnish a person who has clearance and a good background in physics. Would you be willing to train him on the machines and I would volunteer to be the guinea pig?"

"Do you have someone in mind?"

"Yes, I believe so. There is a fellow from the physics lab. I believe he has clearance. His name is Robert Manley."

"You mean Nanoman? Yes, Robert has clearance. In fact, he has very high clearance. He helped us build the IAMGOD machine."

"Let me contact him and see if he might be willing to help me."

"Okay. Let me know what he says, Jacob."

"Yes, I will."

I realized this is my only hope to finish the research before the conference. I put in a call for Nanoman.

"Hello, Robert?"

"Yes, who's calling?"

"This is Jacob Cahn. Do you remember me?"

"Yes, of course. Has the man upstairs given you a message or something?"

"In a way he has, Robert. I have been working at NSA on the IAMGOD machine. I believe you are familiar with it."

"Yes, but how do you know about it?"

"I have been doing my research on human memory using the machine. Here is my problem. I have only four weeks to complete my research before the conference in Bar Harbor. NSA has offered to allow me to use the machine on swing and night shifts for these four weeks. I need a skilled operator to work with me. I can't afford to hire one of their people. So, I wondered if you would be willing to help me out. I could pay you enough to cover your expenses and a little besides. What do you say?”

“I say you are in a big hole and no way to get out of it.”

“Yes, you are right. I am pretty desperate. How about it?”

“Well, I usually work here at the physics lab on swing. You say you only need help for four weeks?”

“Yes, just four weeks, Robert.”

“Has your work on the machine been a success?”

“Yes, it has. NSA offered me a job with all the benefits, including a refuge when the bombs begin to fall.”

“Yeah, they suck you in with stuff like that. Did you refuse to join their black economy?”

“Yes. I told them I did not want to become their slave.”

“Good for you, Jacob. They have given you use of the machine in return for the results of your research of course.”

“Yes, but I won’t need their machine once I determine the neural-glia pathways involved with memory.”

“There has been a switch in your topic. How did you decide to go for the memory thing?”

“Dr. Majors offered me the use of his MEG lab if I served on his research project as his geneticist.”

“I see. Well, Jacob, I like working on those two machines. I might have to squeeze a few tests of my own, say late at night when no one else is around.”

“Just as long as you don’t blow up the machine doing something weird.”

“Okay. I’ll sign on. We can work out the pay scale later. When do you want to start?”

“I’ll get back to you. They want me to sign an agreement before I do any more research in their lab. I believe I will see the paperwork soon, because they are anxious to get the data.”

“Let me know when you have the green light, Jacob.”

“I certainly will. You will be the first to know. Thanks, Nanoman, you are a Godsend.”

“Sure, I know. Sometimes God’s gifts come in strange packages.”

We hung up. I called General Weaver right away and told him Nanoman was on-board. He said the agreement including the four weeks of research would be ready to sign later today or tomorrow. I received a call back from General Weaver only two hours later. He asked me to take a look at my email and to let him know once I signed it. He said he would send a courier to pick it up. I signed it and called him right back. The courier arrived at my apartment an hour later. The General called me back again and said I am free to continue my research. The lab is mine from six to four each weekday evening. I called Nanoman to tell him we can begin this very evening if he wants to. He agreed to go. I told him I will pick him up in front of the physics building at five o'clock. Before leaving to pick up Robert, I went to the bank and withdrew five hundred dollars. I will give the money to Robert when I pick him up.

When I arrived at the physics building, Robert was waiting outside. As he got in the car, I handed the money to him. "Robert, this is what I can afford right now. Will this take care of you for a while?"

"Okay. How much do I have here?"

"You have five hundred."

"Yes, this will get me through the week. I'll let you know when I need some more bread. Of course, this is under the table, Jake."

"Yes, I understand. I'll consider it benevolence."

We have no trouble entering the facility. Robert lost no time in preparing the machines for our first run. I brought along some of the sedatives I had from Dr. Major's lab. I told Robert to generate the low Schulmann resonance. Then I told him to watch for any activity from the cells in the vicinity of my right eye. I gave myself a shot and prepared to enter dreamland.

The shot kept my out for fifteen minutes. When I came fully awake, I ask Robert what he had observed.

"Jake, this is awesome. There definitely is a lot of activity on the glia cells adjacent to the neurons attached to the pupil of your eye. I zoomed in to take a look at the interior of the cells, the nuclei are vibrating in sync with the Schulmann frequency. What was happening with you?"

"I am reliving a memory from last year, playing tennis with Rachel Kaufman, the girl I fell in love with at Emory."

"No wonder the nuclei were excited, heh?"

"Yes, very true. The love we experienced was wonderful, but it had a very sad ending."

"How so?"

"Her father, an Orthodox Jew, refused to sign our ketuba, the

Jewish engagement agreement. Instead, he chose to do the planned marriage thing. She now lives in New York."

"How has the marriage worked out?"

"Terribly! She is lonely and living in a big house on Long Island while her husband works twenty-four-seven at his art gallery in Manhattan."

"Why doesn't she just get rid of the guy?"

"It's complicated. Money is involved."

"I'm sorry Jacob. So, what did you experience in your dream? Have you learned something new?"

"I learned how bad a tennis player I was. Seriously, I will have to look at the recording. Is it possible for you to copy just the portion you zoomed in onto a DVD?"

"Yes. I think I can store at least part of it unto a DVD. I believe there are some blank discs at the drawer below the work station next to the wall."

"This will be very good. I can review the glial activity back in my apartment. Hopefully I will be able to record the location of the memory on the DNA chain. I will most likely find it on a hoch gene on Chromosome 22."

I watch as Robert begins searching for the location of the neuron activity. It takes him a few minutes to find the location to export the data.

"Do you have any ideas what we should look for in our next test, Robert?"

"I am thinking about it. The circuits are hard to trace from the glia to other neurons and other glia. Perhaps we should strengthen those circuits with a heavy dose of copper. No, perhaps with gold."

"Can the human body process gold, Robert?"

"Yes, it can if the gold is very small, at the nano level."

"So, this is where you get your nickname?"

"Yes, it's true. I know a lot about nano particles. I even have some nano gold leftover from a failed experiment last year. Maybe next time we will feed you some of it. Let me first finish exporting the info you need from the Qbit to the disk."

"I'm not sure how we would get the gold to where we need it. The blood stream will not reach past the cell wall."

"I may be able to move it there using radiant energy."

"Radiant energy?"

"Yes. It is very strange stuff, Jake. Tesla came up with it during his experiments when people thought he was completely off his rocker. He was a genius, but he may have been helped by the occult side. Anyway, let me think about this before we try the gold dust."

We worked in the lab until midnight, and then called it a day, but it was nighttime. Several further tests were confirming my theory about Chromosome 22.

The testing went on for the next three weeks permitting me to compile a large amount of valuable information. The nano gold idea paid off. It resulted in a much stronger signature from the glial cells. It allowed us to capture, for the first time, memories stored in my DNA passed down from my ancestors. Then I discovered something odd. I assumed the memories from a parent would only contain memories from their birth up to the conception of the child. I was wrong. I was able to recall memories of my grandmother that dated after my mother was born. How was this possible? Are we all connected possibly by a current such as the Hall current with some kind of mysterious switch board?

This fact was amazing, but also very dangerous to the future of my research. I realized I had no explanation for the source of these ancient memories. If I was to include them in the data, I would be written off like Dr. Tesla. Robert agreed with me. We pondered whether to reveal this ability to recall ancient memory or to deny, or at least postpone the discovery. I realized I had surfaced a monumental problem.

After pondering the issue for over an hour, Robert and I came to the conclusion we needed to eliminate every vestige of pre-birth memory from the data set. Robert was afraid of what NSA might do with it. I needed to protect the credibility of my research. I would be accused of dabbling in the dark arts if this was known.

I made the decision. "Robert, go ahead and delete all of the pre-birth memory records from our data. The NSA must never know about what we have found."

"I absolutely agree with you Jake. We must carefully erase all of our footsteps on this."

I can tell Robert is worried, "Jacob, if the NSA ever finds out we have deleted this data, we will both be viewing life from behind bars, or end up in an insane asylum."

We both swore to never reveal our secret, not to the NSA, or to anyone else, ever.

Robert's worried comments will give me the opportunity to point out to him how this ancient lineage within our brains absolutely confirms the truth of God's Word.

Robert proceeded to completely delete all data revealing prenatal memory, replacing the time gap with duplicated post-natal data. As far as we both were concerned, pre-natal memories never happened.

During the final week in the lab, Robert began testing various crystals and granite rocks. I left him alone to do his own thing. He did not try to explain it to me. All he said was there was enough energy in those crystals to light up the whole world.

Working with Nanoman has been a major blessing. I felt I needed to reach out across the aether, this mysterious plasma substance filling space which is now challenging the theory of dark energy. I wished to touch the soul of this bright young physicist.

"Robert, we have now come across a marvelous discovery, a discovery I feel testifies to the existence of a loving God. Do you agree"

"Jake, I can recall the day you first came asking for my help. I wrote you off as a very naïve screwball. To use your terminology, a real shlemiel. How wrong I was. Yes, I can agree the way this research has been unfolding convinces me I was the shlemiel. I'm going to do some serious praying when I get back to my apartment. I need to follow your example. Thank you for being my friend and showing me the way to live eternally with Yeshua."

"Thank you, Robert, for being there for me. I know the Lord is going to reveal many mysteries of his Creation to you."

It was on Thursday of the last week of classes when I received a call from General Weaver.

"Hello, Jacob. This is General Weaver."

"Hello, General."

"Jake, I note you have not visited our lab for a few days last week. Have you successfully completed your research?"

"Yes Sir, I have. I will not be needing your lab any more, at least for the foreseeable future. I have enough data to share at the conference in Bar Harbor."

"Very good. I am calling you to say I recently held a meeting with Director Hobbs, Drs. Altschul, Majors, Glick, and Adelman, all of your conference and thesis advisors. I made it clear to them you have signed a non-disclosure agreement with us to conceal your research results from the public. Your advisors have agreed to conduct your conference interview in secrecy. They also have agreed to sign non-disclosure agreements. You are probably wondering where this leaves your efforts to receive the degree. Director Hobbs has agreed to confer the Doctoral Degree to you upon your successful completion of your final years of study. How does this sound to you?"

"It sounds great, but also very unusual. I am pleased everyone is on board with it."

"Of course, this means your thesis containing the research will never see the light of day. This will limit your exposure to the academic community. You will not be receiving recognition for the successful completion of your research, except in our eyes."

"Yes, I understand. I won't be receiving any Nobel Prize. I realize this will, in effect, relegate me to the status of a non-person in the academic community. Of course, this does not prevent me from conducting other genetic research. Perhaps I may yet be able to explain the method the body uses to direct precursor cells in the fetus."

"Yes, perhaps so, but you may find it hard to secure a research position with this blank achievement page on your resume. Perhaps I can help you out with this. I am willing to give you my personal recommendation."

"I appreciate your offer, General Weaver. Do you remember me telling you I am not into this genetics field for personal fame and fortune?"

"Yes, I do remember you saying something along those lines."

"I do not need to gain any public acclaim. My overriding goal of my research is to meet the needs of humanity. If not in a lab, then in some other way. Perhaps I will end up working with my friend, Allen in Ethiopia, or Timbuktu? I trust your agreement to protect me from accusations of spying, or whatever will hold true?"

"Yes, we will not interfere with any type of ministry you are describing. You will receive an official document substantiating our agreement. I believe we both are in total agreement with this plan."

"Yes, I certainly am. I will not be revealing any information about your awesome machine. If ever questioned, I can say I was asleep through the whole time this discovery was made."

"Yes, Jacob. You truly were asleep. I have enjoyed your Jewish humor. Don't ever lose it. Thank you for all you have done for us in the lab, and especially for being open to working out this situation in a cooperative manner. I will see you next in Bar Harbor, Maine."

"Thank you, General. See you then."

I am glad the General appreciates my humor and his comment about my attitude. This whole experience at NSA has been a blessing. Robert's conversion is simply awesome. It has made it worth going through this whole process. We both share something incredible, something only few men have every considered, the storage of ancient memories down through the ages. My experience determining the location of genes responsible for memory is truly a gift from God. It will greatly strengthen my faith. Ahead of me are another three years of hard work to complete my doctoral degree. Right now, I am ready for a vacation in Bar Harbor, Maine.

The Conference

I called Joe Tuesday evening to touch base on our trip to the conference.

"Hello, Joe. Jacob here."

"Hi Jake. What's up?"

"Joe, I believe we better firm up our schedule for the trip to Bar Harbor. We will need to make reservations and get tickets to the ball parks. Do you have time to work on it, or do you want me to set something up?"

"Jake, I'm still working on my proposal. Why don't you work up a schedule and make some reservations?"

"Okay, I can work up a schedule. I believe you said you want to take some time getting to Bar Harbor?"

"Yes, I think a week or even a few days more would be good."

"I think I can work out a workable schedule, Joe."

"Jake, do you have a passport?"

"Yes, I have one somewhere in my papers. Why?"

"Perhaps we might want to slip over to Canada."

"I don't think we have time, Joe. Do you like to attend church on Sundays?"

"No, not really, why?"

"I believe you told me your brother Karl lives in Paterson, New

Jersey."

"Yes, he does. My brother Karl is a professor at William Paterson University."

"Well, besides the ball games, I would like to visit a Messianic congregation I heard about located in Wayne. Wayne is very close to Patterson. Do you mind if I schedule a Saturday to attend the Messianic congregation there?"

"I'm fine with whatever you decide to do, Jake. I believe we should leave very early on a Friday morning. We need to avoid the Friday evening crunch in the New York area."

"Okay, I will get back to you, Joe."

I hung up and pulled out my Rand McNally atlas and went to work. We can take the I-95 and Garden State Turnpikes to Paterson and Joe's brother's house. We should be able to make the trip of two hundred miles in four hours. Friday is a night game. I bought tickets on the first base side to the Yanks game with the Oakland A's beginning at six in the evening. We should arrive at Joe's brother's house in plenty of time to get to the ballpark by six. If necessary, we can just go directly to the ballpark before going to Karl's. To get to the Stadium in the Bronx we can take the I-95 across the George Washington Bridge. On Saturday, I want to visit the Messianic Congregation in nearby Wayne. I set aside Monday for a quick tour of Manhattan. I took a look to see where Rachel's home is located on Long Island. It will be just too far out of our way to try to visit her.

Our trip to Boston and Fenway Park will take us three hours via I-87, I-84, the New York State Throughway, and the I-90 Turnpike to reach Boston. Fenway Park is located close to the Boston CBD. I made reservations in the small town of Redding, located near I-90 just north of Boston, for Wednesday. The tickets I ordered is for the Red Sox game with the Orioles is an evening game. This means we will need to get an early start from the Boston area on Thursday morning to arrive at Bar Harbor in one day. It is three hundred miles, and the road from Bangor to our destination is over secondary roads. We will be taking I-95 and the turnpike to Bangor, then south from Bangor on state highways 1 and 3 to Mount Desert Island and the city of Bar Harbor. The Jackson Laboratory is located right downtown, only three blocks from the Harbor. On Friday and the weekend, we will have time to tour the quaint town and the Acadia

National Park on the Island. I need to find out whether the Institute is paying for our lodging. My guess is Dr. Hobbs has taken care of it. Our meetings begin on Monday morning, just two weeks from next Monday. To be safe, I want to leave Baltimore this coming Friday. I called Joe back. He said he is able to leave then.

I took my car into Jiffy Lube. I had them give it a thorough going-over: oil chance, filters, transmission, wiper blades, the works. When I got back, I reviewed and edited my proposal, getting rid of extra spaces, adding commas, changing sentence structures, and getting it as clean as I am able to do without submitting it to a professional editor. I'm sure I have missed finding some syntax problems, but I believe it is acceptable. It is, after all, a proposal, not a published thesis. I eventually will have to submit my final Doctoral Thesis to an editor for review. I called Dad to let the family know about my upcoming trip and withdrew two thousand dollars from my bank account to cover the expenses. By Thursday evening, everything is packed and ready to go. My passport came in the mail yesterday. I gave Joe a call to see how his end is going.

"Hello, Joe. Are you ready to leave?"

"Yes, Jake. I have everything packed. Besides my briefcase, I have one suitcase and a duffle bag. Is there enough room in your car?"

"Yes, there's plenty of room still in the trunk and the backseat is empty. I'll pick you up at seven o'clock."

"Right. Do you have the tickets?"

"Yes, I have them. One thing I forgot to check on. Do you know where we are going to stay in Bar Harbor?"

"Yes. I called Ms. Capoletti and she said we have reservations at the West Street Hotel beginning on Monday the tenth. It is a four-star hotel. I found out the Institute is picking up the tab."

"We will be arriving before the tenth. Should we make reservations there, or at a cheaper hotel or motel, Jake?"

"I think we should find someplace a bit cheaper, Joe. I took a look at the town on Google and there is a Quality Inn on Kebo Street. I'll make some reservations there beginning on the third. We can always cancel them if we find we are arriving later. Do you want to share a room with two beds?"

"Yeah, let's try to keep the costs down as much as we can."

"Okay, Joe. I'll take care of the reservations. Quality Inns do serve breakfast and the rooms are air conditioned. If I have any trouble, I'll let you know."

I hung up with Joe and got right on the computer to make the reservations at the Quality Inn. The reservation clerk said I can cancel up to one day beforehand.

Joe and I departed for the conference early on Friday. I pulled through McDonalds and we bought breakfast. We arrived at his brother's place at half-past eleven. Karl had made arrangements for someone to take over his classes, so he is there when we show up. Karl's two kids are in school, but he introduced me to his lovely wife Patricia. I quickly felt at home with this family.

"Hello. Patricia, Patty from Paterson, no less."

"Yes, they named the city after me, Jacob. I understand you two are going to take in a game at Yankee Stadium this afternoon."

"Yes, that's the plan. Karl, I didn't get you a ticket for the game. I assumed you would be in class. I wonder if we can get another ticket near us?"

"We can try, Jake. We might be able to cancel your two tickets and get three together."

"Yes, let's try to do it."

Karl took us into his study and we tried to make the exchange. There is no problem. We are able to get three seats on the first-base side. The game starts at six o'clock. Karl said it would take us about forty-five minutes to get to the stadium, but he suggested we leave at least an hour and a half before game time to get a decent parking space. We ate lunch and then sat around the den until time to leave. Karl's two kids, Joey and Kristy, came back from school at three-thirty. Joey is ten and Kristy is seven. Joey was disappointed when he found out he would not be going to the game with Uncle Joe. I told Joey I would bring him back a ball. He thanked me and said he was looking forward to it.

Karl told us, "Joey is quite a ball player. Last year he was picked for the league's all-star team. He pitched a scoreless five innings in the playoffs and earned the MVP trophy."

"Wow, Joey, it sounds like you might play on the Yankees team someday."

"Well, maybe. I'm good in math. I could be their statistician."

Patty made some sandwiches and coffee for us and we left for the game. It was not a good night for the Yanks. The Yanks fell behind early and went through most of their pitcher's roster. They could not stop the Oak's hitting. The final score was eight to three. Even so, it was a great evening. We overindulged ourselves on hot dogs, peanuts, and candy corn. We got back late to the Harmon household. The kids were asleep. I gave the ball to Karl to give to Joey in the morning.

I slept in a bit on Saturday morning. Their Messianic services start at ten o'clock. After a breakfast of hot cakes, I was ready to leave.

"Joe, do you want to go with me to the Messianic service?"

"No, Jake. I really want to spend time with the family. Kristy is playing basketball this morning at the grade school. I would like to go and root for her team."

"I understand. I will see you later. I will be back by one or so."

I left, anxious to visit this Messianic congregation led by a very well-known Rabbi David Showers. The service did not disappoint. I arrived a bit early and was able to find a seat close to the front of the room. The service began with music. One of the men came forward and sang a Psalm. The Rabbi ask the people to turn to the book of Shemot, Exodus Twenty-nine and verses thirty-eight and thirty-nine. He first read the verses in Hebrew, the repeated them in English.

Now this is that which thou shalt offer upon the altar; two lambs of the first year, day by day continually. The one lamb shalt thou offer in the morning, and the other lamb thou shalt offer at evening.

The Rabbi carefully explained the rules governing these daily sacrifices: "The first lamb was slain at the third hour of the day, or nine o'clock in the morning. The trumpets were blown and the temple gates were then opened. The last lamb was slain at the ninth hour of the day, at three in the afternoon. Between these two sacrifices, other sacrifices were made.

"My dear mishpocah, at the third hour Yeshua was raised on the cross and crucified. At the ninth hour Yeshua gave up the Ghost. The meaning involved with this ritual should be clear to you. The Lamb of God covers every need, all our sin, every problem, forever. It is the perpetual, eternal remedy for our sin."

The Rabbi saved the best for last. He went on to explain the deep significance of the ritual of these daily sacrifices. He asked, "Do you know what the meaning of the Hebrew word *tamid* means? It is the continual, perpetual, eternal sacrifice by Yeshua for you. It is the Lord's covering for every moment of your life. So, mishpocah, live every moment for him."

Following the message, the people gathered for prayer. As I was leaving, I saw a second group of worshippers gathering in the fellowship hall. I took a look inside the room. The hall was quite large. There were round tables and a long banquet table on one wall. The food is free with is a box for contributions at the far end of the buffet table. The food is very Jewish: bagels, cream cheese, fish, pickles, sauerkraut, and, best of all, cheese blintzes. I just have to stay for lunch.

I sat down with a young couple. The man tells me they are engaged to be married soon. I listened to their plans. The girl has visions of a large family, perhaps eight or ten. The future husband turned to me and said, "Oui vey we will be eating matzah balls the rest of our lives."

Just then the Rabbi came into the room and began visiting with everyone. I took the opportunity to introduce myself to him. "Hello Rabbi Showers. I have so looked forward to meeting you. My name is Jacob Cahn. I have a calling from Yeshua, but first I must finish my degree. I am headed to a conference in Bar Harbor, Maine where I will be presenting my Doctoral Thesis Proposal. I would like you to pray over me."

"Of course, Jacob."

The Rabbi raises his hands and makes the sign of blessing. He prays over me in Hebrew.

"Jacob, just now I received a word from the Lord for you. The Lord says he has a ministry of healing for you. Not only physical healings, but mental and spiritual healings of souls. Then Yeshua said something quite strange. He says, you do not need to use the gold dust. Does this mean anything to you?"

"Yes, Rabbi, it does. The Lord is telling me not to use the very fine nano gold dust to gain better access to the frequencies emitted by the neurons and glial cells in the brain. Truly, the Lord has my life in his hands."

"Yes, I should say he does. I have no understanding of science.

All I know is Yeshua understands it all. He created it.”

“Thank you for sharing this with me, Rabbi. I am very encouraged by this word.”

“Nice to meet you Jacob. Let me say a quick prayer for the success in the meeting you will be attending.”

The Rabbi prays a blessing over me in English. I am blessed twice, one in Hebrew and one in English. I understood only a little of the first one, but plainly understood the second. Then he places his hand on my forehead. I went to sleep as if I was in the lab getting a shot.

When I awoke, I am on the floor. The Rabbi has moved on to other tables. The couple at my table are bending over me. The young girl looks very worried. “Are you alright, Jacob?”

“Yes, I’m fine. How long have I been laying here?”

She says, “You have been asleep only a few minutes. The Rabbi said for us not to be concerned. He is in Yeshua’s hands. I have never seen this before.”

“I am fine. You don’t have to worry.”

I managed to stand up and carefully made my way back to my car. I arrived back at the Harmon household at half-past two. The rest of the day I am still feeling the anointing of the Holy Spirit. I told Patty and Karl I need to get some rest. It isn’t long after laying down before I am asleep.

Joe knocks at my door at seven the following morning. “Jake, are you feeling alright?”

“Yes, I’m fine Joe. This is Sunday morning, right?”

“Yes. We are going to take a look at downtown Manhattan, if you are up to it.”

“Certainly. When do we leave?”

“Right after breakfast. Karl and the family are going to church, so it will be just the two of us.”

“Fine, Joe. I’m sure we can find our way around the city.”

“Karl thinks we should just take a tour. We will not have to drive through the city. I wonder if we can we afford it?”

“Yes, I think we can. Let’s see where the closest one is located.”

Joe found a tour company close by in Paterson. Gardner bus tours is located at 153 Notch Road, just off of the Garden State Freeway and Highway 3. We called and selected the tour of Lower Manhattan with stops at the Empire State Building, Federal Hall, Central Park,

Wall Street, and the 9/11 monument. We made reservations and left for the city. We arrived just before the bus departed at nine o'clock. After the tour, Joe and I took a cab to visit St. Paul's church at the 9/11 site. The hustle and bustle of this city is beginning to unnerve me. I am glad when we arrive back at the Harmon's house again. I am looking forward to some peace and quiet. Karl has steaks on the grill. After dinner, we sat around telling Karl's family about the events of the day. Tomorrow Joe and I will resume the next leg of our journey.

It is raining when I awoke on Monday morning. It's not just sprinkling, it is a real gusher. We left after breakfast and started up the I-287 toward Boston. It rained the entire day. We arrived in the small town of Redding just before dark. We had pizza delivered to avoid the flooded streets. On Tuesday morning, it is still raining heavily. The baseball game is cancelled.

I told Joe, "We might as well go on to Bar Harbor. No use trying to tour Boston with weather like this."

Joe agreed. I cancelled our baseball reservations and we left for Bar Harbor. We arrived in Bangor at one o'clock. The sky is beginning to clear. We decide to continue on our way to Bar Harbor. We finally arrived on the Island at four o'clock Tuesday. The Quality Inn on Kebo Street had a vacancy, so we were able to register a day early. Then we agreed to find a restaurant near the harbor for dinner. We drove the car around the town and settled on a visit to the Bar Harbor Club on West Street. The building is located right on the harbor. There are tennis courts, a golf course, and a huge pool fronting on the harbor. A stone seawall stretches along the harbor. Very ritzy. The management did not seem impressed by our casual dress, but they allowed us to enter the swanky dining room. We both ordered seafood dinners: lobster, clams, a sea greens salad, Kennebec potatoes, and all the trimmings. I am surprised by the reasonable dinner prices. No doubt we are going to gain weight on this trip.

We spent the next few days just exploring the island. A four-masted schooner, named the Margaret Todd, is docked at the wharf. It is low tide, so we walked across the sand bar, picking up sea shells. On Friday, we climbed Cadillac Mountain, the highest point on the Northeastern Seaboard. On Saturday, we spent exploring the Acadia National Park museums and nature trails.

In the middle of the town there is a monument in honor of the sons of Eden who were the defenders of the Union in the Civil War.

The town was originally named and incorporated as Eden. It is not hard to see why they selected Eden as the town's name. The island is extremely beautiful. No wonder so many summer tourists visit the area. During the winter, the number of people on the island dwindles to a few thousand.

On Sunday morning, we relocated to the Grand Hotel. The hotel is an example of classic Victorian architecture. We were both assigned rooms on the third floor I had a view of the harbor. My room is small, but nicely furnished. The walls are decorated with several spectacular paintings, views of the rugged coastland, and the vegetation in the Arcadian Park. Joe and I spent the rest of Sunday exploring the shops in the city. We signed up for a tour of the bay. On the tour we saw puffins, lobsters, seals, pelagic seabirds, and toured the island lighthouse. We encountered quite a few kayakers. We didn't return to the hotel until almost dark. We ate dinner at the hotel and then turned in.

They have scheduled the convention to start at nine o'clock Monday morning. Dr. Altschul called me at seven thirty. He wants to meet with me in the hotel lobby at eight. I quickly dressed and took the elevator to the ground floor.

"Good morning, Jacob. I see you found your way here without any problems."

"Yes. Joe Harmon and I arrived last Tuesday. We have been touring the island. This is a great place for the conference."

"Yes, it is. I just want to touch base with you. Today the meeting here at the hotel will be taken up with a general orientation to the conference. Our special meeting with you is scheduled for Wednesday morning in room 205 of the Jackson Laboratory. Do you know where it is located?"

"Yes, it is located on Mount Desert Street."

"That's right. The meeting is limited to invited participants only. General Weaver will attend along with two others from his staff at the NSA. I believe you have already met Warrant Officer Dan Jenkins, and Colonel Ben Jones. Also, the members of your advisory committee will be present, Drs. Hobbs, Majors, Glick, Adelman, and myself. I assume you have printed copies of your proposal?"

"Yes, I have ten copies, plus the original."

"I want you to destroy two of them."

"I thought perhaps Dr. Hobbs or yourself may need an extra copy?"

"Perhaps so. Go ahead and keep the extra copies. I want you to place each of them in a large envelope and write the names of each of the people invited on them. Be sure you can account for all of your copies. I want you to have each person receiving a copy to sign a receipt. Tell them they are not to make any copies of your proposal. Do you understand?"

"Yes Doctor, I understand. I will have them sign a statement to protect the confidentiality of my proposal."

"Very good. After today's opening meeting, our team is scheduled to have dinner with the NSA people. It will be a private closed-door meeting. You must not discuss these issues with anyone outside of our group. I mean everyone, especially hotel staff or others close by when you are with your advisors."

"Okay, I will watch my tongue around any outsiders."

"Well, I guess this is all we have to discuss at this point, Jacob."

"Yes, Sir. I appreciate the oversight you are giving to me."

"Thank you. We need to handle this whole matter very discreetly. Well, I'll see you in this morning's orientation meeting. They have planned to serve lunch here at the hotel after the meeting."

"Very good. I'll see you there."

I had just enough time before the meeting to place the copies of my proposals in the envelopes I brought with me. I printed the name of each of our group attending our special meeting on the envelopes and placed them in my briefcase. I placed my two extra copies in an unnamed envelope. I locked the briefcase securely and hid it beneath the two large pillows on the bed, completely out of sight. Then I left for the orientation meeting.

After the meeting and luncheon, I returned to my room. I could see the maid had been in to clean the room. I thought, I had better check to be sure my briefcase is still beneath the pillows. It is there. All's well.

Suddenly, I hear the Lord telling me in a still small voice, "Jacob, count the proposals."

I took my briefcase out from under the pillows. The first thing I see is the lock on the briefcase has been jimmied. Shaking like a leaf, I quickly open the briefcase and count the envelopes. One envelope is missing, the one addressed to General Weaver. A cold chill runs up my backbone. I grab the house phone and ask to be connected to General Weaver.

"Hello?"

"Hello General. This is Jacob Cahn. We have a problem."

"How so, Jacob?"

"Someone has stolen a copy of my proposal."

"Are you sure?"

"Yes, I'm sure. When I returned from lunch, I checked my briefcase. The lock has been broken and one envelope is missing. It is the envelope I addressed to you."

"Jacob, do not touch anything. We will be right there."

General Weaver, his two assistants, and the hotel manager arrived at my door in a matter of a few minutes. I explain how I had hidden my locked briefcase beneath my pillows, and how the Lord prompted me to count the contents of my briefcase when I returned from lunch. The hotel manager is on the phone determining the maid who earlier had serviced my room. Colonel Ben Jones is dusting my briefcase, looking for prints. He asks me to pick up a glass from the kitchen and takes my prints off of the glass.

General Weaver is on the phone with Fort Meade. Apparently, the NSA has the ability to observe any location in the country in real time. When Joe saw the commotion down the hall he entered my room, Warrant Officer Jenkins asked him to leave. There is a knock on the door. It is the maid. General Weaver asks the maid if she remembers seeing the briefcase under my pillow. She said she had seen it, but she just left it where it was. She denied trying to open it. Colonel Jones took her fingerprints as well.

General Weaver asks all of us to gather together. He then asks Colonel Jones for the results of his fingerprint tests.

Colonel Jones says, "The only prints I found are Jacobs and the maids. Some of the prints have been partly erased. Probably done by the gloves of the thief. This appears to be a professional hit. Someone has been tipped off to this operation, Sir."

Just then, the phone rings. General Weaver answered it. He listens for several minutes, and then hangs up the phone.

"Well people, we have experienced a theft by a professional, possibly by a foreign agent. We are now following a blue Porsche automobile travelling at a high rate of speed on State Highway 3.

The vehicle will be stopped before it reaches Bangor. Thanks to Jacob's quick response, I believe we will successfully recover the document shortly. Please do not share this information with anyone. If you do so, you could be arraigned as an accessory to the scene of a crime."

It seems the NSA likely has a mole. I am sure none of my professors have arranged for such a theft, since they will be receiving their personal copy of my proposal. Of a certain, the maid is not guilty of any involvement. The plan is too carefully laid out, and she could not know anything about the person occupying the room, nor could the manager know. General Weaver apparently agreed with my thoughts. He excused the manager and the maid, giving them no special instructions other than to keep silent about the incident.

We all waited for another phone call. It came thirty minutes later. The document has been recovered. There will be no prosecution. The culprit has political immunity. I will not have an opportunity to subject him to my IAMGOD machine, the Lord. This probably means the mole will remain hidden within the ranks of the NSA. I recalled again Dr. Glick's word to me a few weeks ago when he commented, "Welcome to the black economy, Jacob."

On Tuesday Joe and I had some free time. The advisors were busy interviewing three other proposals. We decide to take a look at the old schooner tied up at the pier. We ran into General Weaver there. I was curious about the details of the theft.

"Joe, I need to speak with the General privately for a few minutes. Why don't you take another look at the schooner again?"

"General, I believe you shared one fact you probably should not have. You told us the car was a Porsche."

"Yes, I did say it was a Porsche."

"And the thief had immunity from some foreign country?"

"Yes, he had immunity, Jacob. But I didn't say what other country."

"I am guessing this person with immunity might have connections with certain persons in Italy. Perhaps he might even have ties to the powerful Mafia, or even the Vatican?"

"Jacob, you are a very astute person. I will have to be more careful what I say around you in the future. Often, we know more than we would like to know. It is the world we now are living in, Jacob."

"Yes, most uninformed people call this human process natural history. Decisions may result from moral or immoral objectives, but most commonly the result of a godless, immoral pursuit. The Machiavellian battle for power. There exists another level, another paradigm, if you will."

"And what is this other paradigm, Jacob?"

"There is a spiritual dimension to history. Many describe it as conspiratorial history. This view understands there is an unseen battle for supremacy occurring in the heavens. Satan is playing a game of chess with the Almighty. This game has been going on since the very beginning when the anointed Cherubim attempted to take the place of God. The Almighty threw him down upon one small planet in this huge universe. He is the great deceiver of mankind."

"A little more of this and you would have me become a Christian, Jacob."

"Those words were voiced long ago by another powerful person, General. Two thousand years ago King Agrippa said to the Apostle Paul: 'Almost thou persuadest me to be a Christian.' Paul responded by saying: 'I would to God, that not only thou, but also all that hear me this day are both almost, and altogether such as I am, except for these bonds.' The king and his wife failed to receive the message, for their hearts were hardened. No doubt they were hardened by the life of luxury and access to power. The same holds true today, Sir."

"Jacob, I admit this vocation of mine does harden my ability to appreciate the lives of the common man. They are blessed to be less burdened by the heavy responsibilities of leadership in an ever-changing, ever more dangerous world. I will seriously consider what you have shared with me today. Pray for me. I need God's help."

"General, you will find a Gideon Bible in the nightstand next to your bed. It contains the word of life and there is a section to help a person come to faith in Jesus."

"My family are all Catholic. Would you recommend the Catholic Church, Jacob?"

"No Sir, I would not."

"I didn't think you would."

"Jacob, I appreciate you sharing your faith with me. I have to tend to some business now."

As the General is leaving, a black helicopter is landing on the lawn in front of the hotel. I watch as a soldier brings the General a familiar-looking white envelope.

Joe comes back quickly and asks me, "Jake, what in the world is going on?"

"Joe, just forget what you have seen. I can't discuss it."

We watched the copter until it left the island.

The following day, I am walking the two and a half blocks to the Jackson Lab on Mount Desert Street when I see General Weaver walking toward me. He escorts me into the conference room. Everyone invited is present. Dr. Hobbs is seated at the end of the table, plainly taking charge of the meeting. I distributed my proposal to everyone, asking them to sign a sheet verifying they have received their copy of my proposal. Then Dr. Hobbs calls the meeting to order. There is a flag standing in the corner of the room. General Weaver suggests to Dr. Hobbs we begin our meeting by pledging our allegiance to the flag. We all stand. The military people are saluting as we ordinary citizens of our Nation place our hands over our hearts. We all pledge our allegiance to the Nation. Everyone is giving the pledge with a noticeably greater attention. I notice the General emphasized the words "under God." Perhaps he has taken time to read the Gideon Bible in his bedside table. Everyone sits down and Dr. Hobbs begins to address the group.

"Gentlemen, we are gathered here this morning to discuss a very unusual situation. I have never encountered anything even remotely similar to it in my many years at the Institute. Let me begin by saying this young man, Jacob Cahn, impressed me from the very first time I met him. He has proved himself to be extremely capable and has been gifted with a keen intellect, plus an intense interest in the field of human genetics. He also has been a source of considerable controversy from the very beginning, as Dr. Altschul can attest to.

He is outspoken in his opinions, yet he responds to criticism in a conciliatory manor. His research has gone well beyond our early expectations for a first-year graduate student. These meetings here at Bar Harbor are usually to approve a proposed research topic. Jacob has gone far beyond a simple proposal. He has achieved a major scientific advance in our understanding of how the human brain stores our memories, where we find our creative gifts of contemplation and understanding. I think back to Copernicus and Galileo, and their discoveries of the workings of our solar system. They were rejected by the powers that be because the discoveries turned the prevailing paradigms upside down. Today, we have a choice to make. It is complicated by forces beyond the realm of pure science. We are living in a time when scientific truth is in conflict with the desire to divulge truth to the world, a world divided by conflicting and powerful adversaries. This world condition clearly demands we manage the body of new knowledge well if our civilization is to survive. We have given this difficult task into the hands of our government. We hope and pray our Nation's government will honor the trust we have placed in it and will use this new knowledge in a proper way to benefit human kind while protecting us from those who would seek to dominate the whole world. Today, we must come to a decision to protect our National interest while at the same time, protecting the individual freedom to follow his God-given destiny, free from the dictates of powerful elites. With these thought in mind, I will now ask each of you to begin reading Jacob's research proposal. Take your time to do so. We will be serving coffee and donuts. Following your examination of the proposal, there will be an opportunity for each of you to ask Jacob questions. Unfortunately, Jacob will not be able to discuss the specific details of the NSA's new IAMGOD machine mentioned in his proposal. This new machine is classified top secret. You are among the few people who will even know of its existence. Let us begin."

I opened my envelope and began to read my proposal for the umpteenth time. It took an hour and a half, plus my five donuts and three cups of coffee before everyone completed their reading of the proposal.

When the last person closed their folder, Dr. Hobbs asked, "Has everyone finished reading the proposal? Fine. Then we are ready for questions."

The first question came from Dr. Glick. "Jacob, I believe you first planned to pursue a different topic, namely the problem of precursor neurons?"

"Yes, Dr. Glick. I originally wanted to understand how precursor neurons are directed to their positions in the cerebellum."

"Why did you abandon the effort?"

"After some preliminary study, I came to the conclusion there was only a slim chance of solving the question apart from causing harm to the unborn child."

"So, you were not willing to use aborted fetuses in your research?"

"No, I was not. I believe in the sacredness of the unborn, Doctor."

"Thank you, Jacob. I also hold to the same position."

The next question came from Dr. Majors. "Jacob, I assume you realize you have put each of us in a difficult position by utilizing a classified device?"

"Yes, Doctor Majors. As you know, I attempted to use the MEG machines in your own laboratory, but found they did not have the ability to access the cell, its nucleus, or the genes themselves. It was at your suggestion I agreed to use the classified machine."

"Did you understand when you agreed to use the machine it meant the results of your research must remain in the black economy?"

"No, I only recently realized my work at the NSA might interfere with my program at the Institute. Can I ask you a question Dr. Majors?"

"Yes, of course."

"Was it ever your intention to recruit me to join the NSA?"

"No, Jacob, I did not have any such intention. As you, I failed to realize the extent of the NSA's constraints on the use of the new machine. I actually expected NSA to make the machine available for use by public and private research facilities. In my particular research program, I was motivated by the desire to incorporate a fully rounded approach to the overall project, to extend it to include the genetic aspects. I'm sorry if I gave you such a false impression."

"I accept your answer, Doctor. I am not seeking to accuse you of any malfeasance or deception. My question was just meant to clarify the way this whole matter developed. Dr. Hobbs, this catch 22 we find ourselves in was not foreseen by myself, Dr. Majors, the NSA, or anyone else."

Dr. Hobbs is taking time to write down the words of our discussion before continuing. "I understand what you are saying, Jacob. Do we have another question from the committee?"

The next question came from General Weaver. "Mr. Cahn, do you realize how difficult this situation has become. I would like to say this all could be solved had you accepted our offer of employment. Once again, I want to offer you the chance to change your mind and accept our prior offer. The decision would be helpful to all of us here. What do you say? Can we proceed along this alternative I am offering to you?"

"General, I appreciate your desire to offer me a position with the NSA. I did not expect you to offer me a position once again. As you may recall, I earlier stated my reasons for not accepting your generous offer. I have not changed my mind. I feel it will not be helpful to repeat my reasons for not accepting your offer to this group. It is my opinion it might end up raising unneeded emotional responses. I realize this has been an onerous situation for everyone concerned."

Dr. Hobbs is nodding his head. "Gentlemen, I believe we should not stray from the path we are now on. Let's just accept Jacob's desire to forego any reconsideration of the NSA's offer. Are there any further questions?"

General Weaver is looking very frustrated, but there are no other questions. Dr. Hobbs then continues, "Since we have no questions, let me suggest how we can proceed. I have met with the NSA and we have developed a plan for handling this matter. In my possession is a written agreement we jointly have agreed upon. I will pass it along for each of you to read. Let me quickly summarize what the document says. Jacob Cahn agrees to provide all of his research results to the Agency. He also agrees never to release this data, nor to publish the results of his research at NSA to the general public. Jacob agrees to no longer make use of the NSA's facilities unless NSA extends an offer to do so with terms mutually agreed upon. Jacob agrees not to share with the public the capabilities, or even the existence of the IAMGOD machine. In return, the NSA has agreed not to pursue any legal or harmful actions against Jacob Cahn pursuant to his now completed research. The NSA will agree to recognize internally his research achievements and will not obstruct in any way his efforts to receive a doctoral degree from the Institute or any other similar institution. In the agreement, the Institute agrees to recognize Jacob's research proposal and research results as satisfying the research requirements for the Doctoral Degree in Genetics. Jacob's completed thesis based upon the research at NSA, will be held by the NSA only. No copy of the dissertation will be held by the Institute, or by Jacob Cahn. The Institute agrees to grant the Doctor of Microbiology degree with a concentration in human genetics to Jacob once he successfully completes his additional three years of study.

"I believe my summary of this agreement faithfully describes the document developed by our legal advisors. Are there any questions or observations about the agreement I have summarized?"

The room is very quiet. Dr. Hobbs then passes the document around the table. As each person finishes reading it, I try to read their body language. Drs. Glick and Majors seem to be in favor of the agreement. Dr. Altschul plainly is not pleased with it. The General has now become calm. He has resigned himself to favor the agreement. The other NSA people seem a bit frustrated by the agreement, but do not seem overly hostile to it. Dr. Hobbs' demeanor is hard for me to interpret. I believe he is trying hard not to influence the opinions of the others. Dr. Hobbs waits until everyone has read the document. Then he prepares for the vote.

"Gentlemen, since this is a sensitive issue, we are going to decide Jacob's fate by secret ballot. I am asking Dr. Glick to assist me in tallying the votes. I have eight slips of paper here. I am going to permit Jacob to vote also. I will not vote unless there is a tie. I believe everyone has a pen in front of them. Please take one slip. Fold your slips of paper after voting and do not sign your name on your ballot. Unless there are questions, I now will call for the vote. Do I have any questions? Let the record show there are none. We will now proceed to vote on the agreement. Do I have a second?"

Dr. Majors declares, "I second the measure, Doctor Hobbs."

"Very well. All those in favor please mark your ballot 'Yes.' All those opposed please mark your ballot 'No.' You may also choose to abstain. When you have finished marking your ballot, please fold it. When I see all of the ballots folded, I will ask Doctor Glick to collect them. Then I will read the votes. I will not participate in the vote unless there is a tie."

Dr. Glick quickly collected the ballots and handed them to the Director. My forehead is dripping with sweat as Dr. Hobbs reads off the votes one by one, He then announces the results.

"The tally is seven in favor, one abstention. The agreement is passed. I will now ask all of you to place your signature above your name. To protect your vote, I am asking everyone to sign the measure, even if your vote was not in favor."

The Director passes the document down the table for signatures.

Dr. Glick raises his hand to speak. "Gentlemen, I for one want to commend everyone at this table for the way this matter has been handled. We have all learned valuable lessons through this unusual circumstance. We have learned to resolve a difficult matter with grace and consideration for all parties involved, our Government, the Institute, and a young and talented scholar. Our actions today are honoring the high traditions of truthful science. Jacob, I want to say, mazel tov, mazel tov!"

The group instantly responded, clapping their hands and repeating mazel tov over and over. I began crying like a baby. Everyone shook my hand as they departed. The ordeal is over. I have been blessed yet again by my Lord Yeshua.

When I arrived back in my room, I called Dad. "Hey Pop, I have some good news for you."

"What is good news for me is hearing your happy voice, not your worried one."

"Dad, I have passed my research requirement. Not only do I have permission to do the research, but they have accepted my research."

"What? Are you serious?"

"Yes, I'm serious."

"What have you done to deserve this?"

"To put it simply, I have discovered where our memories reside. I have uncovered where we think and how to recall our entire history."

"Really! So, I won't have to worry about forgetting where I left my stethoscope?"

"No, you won't Dad. I'll fill more in for you later. Goodbye."

"Goodbye, Son."

23

A Year of Decisions

"Hast thou given the horse strength? Hast thou clothed his neck with thunder? Canst thou make him afraid as a grasshopper? The glory of his nostrils is terrible. He paweth in the valley, and rejoiceth in his strength. He goeth on to meet the armed men. He mocketh at fear, and is not affrighted, neither turneth he back from the sword."
(Job 39: 19–22)

Rachel's Journey

Joe's proposal was approved on Wednesday as well as my own. We both agreed to celebrate our success with another dinner at the Bar Harbor Club on West Street. Joe ordered a bottle of Champaign, then a second bottle. By the time the dinners arrived Joe is feeling very mellow. I honored my Nazarite vow and drank tea. We consumed lobster, clams, mussels, and crab cakes. We finished the dinner with a bread pudding dessert, smothered in cream sauce. Although the Bar Harbor area is beautiful, I am anxious to get back to my apartment.

"Joe, what do you want to do, stay here until the end of the week, or head home?"

"I don't know, Jake. What do you want to do?"

"I think I would like to leave Bar Harbor and go back to Baltimore.

We can take advantage of those unused tickets at Boston tomorrow. It is a night game. Then we can stop at your brother's house and take in another game at Yankee Stadium or take another tour. What do you think?"

"It sounds good to me, Jake. We will have to leave very early to get to the ballpark by six or so."

"True. I believe I can put the pedal to the metal and make it by the time the game starts."

"Okay, Jake, let's do it."

I paid the bill for dinner over Joe's objections. Then we returned to our rooms. On Thursday morning, we ate breakfast at the hotel. Joe called Fenway to check on our rained-out tickets. He was able to reserve two seats in the fourth row on the third base side. The Sox are playing the Orioles, our home team. The ticket lady said there was a zero chance of rain for today and Friday. We left Bar Harbor by half-past eight. We plan to spend the night someplace just south of Boston. Joe found a Motel Six off of I-93 at Intestate exit 15. We stopped for lunch at Portland and managed to arrive at Fenway a few minutes before four o'clock. It was a magical evening. The Orioles won the game by a single run, a homer over the green wall. Joe managed to grab a foul ball in the third inning off the bat of Chris Dayton, the Oriole's young first baseman. We arrived at the motel very late. On Friday, we surprised the Harmon family by arriving on their doorstep at noontime. Mrs. Harmon made lunch for us and we spent the afternoon playing pool and board games with the kids. Karl arrived back from the University at five. We shared our experiences with them after dinner and turned in about ten.

Saturday morning, we both rose early. I found myself thinking about Rachel. I gathered my nerves and gave her a call.

"Hello?"

"Hello, Rachel. Do you know who this is?"

"Yes, of course I do, Jacob. Are you calling me from a New York number?"

"No, I am calling you from Paterson, New Jersey. Joe Harmon and I are just now returning from Bar Harbor. We are visiting Joe's brother here in Paterson. You have been on my mind. I know I probably shouldn't be contacting you, but I just wanted to hear your voice. How are things going with you?"

"Things? You mean how am I coping with my marriage?"

"Well, yes. I hope you are finding a way to cope with it."

"I am, Jacob. I received your wonderful letter. I have come to the conclusion my life is too short to waste it moaning about my failed marriage. I assume Dr. Mike told you I cannot divorce my husband?"

"Yes, he explained the financial problem. I take it your marriage has no chance for a reconciliation?"

"None, Jake."

"It sounds like you are dealing with this well. I have been so concerned about you."

"Well, I have been thinking about it a lot lately. I also have been reprocessing my argument with Alice."

"Really?"

"Yes, really. When all of this came tumbling down on me, I started to drink. I soon realized alcohol was not going to take me anywhere but down. I went to a psychologist, but she wasn't on my wave length. In desperation, I prayed to Ha Shem. Then it happened. Yeshua came to me in a dream. It was a miracle. When I awoke from the dream, I knelt down and prayed. I told him I would accept the grace he offers if he would heal my broken heart. Jacob, Yeshua is real. He answered my prayer. I have been totally transformed by His love. Jake, the love I am experiencing is so awesome. I thought I would never be happy again. I was wrong. Yeshua has filled me with his love, peace, and joy. What is more, he has given me a new assignment. Do you know what I am involved with now?"

"No, I don't."

"I am in training to go to the mission field. When Rabbi Stein found out I am an archaeologist, he drafted me. I will be taking people from our congregation to minister in Mid-Eastern Arabic countries. I am taking twelve women to Egypt next month. We will be helping leaders of a Coptic Church in the desert city of Siwa in western Egypt. I am also scheduled to teach a short course with the Egyptian Archaeological Society in Siwa."

"I am so happy for you, Rachel. I know you and your staff are going to do a fantastic job ministering to the Egyptian people."

"Why don't you join us for a week or so?"

"I cannot. I still have three years of studies before I can put the Doctoral certificate on the wall."

"Yes, of course. Are things going well with you?"

"Yes, very well. I have already completed my research project and I am looking forward to completing my degree. Once I finish the degree, I plan on applying the new knowledge I have gained, perhaps in Omaha with Dad. That is, if the Lord doesn't take me elsewhere."

"What kind of new knowledge are you talking about?"

"Rachel, the Lord has given me a gift. I have found the key to unlocking memories stored in the human brain. I believe this may lead to a great awakening. Scripture describes God putting the words of Scripture into our mouths during the persecution of the End Times. I believe this awakening is coming soon."

"Awesome, Jake. I will pray for you. The people of the world certainly need waking up. The time for Yeshua's return is drawing very close."

"Yes, we are seeing new signs of the End Times almost every day. Rachel, it has been a real blessing talking to you this morning. We must keep in touch with one another."

"I would ask you to pay me a visit, but I don't want to open old wounds. I hope you understand, Jacob."

"Yes, I understand. You know I will always cherish the times we spent together. Now I can look forward to being together with you and Yeshua in heaven for eternity."

"Yes, we will be together there forever. Please pray for Papa. He needs Yeshua also. I love you dearly, Jacob. I will always be your Rachel."

"Yes, you truly are my Rachel. My namesake was helplessly in love with Rachel the moment he set eyes on her. I inherited his DNA. May Yeshua richly bless you. Goodbye Rachel."

"Goodbye, Jacob."

I gently put down the phone and stretched out on the easy chair, just staring at the ceiling I was lost in thought when Joe knocked on my door. I invited him in.

"Jake, are you reliving yesterday again?"

"Joe, I am reliving a lot of yesterdays."

Joe looked a bit puzzled. I didn't explain it to him. "What did you want to ask me about, Joe?"

"I wanted to ask you if you want to leave today or tomorrow morning?"

"We might as well leave this afternoon if you are ready to, Joe."

"Yes, I'm ready to leave."

"Okay, let's leave right after lunch."

We quickly gathered our things and left for Baltimore. We arrived back at four. After dropping Joe off at his apartment. I unloaded the car and carried everything up to my apartment, stopping on the way to pick up my mail. There were several letters in the mail. I put the letters on the kitchen table and then dumped my stuff onto the living room floor. I will sort things out later. I want first to go through the mail. I threw away about a dozen ads from car dealers, insurance companies, and sales catalogues. I threw the ads away and began opening my mail. The first letter was from General Weaver at the NSA. He sent me a very nice letter, thanking me for the work I had done and for my willingness to cooperate with them. He also included a check in payment for my research at the lab. My hand is shaking as I carefully unfolded the check. It is made out in the amount of twenty thousand dollars. I carefully put it in my wallet. I also received a letter from Dr. Hobbs. He said he is looking forward to next year and wishes me the best. He included next year's class schedule. The next letter is from Dr. Mike. He said Sam Kaufman has seen a change for the better in Rachel. He also said he didn't understand how she now is able to cope with the marriage. Rachel is telling him she has found a new life in Yeshua. Dr. Mike says Sam is now asking him to explain Messianic Judaism to him. Perhaps my prayer for Samuel is being answered. There is another letter from Dr. Mike. The large envelope contains his translation of a Dead Sea scroll recently discovered in cave twelve. He says the original was written in Hebrew. The scroll is entitled The Humane Gospel, The Gospel of the Holy Twelve. Dr. Mike has dated the scroll to the first century AD. He tells me I will find Chapter Sixty-one very interesting. I turned to the Chapter and began to read. I couldn't put it down. Chapter Sixty-one is a parallel to Chapter Twenty-four from the Gospel of Matthew almost word for word. And there is more. Chapter Sixty-one mentions several details of the End Times not found in Matthew's Gospel. I put everything else on hold to study the document, slowly reading each of the nine verses. I scanned the verses into my computer and read them over and over again.

This scroll is truly amazing. Verse one is almost word for word from those of the Gospel of Matthew. Verse two says there will be many false christs who will deceive mankind by applying the name of God to themselves. These false prophets will confuse the people and mislead them. Also, there will be things taking place on the earth having never taken place before, never seen by any other generation. Verse three reads like the article in a current newspaper, describing a time of wars and rumor of wars, but the end is not yet. The scroll claims when the end does come, we will enter the "Great Rest." This must refer to the Millennial Kingdom.

Before the Great Rest, the scroll describes a time in which the powerful shall gather up the lands and riches of the earth for themselves and shall oppress the greater number who have nothing. Do we see this coming now? Yes, we do.

In verse four there is an interesting wording. It reads "People will be held in bondage, but not in prison." They will be used as slaves by the rich. No doubt this refers to a world-wide economic system described in the Book of Revelation as the mark of the beast. The evil and cruel rulers even lust after the taste of flesh and blood as in the time of the Nephilim. The blood will flow.

Verse five states this cruelty shall be so great no man or beast can escape from it, save the holy angels and the elect of God. A strange savior shall rule the minds of many and this generation shall think evil is good and good is evil. Could this strange savior be a Roman Pope?

In verse six the scroll says this strange god will work many miracles and deceive the people of every nation into worshipping him, but he shall not succeed, for in verse seven. It states: "The Eternal Spirit of All, the Holy Spirit, shall send forth his holy messengers to restore the Holy Law hidden by wicked men through their vain traditions."

All who do not believe shall perish. Those who do keep the restored law will be hated by all nations. Many who are offended by the Law of God shall betray their friends, listening to false prophets.

In verse eight the document describes a horrible time when evil shall be so rampant God's name shall be blasphemed like never before in the history of the world, more often than the stars of heaven.

Verse nine describes the shedding of the blood of God's innocent believers and a time when humanity will be deceived. How man will be deceived! I remember reading the passage in Revelation, Chapter Eleven where it says the Evil Blasphemer will apply God's holy name to himself. The similarity of the words of this scroll to our present age is striking.

I know the Lord's prophecy received by Pastor McGuire so many months ago is soon to be fulfilled. I am called, called to minister along the side of the two witnesses, one of a hundred and forty-four thousand ministering evangelists in the Last Days. Microbiology is a worthwhile endeavor, but this is my true calling.

On the following Monday, I put in a call to Dr. Adelman. Perhaps he may want my services again this summer.

"Hello, Doctor. Jacob Cahn here."

"Hello, Jacob. I tried to find you after the conference. I wanted to talk privately with you, but you already had left."

"Joe Harmon and I both wanted to leave early. As far as I knew, we weren't under orders to remain to the very end of the conference."

"Yes, you were free to leave. The majority did leave after receiving the results of their interviews."

"Are you possibly calling me about working this summer for you?"

"Yes, I wanted to check with you to see if you wanted me to work the summer sessions again."

"I'm fine with it."

"Consider yourself hired, Jacob. We will be offering the advanced course in virology for both sessions. I will send you copies of the text and the lab books. Will you be trying to continue your research at the same time?"

"No, I will put my genetic research on hold for the summer."

"I appreciate your willingness to set it aside for the summer sessions, Jacob."

"These fields constantly are exploding with new knowledge. I'm sure I will be learning more about virology this summer, Doctor."

"Yes, you are right. Very well then, I will give you a chance to look over the new materials and I will give you a call to set up a meeting to go over the course."

"Thank you, Doctor. I appreciate the opportunity."

"I know it is nice to receive a paycheck now and again. Talk to you soon, Jacob."

"Goodbye, Doctor."

Even though I have the NSA check, I felt relieved to know the summer work will put food on the table. I perhaps will have another use for the check. I gave Robert Manley a call.

"Hello, Robert. Jacob Cahn here."

"Hello, Jake. I understand your research has been approved by the Institute."

"Yes, it has Robert. For your information, the nano gold did improve the magnetic signal from the neurons."

"Yeah, the colloidal gold really improves the conductivity of low voltage electrical circuits. If you want to purchase some nano gold, I get mine from Delta Refining Company."

"How much does it cost?"

"That depends upon the size of the colloidal particles and how they are suspended. The price for a single gram of one-micron of gold is about $350.00. The larger quantity you buy, the less the cost, but you won't find any priced below a hundred dollars or so."

"What size did we use at NSA?"

"We used five grams of one-micron gold."

"So, if I wanted to use gold in a future test, it would cost me about fifteen hundred dollars just to make one test?"

"Yes, Jake. It is expensive."

I calculated my new-found wealth would only support thirteen tests using gold very sparingly.

"Robert, the Lord told me I will not have to purchase any. Thank goodness."

"Okay, Jake. That's cool."

I hung up with Robert. The next thing on my agenda was to do my washing. The apartment machines are on the second floor. I gathered up my dirty clothes, some quarters, and made my way downstairs. Once I finished the washing I made dinner and then watched the news until bedtime. The next morning, a courier dropped off the new text and lab workbook. I got right to work reading the text. The first summer session begins next Monday.

This summer's days were unusually hot and humid. My apartment air conditioner did not work well. I spent a lot of evenings in the Student Union, studying Hebrew. I felt the need to learn the language. I bought a complete set of DVDs teaching the pronunciation and meanings of some three thousand words.

A Hebrew Ivrim

Before the summer sessions finished, I received a call from Director Hobbs. He asked me to meet with him in his office at three o'clock on Thursday. I have no idea what he might want to discuss with me, unless it is to offer me a job of some kind. I arrived right at 3:00. Robbie is there to welcome me as I entered.

"Hello Jacob. I haven't seen your face for too long. Where have you have been hiding?"

"I have been working for Dr. Adelman again this summer."

"Oh, I didn't know you were helping him out once again. Dr. Hobbs just stepped out. He said he would return shortly. Have a seat."

I sat down and waited for the Director. Five minutes later he arrived.

"Hello, Jacob. On time again, I see. Come into my office. I want to share something with you. Robbie, we don't want to be disturbed."

"Alright, Doctor."

I followed Dr. Hobbs into his office and sat down. As usual, he did not waste any time in getting to the subject of our meeting.

"Jacob, things here at the Institute are soon going to change. Effective the first of September I am going to take a sabbatical. Marge and I want to take some time to travel, take in the sights. When my sabbatical is over the following year, I plan to retire."

"I hope the things I have put you through are not what has caused you to bow out of the Directorship, Sir?"

"No, not at all, Jacob. Times are changing and some of the changes I am not in agreement with. The Institute is feeling a lot of pressure to use the CRISPR technology across species. As you know, I have been very critical of crossing species, even when working with plants. The major corporations whom will be nameless at this time, are heavily involved. Very soon, these copyrighted higher yielding genetically modified seeds will only be available from them. These corporations will totally control the world's food supply. This is putting many plant varieties at risk of extinction. We also are hearing of labs using CRISPR to introduce animal DNA into human DNA at the germ level, permanently altering the world's progeny."

"Doctor, the recent Microbiology Ethics Conference has failed to come up with any firm guidelines."

"Yes, it is true. Jacob, and it is a sad situation. My heart is not in this. The Institute will need new leadership in tune with the times. Dr. Altschul will be Acting Director of the Institute during my absence. I will not be able to shield you and others from decisions he may wish to make regarding this CRISPR technology."

"If I understand you correctly, you are suggesting I may experience new challenges to my research next year?"

"I doubt if you will experience any serious difficulties in finding research opportunities. Three research groups already have asked me to have you contact them. However, some of the values you hold will certainly be out of favor."

"I see. I want to share something with you also, Dr. Hobbs. As you know, I am a Jewish mensch. We Jews trace our lineage back to the ancient nomadic Hebrews. These Hebrews were known as the Irvi, the singular form of the word is Irvim. The word means to pass over.

Doctor, I have made the decision to become a modern day Irvim. I am going to pass over the opportunity to work as a geneticist here at JH next year. I have decided to take another path as I seek to find my ultimate calling. The Lord has had to warn me several times to avoid seeking worldly things, fame, wealth, and authority. He has told me I am to remain celibate. So, today I am passing over to another pursuit in my life, Sir. I believe we are, in a sense, at the same crossroads in our lives now. You are passing your leadership baton to others. I am passing my genetics research opportunities to others as well. In my opinion, we are both moving on to a new and greater calling."

"Can you expand on this greater calling of yours, Jacob?"

"I have been called to evangelize for the Lord. It has been a long process for me to arrive at this point. Pastor Bill McGuire in Atlanta said he believes I am called to be one of the twelve thousand evangelists from the Tribe of Judah to serve in the End Times."

"This is quite astonishing, Jacob. I'm certainly aware of the end time signs mentioned in Scripture. They are now occurring at an accelerating rate, almost daily. Still, you have a lot to offer to the field of Microbiology."

"I understand what you are saying. I believe God will use my training at the Institute in some way. Anyway, my mind is made up. I will be crossing over Jordan soon. I want to thank you for sharing your future plans with me. I felt compelled to share mine with you as well."

"This decision is quite a surprise, Jacob. I had no idea you were contemplating such a decision."

"Yes, I realize my decision is utterly unexpected. I hope you will forgive me for keeping the lid on it until now. I am looking at this change with a considerable amount of anxiety."

"Well, Jacob, I suppose I must allow you to pass over to the other side. Please keep me informed of your whereabouts and how the Lord is using you. I am sure he will be providing a way where there appears to be no way, Jacob."

"I promise to keep in touch with you."

"Do you know where you will be headed out from Baltimore?"

"No. It is a mystery. The Lord hasn't revealed it yet. I know he will do so when it is time. He is never early, and never late. The word will come right when it is supposed to come."

At this point, Dr. Hobbs stood up from his desk and escorted me through the office door and into the hallway. We shook hands. Looking into his eyes, I realized I am more than a student microbiologist-in-training to him. We have forged a deep bond together. I have found another brother in the Lord.

It was ten minutes past four when I left his office. I didn't know what to do, or where to go. I needed to vent my emotions, but there is no one to turn to. Terry and Ramona have gone to California to watch one of their horses run in the Santa Anita Futurity. I finally just returned to my apartment. I pulled a pizza out of the refrigerator and made dinner. Then I sorted through my movie collection. I needed to get lost in something. I decide to watch the movie *Secretariat*, the story of the greatest race horse ever to step onto a race track. The true story is a grand example of a woman's desire to succeed against overpowering difficulties. The opening scene shows this super horse of all time, pawing the ground in the starting gate, taken from verses nineteen through thirty-five of Chapter Thirty-nine in the Book of Job. The Christian music playing in the background is, even now, feeding my soul. With the Lord's help, I will overcome every obstacle standing in my way. Tomorrow, Friday, will be the first day of a new path in my life. I will be dipping my feet into the Jordan.

I called Dr. Hobbs at nine o'clock in the morning. Robbie put me right through to him.

"Good morning, Doctor. This is Jacob again. I need to start the ball rolling, Sir."

"You are deciding to leave so soon?"

"Yes, I am. I want to take my leave of the Institute effective after the summer sessions."

"Are you sure, Jacob?"

"Yes, I am very sure. I have no doubt about this decision. I will complete my work with Dr. Adelman this summer and then fold up my tent and move on into the wilderness."

"The Institute is losing a talented Microbiologist, Jacob."

"Perhaps so, but I'm not indispensable. Joe Harmon will still be here and he is a very good researcher."

"Yes, Joe is a good, but he lacks the intuitive understanding you have, Jacob."

"Thank you, Doctor. Do I need to give you a letter to seal my decision with the Institute?"

"Yes, you will have to send me a letter and then obtain a clearance from the admin office. Let me know if you have any problems, and keep me posted."

"I certainly will, Doctor. Goodbye."

"Goodbye Jacob, and the Lord watch over you."

I felt relieved after the phone call. Now is as good a time as any. I picked up the phone and called Dad.

"Hello, Dad."

"Hello, Son. Nice of you to call. Are things going well with you?"

"Yes, but I have something important to share with you."

"What is it this time? Are you out of gelt?"

"No. In fact NSA paid me twenty thousand dollars for my research with them. I have made another hard decision. Things are not going well politically here at the Institute. The director, Dr. Hobbs, is taking a sabbatical and he shared he will be retiring a year from September."

"So, how is this bad? There certainly must be others qualified to take over?"

"Yes, they are technically, but not morally qualified. Many in the Institute feel they are being left behind in the application of CRISPR technology."

"How so?"

"Many labs are now crossing human DNA with animal DNA, even at the germ level. I have heard the Military is producing super soldiers using the technology. Several labs around the world are trying to produce human organs by introducing human DNA into pigs."

"Do you think the Institute will change their policy outlawing this practice?"

"Yes, I do. Dr. Hobbs believes they will do so just as soon as he removes himself as Director."

"So, what are you going to do, finish your degree elsewhere?"

"I'm not going to finish my degree ever, Dad."

"What? You are going to leave after spending all this time, money, and effort without a degree?"

"Yes, I am. I am going to become an evangelist, perhaps in Israel."

"No, not an evangelist to convert Jews to Christianity. You can go back to medical school and come work with me here in Omaha. I can even put you to work coding medical records right now, today."

"I cannot, Dad. The Lord is telling me to be a good Hebrew."

"What do you mean?"

"Dad, I have been studying the Hebrew language. The meaning of the word Ivrim is one who crosses over. Dad, we Hebrews had to leave Egypt before entering the Promised Land, and we crossed over the Red Sea before crossing the Jordan River. When I left Emory, I left slavery and became a free man. Free to battle the giants, if you will. Now I'm using this freedom to cross over into the promised land, to be of service to Ha Shem."

"Jacob, if you must go ahead and do your own thing I hope you quickly get lost in your promised land and come back home. Goodbye Son."

I felt relieved after the phone call. He at least called me son before hanging up the phone. I spent the rest of Friday preparing for the advanced course in virology.

24

End of the Age

"And I saw another angel ascending from the east,
having the seal of the living God.
And he cried with a loud voice to the four angels,
to whom it was given to hurt the earth and the sea,
saying, hurt not the earth, neither the sea, nor the trees,
till we have sealed the servants of our God in their foreheads."
(Revelation 7:2–3)

Persecution of the Truth

On the following Saturday Dr. Adelman called and asked me to meet with him at seven o'clock on Monday, the first day of the summer session. He wanted to go over the schedule for the class in Advanced Virology. The class will begin at nine o'clock. I met with him in the Student Union over breakfast.

"Jacob, are you ready for another busy summer session?"

"Yes, I am. I have read the text over twice. I like the way it reads. I have something to tell you about that will surprise you greatly."

"What do you mean?"

"Doctor, I will not be completing my degree, but I will not be leaving until after the summer sessions."

"I don't understand, Jacob. Why have you made such an unusual decision? Is someone ill in your family?"

"No, my family is just fine. There are reasons I am not free to discuss at this time. You will soon understand why. All I can say is that the Institute is changing."

"I guess I am still out of the loop. Has Dr. Hobbs shared something with you?"

"I'm sorry, but I can't comment, Doctor."

"Well then, let's make your last summer a good one, Jacob."

"Yes, we certainly will. I am looking forward to it."

"I will see you later this morning. I'm picking up the tab, Jacob."

"If you insist. I will be there a few minutes before nine o'clock."

I have an hour before the class is scheduled to begin. I went to the Rec Center and took a run on the indoor track. Then I took a quick shower and left for the class. As last summer, Doctor Adelman is well prepared and gave a stern, but friendly overview of the class to the students. I recognized many of the students. They were in the introductory course last year with the Doctor and myself.

I forgot just how intensive the summer sessions are. Before I knew it, the first session is over. Then it is right on, non-stop into the second session. The limited time I have away from the classes I spend working on my Hebrew. Dr. Glick and Rabbi Epstein are spending time with me, explaining the syntax of this interesting language. A single word often has several meanings, depending upon the context. Many times, the language reveals deeper levels of meaning framed in the terms of the agricultural economy of those times. Gradually I found myself becoming more and more fluent in the language. Rabbi Epstein took me aside one Sunday and asked me to give a message to his home group in Hebrew. I gathered my chutzpah and took it on. It is evident the Lord is in this. What he has in mind for me next, I am not sure.

The week before the end of the second summer session, I received a call from Dr. Altschul, requesting I meet with him at nine o'clock on Friday at his office. I have learned nine o'clock meetings are always important. I put on my suit and polished my shoes. I arrived at his office at five minutes to nine.

"Hello, Jacob. I'm so glad you came. You are probably wondering what this meeting is all about."

"Yes, I am. It's one of those nine o'clock meetings. It must be important."

"Yes, it is one of those meetings. Jacob, I believe Doctor Hobbs informed you he intends to take a sabbatical starting September first."

"Yes, Sir. I believe he and his wife want to take some sightseeing trips."

"He may also have told you I will be taking over his directorship during his absence. It is my desire to continue the excellent way Dr. Hobbs has led the Institute. I will be making some major changes. This is why I have asked you to meet with me today. One of the changes I intend to make is to expand our CRISPR research capabilities. I am initiating several new research efforts in cooperation with the Pentagon and NSA. I have been in contact with General Weaver at NSA. He mentioned your name. He is very impressed with your work and your creative abilities. I know we have encountered differences in the past, but I would like you to consider joining me in developing a research program with NSA and the Military. The remuneration will be very substantial. How does this sound to you Jacob?"

"Dr. Altschul, I wish you well with this new undertaking, but I cannot agree to participate in these new CRISPR research programs. I made the decision to move on in another direction. I will not be continuing my studies here."

"Are you serious, Jacob? Do you intend to finish your degree elsewhere? Perhaps I should tell you the position will mean you will be part of our permanent staff. The salary will start at two hundred thousand dollars a year. They are willing to pay you this salary even before you receive your degree."

"I'm sorry Doctor, but I cannot accept your gracious offer. I have ethical concerns with the application of CRISPR technology. I cannot agree to inserting animal DNA into humans, altering the handiwork of my God."

"Humans are animals too, Jacob. We share the basic cells of life with them."

"Yes, but we have something they do not have, a living soul. Our souls must be protected against corruption by foolish men of science who fail to see the ultimate outcome of these efforts."

"Am I up against this God subject with you once again?"

"Yes, it seems you are, Sir. To change humanity's DNA is an abomination to God Almighty. Scripture speaks about the sacredness of our human lineage, and all of God's unique species. I can never agree to commit to a practice assigning me to the fires of Hell."

"You Jews and Christians are out of your minds. The new technologies of AI, and Qbit computers will put an end to human disease and ultimately death itself. Science soon will create superior beings, life forms capable of living forever."

"Doctor, transhumance is the greatest sin the Devil has ever come up with. I only wish you would set your world view aside long enough to rationally consider the teachings of the Scriptures. The only way you will ever experience an eternal life filled with faith, hope, and love is though Yeshua Ha Mashiach, Jesus Christ the Messiah."

"Change my world view? Jacob, I curse the day you every applied for admittance to this University, and I curse the God you serve. You religious zealots are a disgrace to the world of science."

"Perhaps we are to your science, but the science I study and love comes from the Creator of the Earth, the Universe, and the heavens above. I do not hold any anger against you, Doctor. In fact, I often pray for you. My God still loves you, even when you dishonor his holy name as you have done this morning."

"Jacob, you will soon see. Science will eradicate your God from the face of this planet, and the ends of this universe. Get out of my office. I don't want to see your ugly Jewish face ever again."

I left Dr. Altschul's office without slamming the door, although I felt like doing so. Actually, I felt like tearing it off its hinges. There is no way I can be reconciled with this man who denies the existence of the Creator unless God personally zaps him. The conflict left me feeling very upset. I changed my clothes and went to the Rec Center for a run. I ran for two hours straight. Then I went to the Student Union to have lunch, still upset. I ate only half of my hamburger and fries. There is no one to confide in. Terry is still not back to the farm. No doubt Dr. Glick is busy preparing for the start of the Fall semester. I called Rabbi Epstein. Perhaps he might have a good word for me.

"Hello, Rabbi Epstein. This is Jacob, Jacob Cahn."

"Hello Jacob. What is going on with you? You sound upset."

"Yes, I am upset."

"Do you have time to pay me a visit today?"

"Yes, I don't have anything scheduled."

"Okay, I'll meet you at the church at three o'clock."

"I'll be there. Thank you so much."

"Take your time. I don't want you to add a ticket to your troubles."

I arrived at the church five minutes after three o'clock. Rabbi Epstein is waiting in the parking lot for me. He took me to his office in the basement of the church building.

"Jacob, welcome to my holy of holies. There are holes in the rug and holes in the ceiling, but God does provide."

"Yes, he does. I can confirm it. My problem is not a holy financial problem. I just experienced a very bad meeting with Dr. Altschul at the Institute. We had differences in the past, but this meeting was straight out of hell itself."

"Did you feel like you were being persecuted for sharing the Word of God with this man, Jacob?"

"Well, I was treated very abusively. He ended up losing his temper and cursing me and God."

"Cursing you say?"

"Yes, the man cursed God and said he didn't want to see my ugly Jewish face ever again."

"And how did you respond to his rantings, Jacob?"

"Somehow I managed to tell him I often pray for him. I told the man God still loves him, even when he dishonors his holy name."

"Bravo, Jacob! Let me quote the first verse of Chapter One in the Epistle of James: '*My brethren, count it all joy when ye fall into diverse temptations, knowing this, that the trying of your faith worketh patience.*'

Persecution seems strange to us in this country of ours, because we have been blessed to live in a nation covered by God's grace and protection. In these end times, we are encountering ever greater persecution. The enemy knows his days are numbered, and he is going around like a roaring lion, attempting to destroy all goodness and unity. You have been singled out for some of the lion's attention; only God knows why. Let's pray and lift it up to the Lord."

"I am ready for prayer, Rabbi. I need to know where I am

headed.”

The Rabbi put his tallit, his prayer shawl, over his head and began praying in the Spirit. I bowed my head, but no words came out of my mouth. He must have prayed for ten minutes or more. Then he quietly removed his tallit, carefully folded it, and placed in on the shelf behind his desk.

“Jacob, I have a word for you from the Lord. Are you ready to hear it?”

“Yes, of course. I’m very ready.”

“The Lord says you will receive a call, a phone call actually. It will be from someone you know, but not very well. This person will ask you to go on a journey, a long journey. You do not need to worry about how you can arrange to go. It will be provided for you. God says he will lead you to a stranger, but not right away. This stranger’s name is Zvi.”

“Rabbi, this is incredible. I can confirm the Lord’s message. Quite a few months ago, the Lord spoke to me. He said I would meet a stranger who would become a mentor to me. Guess what the man’s name is?”

“Could it be Zvi?”

“Yes, it is Zvi. What kind of a name is this?”

“I’m not sure, Jacob. It may be a nickname. It sounds perhaps Polish. One thing we do know, Jacob. You will be receiving a phone call.”

“I have not told you about a decision I have just made. I am going to leave the University before the Fall semester begins. I know I am a Hebrew, an Irvim. I do not know where I am going, but I will follow the Lord into the wilderness if necessary. I am willing to make Ivrim, to cross over into a new land.”

“Marvelous, Jacob. Just as soon as you receive the phone call, I will ask you to share with the entire congregation. God bless you.”

“Thank you, Rabbi. I will keep you posted.”

The Rabbi walked with me to my car and he gave me a big hug. My heart is full of joy as I make my way back to my apartment. I arrived back by five-thirty, still feeling the elation of our meeting. I was hungry. I found half of a pizza in the refrigerator, plus some left-over Dinty Moore stew. After dinner, I settled down and read the Epistle of James from start to finish. I couldn’t help wondering who I would receive a call from. Perhaps Pastor Bill, or Allen? No, the Lord said it would be from someone I barely knew. I spent the evening continuing to read with one ear open for a telephone call. None came. I finally turned in at eleven thirty.

The Ministry Begins

On Saturday, I was up and ready to attend the church service. When I arrive, I learned Rabbi Epstein had invited Rabbi Weiss to give the message. Rabbi Weiss has just returned from another trip to Israel. He reports many Jews in Israel are waking up. They are turning to Yeshua. He believes a very significant move of the Holy Spirit is going on there, but there is resistance from the Orthodox, the current-day Pharisees.

Where have we read this before? Rabbi Weiss is very concerned about the Pope of Rome. The Pope has seduced many religious groups into joining with the Papacy to form a new universal religion of peace. The Rabbi said the Lord told him there soon will emerge a mighty army to oppose this false doctrine. Then the Rabbi shared something remarkable. He said several respected rabbis are claiming the Two Witnesses will soon appear in Jerusalem. Rabbi Weiss turned to the Book of Revelation and read verses three and four from Chapter Eleven:

And I will give power unto my two witnesses, and they shall prophesy a thousand two hundred and threescore days, clothed in sackcloth. These are the two olive trees, and the two candlesticks standing before the God of the earth.

The room erupts in praise. The people are clapping and shouting hallelujahs. I quickly turned to Revelation Chapter Seven in my Bible where it describes the one-hundred and forty-four Hebrews, twelve thousand from each tribe, sealed by God. They are called to preach the Word of God to every living creature on the earth. This sealing of the tribes seems to occur just before the rapture of the believers and the appearance of the two witnesses. It is the great awakening I have been striving to understand. I began to pray silently in the Spirit. Before I realized it, I was preaching very loudly in Hebrew and in English. The congregation was stunned.

"In the last days, a spirit of deception will draw away many from the Word of the Almighty. The sun will turn black like sackcloth, and the moon will fail to shed its light. The seas will roar and mountains will pour forth fire and smoke. Men will seek to hide in the holes of the earth, saying hide us from the God of heaven. Yet I will not forsake my own. Those who have stood against this evil time. Rejoice, my people. Your deliverance is at hand."

As I stopping preaching there fell, what I can only call, a holy hush. Some are staring at me, as if I have lost my mind. Others are silently praying in the Spirit. My eyes turn to Rabbi Weiss. He is walking off the platform toward me. He places his right hand on my forehead. I feel myself falling under the power of the Holy Spirit. As I am falling, I am leaving my body. I am lifted up to a high mountain. On the mountain top, I see a city. It is the New Jerusalem. Where once there was a Dead Sea there is now a wide river flowing from the City of God. One branch flows to the west, down to the Mediterranean Sea. A second branch is flowing eastward into the Araba, now covered with green vegetation. There are fruit trees stretching as far as I can see along the shores of the two rivers. Everything is covered with a bright light, yet there is no sun or moon. The light emanated from the center of the city. Then a voice calls out in Hebrew: "Ehayah Asher Ehayah." I understood the words. It is the name of God, the "I Am that I Am."

When I awoke from my vision, the congregation had left. Only Dr. Glick and the two rabbis remained. They helped me up and into the pew. Rabbi Weiss began to quote a Scripture over me in Hebrew. I understood every word, even the place where the text is found. The passage is from the Book of First Kings. Chapter Nineteen, and verses eighteen and nineteen:

Yet I have left me seven thousand in Israel, all the knees which have not bowed unto Baal, and every mouth which hath not kissed him. So, he departed thence and found Elisha, the son of Shaphat, who was plowing with twelve yokes of oxen before him, and he with the twelfth, and Elijah passed by him, and cast his mantle upon him.

Then Rabbi Weiss took his prayer shawl and placed it over my head, saying, "Jacob, I am anointing you with this shroud of ministry. The angels even now are sealing you with the protection from the

Lord. God has revealed to me your destiny. It is to usher in the great day, the Day of the Lord."

I was overcome by the whole experience. All I could manage to say was: "Let it be, Lord. Let it be."

I returned to my apartment. The first thing I did was to begin packing everything I owned in boxes, leaving out only the bare essentials. The next thing I did was to put in a call for Dr. Adelman. I left a message for him to call me as soon as possible. I considered the packing and the phone call an act of faith. Then I waited for the phone call. None came. I ate a peanut butter sandwich and drank some coffee. Still no phone call. I gave up waiting and went to bed. On Sunday, I was still anxiously waiting for the phone to ring. I finished packing my belongings. I did not leave the apartment for fear I would miss the call. Then I realized, this is ridiculous. God has scheduled it. This call will arrive when I am here. Once I had this firmly in my head, I changed into my running clothes and left for the Rec Center. Joe Harmon was working out on the weights when I arrived. I took a few laps on the track, then asked him to go to lunch. I needed to share my experience with him, to bring him up to date.

"Joe, I will be leaving the Institute soon. I am being called to serve the Lord. He has revealed this to me last week."

"I don't understand it. Are you leaving without finishing your degree, Jake?"

"Yes, I am. The Lord told Rabbi Epstein I will be receiving a phone call. He said it will point out my next step on this road."

"I take it you haven't received the call yet?"

"No, I haven't. I have been expecting the call, but the phone hasn't rung."

"Why would the Lord wait before giving you the call?"

"I don't know. I went ahead and packed all my things to show him I am willing to follow, but so far, no call."

"Perhaps he wants you to do something else, something other than packing up your things?"

"I don't know what it would be. I am ready to leave. I have enough money to go wherever he tells me."

"You have plenty of money? Doesn't the Lord often ask people to trust him to supply all their needs?"

"Yes, it's true. I am now pretty well fixed financially. Perhaps this may be a problem instead of a blessing? Can I ask you a personal question, Joe?"

"Sure."

"Are you financially able to meet your expenses next year at the Institute?"

"Not really. I will have to take out a student loan this year. I was not able to find a job this summer and the trip to Bar Harbor was a bit expensive."

"Yes, we did spend a lot on the trip. You know, I am scheduled to proctor again this summer with Dr. Adelman."

"Yes, I remember you saying he was eager to get you back again."

"Joe, I believe you have hit on something. I left a message with the Doctor to give me a call. Have you taken the course in advanced virology?"

"Yes, I took it during my last undergraduate year."

"Very good. You know the NSA paid me twenty thousand dollars for my research. If you were able to take my place and proctor Adelman's course in the Fall, and I gave you the NSA money, would it be enough to meet your expenses next year?"

"You can't be serious."

"Yes Joe, I am serious. I am thinking my financial well-being may actually be blocking this phone call. Rabbi Epstein told me the Lord would be giving me everything I need for the journey. At the time, I felt God had already supplied my need. Perhaps this is not where my provision is to come from. Perhaps I need to settle this money thing right away."

"Do you think NSA will take the money back?"

"Well, I don't see why they would not take it back. Let's find out right now."

Even though it is Sunday, I dialed the emergency phone number General Weaver gave me.

"Hello, General Weaver."

The General's voice answered "Yes, this is General Weaver. Do we have a problem?"

"Not a serious problem, General. This is Jacob Cahn. I want to apologize for this call on a Sunday, but I need to ask you a question."

"Well, let's hear it Jacob."

"Sir, I am wondering, is it a problem for you if I send the money you paid me back to the NSA?"

"What? I have never, in all of my thirty years of service, been asked to return a paycheck. I do not know how to do it, and I don't even want to try. Do you know how hard it is to deal with these accountants here? I guess you do not. Jacob, whatever you do, don't send the money back to me. If you haven't cashed the check, please do so immediately. Cash the thing and give it to your aging grandmother or something. Do you understand?"

"Yes, Sir, I understand, but I already have deposited it in my bank account."

"Spend the money, Jacob. Spend the money on your girlfriend. As I remember, your car is old and getting ready to die. Spend it on a new car. Otherwise I will have to send Warrant Officer Jenkins to arrest you."

"Thank you, General. You have solved my problem. Have a nice day."

"Joe, after we finish eating, let's take a trip to my apartment. I think you have solved both of our problems."

Joe and I quickly finished eating and walked to my apartment. I opened my dresser where I had stashed my check book and wrote out a check for twenty thousand dollars. I handed it to Joe. Joe is thanking and hugging me when the phone rings. It is Dr. Adelman returning my call.

"Hello, Jacob. What is up?"

"Doctor, something has come up. I will not be able to help you with the Fall Advanced Virology course. I do have a solution. Joe Harmon has taken the course a year ago and I can vouch for him. He is an excellent, student, but needs some gelt. I wonder if you would consider hiring him in my place?"

"Well, perhaps. If you don't mind telling me, why are you not going to be able available?"

"Doctor, I will not be on campus. I have been called to the ministry."

"Really? I will miss you, Jacob. Tell your friend Joe to contact me as soon as possible. I will discuss your idea with him."

"Thank you, Doctor. I really appreciate your help."

"Keep in touch, Jacob."

"Yes, I will keep you informed. Goodbye, Sir."

As I close my cell phone, the phone rings again. "Hello, Jacob Cahn here."

"Hello, Jakub. Do you know who this is talking to you?"

"I am not sure, but it sounds like Yehuda Sorenson."

"It is! This is Yudi."

"My goodness. It is good to hear your voice, Yudi."

"I want to know, Jakub, are you married yet?"

"No, Yudi, I'm still single."

"Are you even engaged?"

"No, I haven't signed any ketuba. I won't be renting a tuxedo any time soon."

"*Ganz gut!* This isn't a sales call, Jakub. At least not a call to rent you a tuxedo."

"Yudi, are you matchmaking for me again?"

"No, I've given up the ketuba idea, Jakub. I want you to come and talk with me. Can you come to my shop today?"

"I suppose I can meet you there. What time?"

"I will open the shop special at two o'clock."

"Okay, Yudi. I will see you at two. Goodbye."

Joe is amazed. "Jake, is this a God thing I'm hearing?'

"Yes, it is Joe. I haven't heard from Yudi for months. I was told the phone call would be from someone I knew, but not a close friend. It certainly is from God. This check will ease the pain this Fall, and I believe you will get hired by Dr. Adelman also."

"This is wonderful. I can't thank you enough."

"Just thank the Lord, Joe. He is taking care of both of us."

"Yes, he certainly is. This is unbelievable, a true miracle. I am going to have to tell my folks about this."

"Okay, Joe. Also, tell them you are growing closer to the Lord. I can see you are doing exactly what the Lord wants of you."

I watch as Joe leaves the apartment house. He breaks out in a run across the campus. I left a few minutes later to visit my Yiddish friend Yudi. I arrived at his shop at five minutes past two. Yudi and his two daughters are standing in the open doorway, waiting for me. Yudi shows me into his office and introduces his girls.

"Jakub, this older tokhter of mine is Reba, and this one is my impish youngest daughter, Dinah. Reba is now twenty-one, and Dinah has just turned eighteen."

"Yudi, I am wondering why you have asked me to meet with you and your daughters?"

"Oy, I can tell you if you wait a moment. My daughters want to take a *hilekh*, a trip to Israel. They want to make an *Aliyah*. I need someone to go with them, to be their *shomer*. You know, to watch over them.”

“You want me to look after them?”

“Yes, look after them to be sure they get settled in a good place. They want to go to a *kibbutz*, you know, a farm commune.”

“Oh, now I understand. You want me to be their chaperone?”

“Yes. I need to stay here at the shop and take care of Mame, my wife. Can you do this for me?”

“Yes, Yudi, it so happens I can do this. But there is a problem of money. I want to do this for you for free, but I don’t have enough gelt to buy the airplane ticket or cover my basic expenses.”

“Not a problem! I can pay for the tickets and for the kibbutz *ding*, the rent. The kibbutz covers the food. I can give you some *kashene* gelt, pocket money.”

“Yudi, I need to share something with you. I am going to Israel to serve Ha Shem. I have resigned from the University to do this. I have even been studying Hebrew, and I am fairly fluent in it now, but I don’t know enough words in Yiddish.”

“Yes, Hebrew is not Yiddish.”

“It won’t be a problem since I’m not taking you along. I will take a Yiddish dictionary along with me. I do have another question for you. How long do you need me to watch over your kinden, your kids?”

“Just until they get settled in a safe place. Then you are free to come home or stay in Israel, whatever you want. I will pay for your trip back if you come home.”

“Yudi. I will do it!”

“Oh, Mazel tov! Mazel tov! You are truly sent from Ha Shem to help us.”

“Yes, I am your *letsgelt*, a going-away gift from Ha Shem. When do we leave?”

“I will make the reservations and call you back soon, Jakub”

“Okay, Yudi. I will wait for your call.”

The girls are ecstatic. They are twirling, jumping up and down, and turning cartwheels as I leave. I am dancing inside as well as I made my way back to my apartment.

I did not have long to wait before I received a call from Yudi. He has booked us on a shuttle out of BWI at 7:00 Thursday morning connecting with an El Al flight leaving from LaGuardia. We will be staying at the Kibbutz David Judea. It is located on the outskirts of the city of Afula, just south of Nazareth.

I have everything boxed and ready for the Salvation Army. There is one other item I need to attend to, Rachel's ketuba. I wrote a letter explaining I am leaving for Israel. Then I carefully wrapped it, placed it in a box, and sent it to Rachel. I sent an email to all my friends in Atlanta and to the Johnsons explaining my departure on my new path. Everything I own is boxed except for my good suit, a few changes of clothes, my important papers, my Bibles, and my Hebrew DVD and dictionary. I managed to squeeze all of them into my old suitcase. I loaded the car with my boxed-up belongings and dropped them off at the Salvation Army store in Timonium. I finally am ready for my Aliyah to Israel. Joe Harmon agreed to take the girls and I to the airport.

As we pulled up in front of the terminal, Joe asked "Jake, what shall I do with Old Blue?"

I handed him the signed title, and said, "Joe, Old Blue is yours. Be sure to check the oil from time to time."

He thanked me and we said goodbye. I retrieved a baggage cart and loaded it with our luggage. After passing through security, we proceeded to walk to Gate 5. I bought the girls and myself some magazines, sandwiches, and coffee. We waited for an hour, then boarded the New York shuttle. We arrived in LaGuardia an hour later. I took the girls to Gate 24 where our flight is waiting. We have seats in the upper cabin. We departed the gate at 7:05 on Thursday. I counted the seconds as the Jumbo Boeing 787 with all four jet engines screaming, began its roll down the runway. We are airborne in thirty-two seconds. This is a direct flight to Ben Gurion Airport. I let the girls take turns at the window, but there wasn't anything to see but clouds and ocean. The first land we saw was off the coast of France the next day. We arrived at Ben Gurion at 12:20 on Friday. Yudi had arranged for a rental car for us. We took a tram to the rental agency and picked up the car. It is a small Fiat, but we managed to squeeze everything in. We arrived at the kibbutz at a quarter to three. We are too tired to do anything except eat a quick dinner. I told the girls to call Yudi and meet me in the dining room in the morning. Breakfast is scheduled from six to eight o'clock. I told them I would meet them at eight. Tomorrow will be a big day. It took me all of two minutes to fall asleep.

I awoke at seven the next morning. It is Shabbat. I walked down to the breakfast area. The girls show up just a few minutes after I did. I realize for the first time this kibbutz is Orthodox. You do not order your breakfast on Shabbat. Everyone just takes what they want from the food on the buffet tables. After breakfast of yogurt, coffee, flat bread with honey and fruit, the girls asked me to take them shopping in Nazareth. I am determined to protect these girls. I told them in no circumstances are we going to Nazareth. It is not safe. I will take them to Tel Aviv, or Haifa, but not to Nazareth. So, we left for the big cities. I felt like a duck out of water, but succeeded in finding a shopping mall in Tel Aviv. I told them to call me on my cell when they are ready to return to the kibbutz. I just wandered around the mall area for a while. I found a coffee shop at the end of the block. There are several older men playing cribbage at a table outside. I bought a cup of coffee and decide to join them. "Good morning. Do any of you speak English?"

One man volunteered to answer. "Yes, we all speak English. Are you a tourist?"

"No, not really. My name is Jacob Cahn. I have just made my Aliyah to Israel. I arrived only yesterday."

"Really? This is unusual. We usually meet old people making Aliyah."

"I am taking care of two of my friend's daughters until they get situated. Right now, we are staying at the Kibbutz David Judea near Nazareth."

"So, you must be Orthodox, but you don't look Orthodox?"

"No. I am a completed Jew. I believe in Yeshua Ha Mashiach."

"Jacob, you won't want to stay very long in Kibbutz David Judea. I believe it is an Orthodox kibbutz."

"I believe you are right about the kibbutz. Only yogurt and matzah with fruit was served on Shabbat."

"You are going to have to find another place to stay. Are the girls Orthodox?"

"Yes, they are. Their papa owns a tuxedo rental store in Baltimore, Maryland."

One of the other men spoke up. "Jacob, my name is Jerry Jenkins. I was raised in Baltimore, Maryland."

"That's interesting. I just left Johns Hopkins."

Jerry asked, "What degree did you receive at JH?"

"I only have a BA, Jerry. I was studying to be a microbiologist in the McKusick-Nathans Institute of Genetic Medicine, but I made the voluntary decision to drop out of the doctoral program."

Jerry wanted to know more. "Too tough? Did they wash you out?"

"No, I wasn't washed out. I voluntarily resigned from the Institute for personal reasons."

He is looking perplexed, still wanting more information.

"It's a long story, Jerry. Basically, I had to leave because of changes being introduced by the new interim director. The interim Director plans to disregard the ethical concerns facing current genetic research. I cannot agree to crossing human DNA with animals, especially when done at the germ cell level. This level means all progeny will retain the inserted DNA, permanently altered human DNA."

The third man introduced himself at this point. "Jacob, my name is Rabbi Joseph Gershom. I also am a converted Jew. I am here playing cribbage with these two friends of mine, trying to convince them in the truth of our Messiah. They like cribbage and I like cribbage, but I like teaching people about Yeshua more."

"I didn't expect to run across completed Jews so quickly. I don't believe I caught the name of your third cribbage partner?"

"Jacob, my name is Zvi."

"Just Zvi?"

"Yes, just Zvi. As you can see, I'm old. I was born in Warsaw, Poland in 1933."

"You lived in the Warsaw ghetto?"

"Yes, I lived there until near the end of the war."

"How did you survive?"

"When my parents were sent to the camps, I stayed with a young couple and their daughter in their home. We lived on the food they grew in their garden."

"Was the garden hidden from the Nazis?"

Yes, it was enclosed by a stone wall?"

"Yes, I know it was. What were the names of the people?"

"It has been so long ago and I was very young. I believe the woman's name was Hilda and the daughter's name was Rose. I can't remember the father's name, or their last names. One day the Nazis came and took the husband and wife to the camps. Rose and I hid in the garden. That night we both packed our things and left. I took Rose through the sewers and we escaped. I got lost and Rose couldn't find me. I never did find out what had happened to her. I made my way to Greece and stowed away on a ship. That is how I arrived in Israel."

"Quite a story. I have a story for you also, Zvi. My Rabbi Jonathan Epstein received a word from Ha Shem for me. The Lord told him I would be taking a trip and I would meet a man named Zvi. The Lord said this man would provide a great help for me."

"This is true. I had a dream telling me to expect a young man named Jacob."

"There is more. What I'm going to tell you next may sound impossible and weird. Zvi I have the ability to awaken memories in people, even memories of both sides of my family lineages, and I can prove it to you."

"Yes, I'm listening."

"At the end of the Second Word War, a woman by the name Hilda Stein and her daughter Rose emigrated to Buffalo, New York. While working at a hospital as a custodian, Hilda introduced her daughter to a Jewish doctor by the name of Joel Cahn. They were married two months later. Hilda died a year after their marriage. Joel and Rose then moved to Omaha where Joel establish his medical practice. I was born there, as was my sister Emily. Zvi, Joel and Rose Cahn are my parents. Hilda Stein was my grandmother, but she died before I was born. Interestingly I have suffered for years by a persistent dream about a garden such as you have described. Zvi, the Lord is in control of all things. He has woven a magnificent tapestry joining our two lives together for this very time."

"Yes, you have blown my mind. This is beyond my wildest dreams. Welcome to the battleground, Jacob. We need to talk more about this later."

Jerry wanted to know about the two daughters who had made the Aliyah.

"Jacob, how long are the two girls going to stay here in Israel?"

"I really don't know. Reba is twenty-one and Dinah is just eighteen. As soon as I know they are settled in, I will be on my own."

Zvi said, "Jacob, you will be staying with me. My apartment is small, but it has a loft. You can sleep in the loft. It is located here in Tel Aviv. Let me give you my card with my phone number and address."

I took the card he handed to me and placed it in my pocket.

"When can we meet to talk, Zvi?"

"How about ten o'clock tomorrow morning?"

"Yes, I can arrange it with the girls. There is a map in the rental car. If I have trouble locating your apartment, I'll give you a call."

Just then, my cell phone rang.

"Guys, it is the girls. They are ready to return to the kibbutz. I must say goodbye to you three cribbage players and make my way back to my car. We will be in touch."

I asked the girls how their shopping trip went on the way back to Kibbutz David Ben-Judah. Reba said, "Jake, we first had to go to a bank in the mall and ex change our dollars for shekels. The exchange rate is 3.7 shekels for every dollar, less the bank's charge for the transaction. There are one hundred agoras to a shekel. The stores are really great, very modern, and the prices are very reasonable. Once we figured out how little we were spending in dollars, we bought more stuff."

"Girls, I have a question for you. Did you know the kibbutz Yudi has selected is Orthodox?"

Dinah said, "Oh yes, Jake. We are comfortable with the kibbutz. We attend an Orthodox synagogue in Baltimore."

"So, you girls are comfortable there. Do you feel safe in the kibbutz?"

Reba answered, "Are you serious? Haven't you seen the guns in the gun case in the lobby, and the number of Israeli soldiers living there? Yes, we are very at home at the kibbutz."

"I am pleased to hear this. While you were shopping this morning, I met three guys in front of a café down the street. One of them, a man named Zvi, has invited me to stay in his apartment free of charge. It has a loft. I have a lot in common with these three guys. Would you be concerned if I left you on your own at this point. If you do, I will remain at the kibbutz awhile longer. If not, I would like to take up my friend's offer and move from the kibbutz to Tel Aviv? So, this is my proposal. What do you think?"

The girls looked at one another and briefly discussed it.

Reba spoke up first. "Jake, we both have decided to accept your marriage proposal. Are you proposing to me or my younger sister Dinah? I'm smarter, but she is better looking?"

They both started laughing.

Dinah then said, "Jake, we actually are glad to be rid of you. How does that make you feel?"

They restart their silly laughter again.

"Okay, girls. I can take a hint. I will leave. Don't think I am going to just disappear, though. I will be checking up on you, now and again. I will keep in touch with Yudi also, so don't let me down. Any problems will find you headed on a plane back to Baltimore. Do you understand me?"

They both chimed in together, "Yes, we understand Uncle Jake."

This is how I became known in Kibbutz David Ben-Judah as "Uncle Jake."

The following morning, I checked out of the kibbutz and made my way back to the airport. I turned in the rental car and took a taxi to Zvi's place. Zvi is waiting on the porch when I arrived. We shook hands and I sat down on the porch with him.

"Jacob, do you know why you are here in Israel?"

"Yes Sir, I believe I do."

"I would like you to tell me."

"Zvi, I am here to usher in the great Day of the Lord. I have been called to serve as one of the anointed and sealed virgin men in the Lord's army. I am a Jew from the Tribe of Levi. I have been given gifts to use to bring healing and restoration to the people here in Israel, whether Jew, Arab, African, or any other ethnic people group."

"Yes, this is your calling. As with Moses, I believe I will soon pass from the scene before this mighty conflict takes place, even though it is coming very soon. We must make hast to prepare you for the time when the Jordan is crossed and the battle lines are drawn. Not only has the Lord revealed your name to me, but also the names of the two witnesses. We will work to secure the necessary technology to enable you to identify the tribal markers of the one hundred and forty-four servants of the Lord. This is simply an amazing time to be alive."

"Yes, it certainly is Zvi. Yeshua is our Savior, our Healer, our Baptizer in the Holy Spirit, and our soon coming King."

End

About the Author

Lee Bush lives in Kalispell Montana with his wife Mildred. They have been married 55 years and are the proud parents of three adult children and seven grandchildren.

Lee earned his Bachelor's Degree in Economics from UCLA in 1966 and his Master's Degree in Geography from the University of Arizona in 1969. He has enjoyed a varied career during his lifetime. He served as an instructor at the University of Nebraska, Omaha Campus for seven years. He left the University and obtained a job as a ge-

ographer with the United States Census Bureau in 1976. He received the bronze medal award from the Department of Commerce in 1993. Lee was offered an early retirement from the Census Bureau in 1994 and moved to California in 1996. Lee served as an administrator at Hope Chapel Foursquare Church in Canoga Park, California for six years during which time he obtained a license as a Foursquare Minister. He studied and became a volunteer apologist with Reasons to Believe ministry in 2005. He is currently an ordained minister with the International Church of the Foursquare Gospel, retired.

Lee has an ongoing interest in science and Biblical Studies, especially as they relate to the times in which we now live. It is his belief we are now experiencing the fulfillment of ancient Biblical prophecies of the end times. Lee accepted the Lord Jesus Christ as his Savior at age 16 and has been an active member in churches during his adult life.

Glossary of Yiddish and German Words

Adonoi ....The Lord
Akhi yedidi ....A good buddy
A klog ist mir ....Woe is me
Aliyah ....Migration to Israel
Apikoros....................A heretic
Bar Mitzvah ....Jewish's boy adulthood celebration
Bemidbar...................Book of Numbers
Bris ....Circumcision
Bubele ....Expression of endearment
Bubkis ....Nothing, or less than nothing
Cepher ....The Word (Bible)
Chavruta...................A joint project
Chutzpah ....Brash courage
Completed Jew ....Jewish believer in Yeshua
Corban.....................A act done for God
Ding........................Rent
Fachadick ....A mixed-up confused person
Ganz gut ....Very good
Gelt........................Money
Goy ....A non-Jewish man
Ha Shem...................The Name (Substitute for God)
Hamisch ....A wise and down to earth person
Hilekh......................A trip (Perhaps to Israel)
Irvim........................People
Kashene....................Kosher person
Ketuba......................Jewish engagement agreement
Khaver......................Friendly person
Khosn ....Groom
Kibbutz.....................A communal farm
Kippah ....Skull cap
Kinden......................Children
Kleyn.......................Small
Klipp ....Gabby woman (Aren't they all?)
Kop.........................Head
Kvetching ....Complaining

Lekhaim! ....To (long) life
Lekherlekh ....Utterly ridiculous
Letsgelt ....Going away gift
Loch in kup.................Hole in the head
Mame........................Mother
Mazel tov...................Congratulations
Mazzaroth ....Mazzaroth (The Way Constellations)
Mekheiyeh ....It's a pleasure
Mensch......................A handsome good guy
Meshugeneh...............A crazy person
Mezuzah ....Small container on doorpost
Midrash ....Rabbinical teachings
Mishpocah ....Family and good friends
Mitzvah ....Doing good deeds
Modeh Ani.................Morning meditation prayer
Oui veh ....Help, it's painful!
Shabbat.....................Saturday, the Seventh Day
Shadchen...................Matchmaker
Saichel ....Common sense
Shema Yisrael.............Hear O Israel, the Lord is one
Shishka......................A non-Jewish woman
Shlemiel.....................A clumsy klutz
Shomer ....A guardian for youth
Schmaltzy ....Exaggerated sentimentality
Schmuck ....A real jerk
Tallit ....Prayer shawl
Talmud.......................Entire Jewish Bible
Tamid ....Perpetual Sacrifice
Tefillin.......................Small scripture box worn on the head
Tokhter.......................Daughter
Tsatskelah ....A beautiful and sexy girl
Tsuris ....Big troubles
Vielen dank ....Many thanks
Yeshiva ....A Jewish school
Vos ist dos? ....What's going on?
Zekukim.....................A bride's dowry
Zerr gut ....Very good
Zerr Schlect...............Very bad situation